Molten Salt
Techniques

Volume 4

Molten Salt Techniques

Volume 4

Edited by

Robert J. Gale
Louisiana State University
Baton Rouge, Louisiana

and

David G. Lovering
The David Graham Consultancy
Swindon, Wiltshire, England

Plenum Press • New York and London

Library of Congress Cataloging in Publication Data

(Revised for vol. 4)

Molten salt techniques.

Includes bibliographies and indexes.
1. Fused salts. I. Lovering, D. G. (David G.) II. Gale, Robert J., 1942–
QD189.M59   1983   546′.343   83-9582

ISBN 0-306-43554-3

TO THE CLASS OF '63, MENTORS & FRIENDS
WHO MADE IT ALL A LOT OF FUN!

Validius est naturae testimonium quam doctrinae argumentum
— *St. Ambrose*

Contributors

Kenneth G. Furton • Department of Chemistry, Wayne State University, Detroit, Michigan 48202

Marcelle Gaune-Escard • S.E.T.T., Université de Provence, 13397 Marseille Cedex 13, France

R. Jeanloz • Department of Geology, University of California, Berkeley, California 94720

Thomas D. Kaun • Argonne National Laboratory, Chemical Technology Division, Argonne, Illinois 60439

Brian R. Kersten • Department of Chemistry, Wayne State University, Detroit, Michigan 48202

R. Narayan • Department of Chemistry, Indian Institute of Technology, Madras, 600 036 India

Paul A. Nelson • Argonne National Laboratory, Chemical Technology Division, Argonne, Illinois 60439

Hiromichi Ohta • Department of Metallurgy, Faculty of Engineering, Ibaraki University, Hitachi 316, Japan

K. L. N. Phani • Department of Chemistry, Indian Institute of Technology, Madras, 600 036 India. *Present address:* Central Electrochemical Research Institute, Karaikudi 623 006, India

Rena M. Pomaville • Department of Chemistry, Wayne State University, Detroit, Michigan 48202

Colin F. Poole • Department of Chemistry, Wayne State University, Detroit, Michigan 48202

Salwa K. Poole • Department of Chemistry, Wayne State University, Detroit, Michigan 48202

W. W. Warren, Jr. • AT&T Bell Laboratories, Murray Hill, New Jersey 07974

Yoshio Waseda • Research Institute of Mineral Dressing and Metallurgy (SENKEN), Tohoku University, Sendai 980, Japan

Q. Williams • Institute of Tectonics, University of California, Santa Cruz, California 95064

Foreword

Once again a group of dedicated authors has succumbed to the entreaties of determined editors to offer another volume in a "how to" and "what is" series in the molten salts area. The breadth of techniques employed is well represented in this present volume, where the molten salts discussed range over about an order of magnitude in temperature, and the techniques represent those at the cutting edge of measurement. Here we find chapters dedicated to one class of melts (amides), a variety of fundamental techniques applied to a host of different molten salt systems, and one, on battery testing, which arises from an area that has been in the forefront of proposed new technological applications of molten salts. The availability of concise, collected literature detailing experimental lore from experts in the areas encompassed by these volumes in invaluable. A single use of one of the practicalities presented in these volumes can, to put it in the crassest possible terms, repay the user for the cost of the entire set.

The worth of these chapters to both aging practitioners and ardent neophytes is huge; in writing a foreword to such a collection I feel it appropriate, though perhaps a bit presumptuous, to offer thanks on behalf of the molten salt community for the dedication of both authors—past, current, and future—and editors to this chore.

Robert A. Osteryoung

Department of Chemistry
State University of New York at Buffalo
Buffalo, New York

Preface

We are glad to report that our optimistic hope, expressed in the last volume, for continued progress, development, and expanding applications of molten salts has been amply borne out in the intervening period. It has been apparent both from the published literature and through personal contacts that the catchment for the use of melts among scientists, technologists, and industrialists is dramatically widening. It would be naive, however, to accuse us proponents of being smug, for what is happening is that ionic liquids are now being assessed and investigated in their rightful role as only one among many media available in the liquid state—i.e., neither with favor nor with prejudice. In short, they have ceased to be seen as "new," "different," "cult," or in any other way that sets them apart from other media. Of course, they bring unique problems—it is the purpose of this series to assist in their solution. It has been the role of our distinguished authors and foreword writers to raise awareness of the possibilities of adopting molten salts for diverse applications and for providing the necessary background knowledge to make such choices possible. It is to them and their contemporaries that we must truly direct our appreciation.

In the academic arena, low-melting mixtures (often complex ions + organic salts) continue to attract considerable attention, with several new systems, including organoalkali metal complexes, being reported. In the industrial sector, power sources based on molten systems have attracted substantial backing worldwide. There is a very real prospect for initial introductions of molten carbonate fuel cells and alkali metal batteries in the not too distant future. New molten salt processing methods, such as fine powder precipitation and compound semiconductor preparation are beginning to appear. Watch this space . . . !

This volume differs from previous ones in a number of respects. Most chapters deal with techniques *per se* rather than melts. Notwithstanding, Narayan and Phani introduce us to amide melts, which have not received much attention previously. Early work on mixed amide–nitrate systems had been conducted by Russian workers seeking concentrated liquid fertilizer formulations; scarcely might they have guessed where their studies would lead.

Poole and his co-workers offer us an unusual insight into the world of chromatography—a field not normally associated with melt technology. However, the connection was always there. Now we can see how the development of new molten/liquid stationary phases can be adapted and deployed for the detection, separation, and analysis of the ever more complicated range of mixtures confronting analysts.

The chapter by Ohta and Waseda makes particularly interesting reading and is a model chapter in many ways. Not everyone is excited by thermal conductivity and diffusivity measurements, but be assured that they underscore very many industrial engineering applications, not least their implications for corrosion of structures. We recommend reading this chapter not only for its obvious content, but also as an exposition of the skillful methods described, including flash laser techniques.

Few workers have attempted to apply magnetic methods to melts, not least due to the high cost of the equipment required. It is entirely appropriate therefore that the chapter on this subject should come from one of its most eminent exponents, working in one of the best equipped and most pioneering laboratories in the world. Warren relates methods of susceptibility, nuclear magnetic resonance, and electron spin resonance in a thoroughly substantial yet clear chapter. Sadly, not many of us will be able to follow in his footsteps! It is important, however, that those techniques be set down for the benefit of the chosen few.

The wealth of detail, the pitfalls, and some expedients in calorimetry are amply explained by Gaune-Escard in her most helpful chapter. Many of us will have recourse to such thermodynamic data at some time and it is reassuring to know how, *in extremis,* we may be able to determine missing data for ourselves. Minority fields always suffer from a paucity of information; it may well be that the information contained in this chapter could be adapted for use at other temperatures, with other liquids, perhaps including, for example, the liquid gases.

The present volume goes to press in the year after California experienced a particularly severe earthquake. Few of us become actively involved in earth sciences, but equally few are not fascinated by the subject. The chapter on ultra-high-pressure techniques by Williams and Jeanloz cannot fail to interest readers—the authors are, of course, Californians. We trust they will not run short of funding for their important work. For the majority of uninvolved readers, it may be worth recalling that familiar chemical reactions taking place at atmospheric pressure can become transformed at geologic pressure, producing unsuspected reaction products. It is usually necessary to resort to original thermodynamic formulas in these situations and to proceed with caution.

The final chapter in this volume comes from Nelson and Kaun at Argonne, a school long associated with advanced battery and fuel cell development. These authors have selected two devices under present development to illustrate the many complexities of materials choice, fabrication technology, and cell testing.

The approaches described will prove invaluable to workers contemplating advanced cell development and show why progress toward a marketable version can be so protracted.

We thank all of the contributors to the present volume, especially for their forbearance in the chores of bringing their works together in this series. As always, we are glad to receive criticisms and comments from wherever they may come.

Robert J. Gale   David G. Lovering
Department of Chemistry   The David Graham Consultancy
Louisiana State University   34 Westland Road
Baton Rouge, Louisiana 70803–1804   Faringdon SN7 7EJ, Oxfordshire
                                                         United Kingdom

Contents

Chapter 2.   Physiochemical Properties of Liquid Organic Salts Using Chromatographic Techniques

Colin F. Poole, Kenneth G. Furton, Rena M. Pomaville, Salwa K. Poole, and Brian R. Kersten

Chapter 3.   Thermal Conductivity and Diffusivity Measurements

Hiromichi Ohta and Yoshio Waseda

Chapter 4.   Magnetic Measurement Techniques

W. W. Warren, Jr.

Chapter 5. Calorimetric Methods

Marcelle Gaune-Escard

Chapter 6. Ultra-High-Pressure Experimental Techniques

Q. Williams and R. Jeanloz

Chapter 7. Battery Construction, Testing, and Materials

Paul A. Nelson and Thomas D. Kaun

1

Amides and Amide Mixtures, with Special Reference to Electrochemical Properties

R. Narayan and K. L. N. Phani

1. Introduction

The amides are fairly stable over a wide range of temperatures, they exhibit low vapor pressures, they can be contained in glass, and they are environmentally safe. Further, the amides offer a wide range of physicochemical characteristics desirable in various reaction media and electrolytic solvents. Melting points of mixtures with various inorganic salts vary from 0 to over 100°C. Other properties such as viscosity, electrical conductivity, and solubility also can be changed by changing the composition of the mixtures and/or the temperature. Applications of amides/amide mixtures include thermal batteries, anodizing and electrodeposition baths, and photoelectrochemical cells. A few typical applications, with the temperature of interest and the amide mixtures employed, are presented in Table I. Preparation, purification, and handling of these melts for research, development, and commercial use in these applications is therefore of interest.

A review entitled "Electrochemistry in Amides," written by Reid and Vincent,[1] appeared 20 years ago. The amides considered were formamide (FA), N-methylacetamide (NMA), N,N-dimethylformamide (DMF), N,N-dimethylacetamide (DMA), N-methylformamide (NMF), and acetamide (AA). A good description of the physical properties, thermochemistry, thermodynamics, and conductometry of amides can be found in this and other reviews (see, e.g., Ref. 2). Since the appearance of this review, considerable work in DMF has been published.[3,4] Acetamide as a single solvent and as part of a eutectic has gained

---

R. Narayan and K. L. N. Phani • Department of Chemistry, Indian Institute of Technology, Madras 600 036, India. K. L. N. Phani's present address is Central Electrochemical Research Institute, Karaikudi 623 006, India.

1

TABLE I. Typical Applications of Amides

Amide/amide mixture	Temp. (C)	Application	Ref.
Urea + MNO_3 (M = K,Na,Li,NH_4^+)	45–80	Synthesis of α–alumina and related oxides	146
		Anodizing of Al, Ti	58
		Fertilizer	26
Urea + NH_4Cl	101.5–400	Hard soldering flux	147
Acetamide + KCl	80–140	Thermal battery	31
Urea + KCNO	135	Reaction media to produce cyanate—nitriding	148
Acetamide + Urea + NH_4NO_3 + $LiNO_3$	Ambient	Electrodeposition	38, 39
		Thermal battery	
		Photoelectrochemistry	78
		Liquid fertilizer	26–29
Urea + nitrates	140	Inhibition of passive layer formation in lithium cells	114, 115
Acetamide + KCNS	Ambient	Coordination chemistry	145
		Photoelectrochemistry	132
Acetamide	80–140	Electrochemical reaction media Electrodeposition	52, 57
		Thermal energy storage	149
N-Methylacetamide + NH_4NO_3	0	Anodizing, batteries, photochemistry	40
Tetramethylurea	Ambient	Polymer solvent	150
		Differentiating solvent for barbiturates	151
$NaNO_3$–KNO_3 + acetamide/urea	63.5	Anodizing	119
DMA, DMF		Lithium batteries	131
Ammonium-formate + formamide	120	Deposition of Cr and Mo from oxides	152
N-Methylacetamide		Deposition of *p*- and *n*- type semiconductors	153

recognition recently.[5–39] With the exception of AA and urea, the amides are liquids at room temperature and can be classified as room temperature molten solvents. A new group of room temperature melts is the binary and ternary eutectics of acetamide and urea with additions of an inorganic salt, and electrochemistry in these melts has been reported recently. Many of the room temperature eutectics are the outcome of a search for a liquid fertilizer by Soviet agricultural scientists.[26–29] A few have been studied for application in thermal batteries,[30–34] while recently the electrochemical behavior of several systems has been reported by McManis *et al.*[35–39] The potential utility of amide melts has also been referred to by Lovering and Gale.[40] Lovering has made some preliminary studies in a mixture of NH_4NO_3 + N-methylacetamide, which appears to form stable liquids, melting below 0°C.[40] These may usefully extend the range of electrolytic devices still further.

2. Properties

Some of the properties of the pure solvents, which will be useful to the electrochemist who is planning to do research in these solvents, are given in Table II. All amides have large dipole moments and dielectric constants. Substitution of both hydrogens on the nitrogen atom by methyl groups tends to decrease the dielectric constants. The acetamides are more stable thermally than the corresponding formamides.

The orientation polarization through hydrogen bonding results in a high dielectric constant, and this is unusually high for NMA.[41–43] The hydrogen bonding is believed to occur through the carbonyl oxygen. Raman spectroscopic studies[44] show that the NMA molecule occurs either in the *cis* or the *trans* form, there being no rotation about the C—N bond. The *cis* form would lead to dimerization, while a linear polymeric species can be obtained from the *trans* form. Since two hydrogen atoms are available in the parent amides, they are able to form linear chain polymers, dimers, and a number of intermediate forms. Acetamide is inferred to have a high proportion of dimers. Substitution of both hydrogen atoms rules out hydrogen bonding and hence gives rise to relatively low dielectric constant.

Tetramethylurea (TMU) has a wide liquid temperature range. The strong association due to hydrogen bonding usually observed in the amides is absent in TMU. The high boiling point is indicative of a high dipole moment (3.37 D). The moderate dielectric constant (23.45 at 25°C), low viscosity, and low specific conductance indicate TMU to be a useful solvent for studying the behavior of electrolytes. TMU has the same capacity as acetone for dissolving strongly ionized salts. Iodides are soluble while bromides dissolve to some extent in TMU, but alkali and alkaline earth chlorides are sparingly soluble. Solutions of iodides tend to become yellow, indicating possible side reactions. Perchlorates and tetraphenylborates show good solubilities in TMU. TMU was marketed as a polymer solvent for a number of years.

The amides are of particular interest to electrochemists because they are good solvents for both inorganic and organic species. In an investigation,[45] only a few of the 50 odd salts studied were found insoluble in NMA, while many organic compounds were easily dissolved. When no specific interactions are involved, solvation energy can be related directly to the dipole moments and hence to the dielectric constant. However, prediction of solvation characteristics based on bulk dielectric constant is not reliable since solvation depends as well on intramolecular attraction and association. The effects of the high electric field

TABLE II. Physical Properties of Formamide and Acetamides[a]

Amide	mp/fp (°C)	bp (°C)	density (g/cm^3)	Viscosity (cP)	Dielectric constant	Refractive index	Minimum specific conductance observed (r^{-1} cm^{-1})
Formamide	2.51–2.57	218.0	1.1292 (25°)	3.30 (25°)	109.0 (25°)	1.4475 (20°)	2.0.10^{-7} (25°)
Acetamide	79–81	221.5	0.990 (86°)	2.01 (86°)	62.5 (86°)	1.4274 (78°)	2.6.10^{-6} (94°)
			0.986 (94°)	1.63 (94°)	60.6 (94°)	1.4160 (110°)	
			0.980 (105°)[b]	1.32 (105°)[b]	56.3 (105°)[b]	1.4079 (130°)	8.8.10^{-7} (87.5°)
			0.967 (120°)	1.06 (120°)	74.0 (25°)[c]		
N-Methyl-acetamide	29.5	206.0	0.9503 (30°)	3.885 (30°)	186.0 (25°)[c]	1.4277 (30°)	1.0.10^{-7} (40°)
			0.9421 (40°)	3.019 (40°)			5.0.10^{-8} (30°)
N,N-Dimethyl-acetamide	−20.0	165.0	0.9366 (25°)	0.919 (25°)	37.8 (25°)	1.4351 (25°)	0.8.10^{-7} (25°)
			0.9232 (40°)	0.766 (40°)			
Tetramethyl-urea	1–2	175.2 176.5	0.9619	1.4	23.45		2.0.10^{-8}
Urea	132.8		1.323 (20°)			1.4840 (20°)	

[a]References 1, 2, and 59.
[b]Reference 35.
[c]Extrapolated.

due to an ion on the solvent behavior are also to be taken into account, and there may be specific ion–solvent interactions such as acid–base (Lewis) interactions.

Parker[45] has suggested that, in protic solvents, hydrogen bonding may be superimposed on ion–dipole interaction, particularly when anions are considered. It has been reported that, in water, anions have much greater solvation energies than cations of similar size.[46] In protic solvents, solvation energies decrease in the order $OH^- \simeq F^- >> Cl^- > Br^- > I^- > SCN^-$. This order is somewhat revised in aprotic dipolar solvents, since anion solvation involves only ion–dipole interaction. It may be pointed out that DMA is a better solvent than nitrobenzene, which has the same dipole moment. This is attributed to a delocalization of charges on the oxygen atom and a consequent high π bond order. In the amide group, there is a considerable overlap between the lone pair of the *p* orbital of the nitrogen and the carbonyl π orbital. The π bond order in the C—O bond is decreased with a corresponding increase in that of the C—N bond. In nitrobenzene, in spite of π delocalization, the charge is not concentrated near the oxygen atom. Alternative approaches to solvation structure characteristics and enthalpy of solvation can be found in several articles.[47–51]

An interesting observation made in the study of the reduction of benzaldehydes and nitrobenzaldehyde is that acetamide can solvate the —CHO group as water does.[52] The aldehyde is inferred to exist as a *gem*-diol in aqueous media according to the following equilibrium[53–55]:

$$
\begin{array}{ccccc}
\text{H} & & & & \text{H} \\
| & & & & / \\
-\text{C}=\text{O} & + & \text{H}_2\text{O} & \rightleftharpoons & -\text{C} \begin{array}{l} \diagup \text{OH} \\ \diagdown \text{OH} \end{array}
\end{array}
$$

Polarographic and cyclic voltammetric studies confirm the existence of two forms of the aldehyde, both in water and in acetamide.[52] The corresponding *gem*-diol in acetamide is

$$
\begin{array}{c}
\text{H} \\
| \\
-\text{C}-\text{OH} \\
| \\
\text{NHCOCH}_3
\end{array}
$$

3. Purity of Amide Solvents

3.1. Hygroscopicity

High hygroscopicity is common to most amides. In storing amides, wet floors, wet roofs, and moist storage conditions must be avoided. Storage where contact with acids or strong bases can occur is to be avoided. Water may be considered as one of the impurities in amide solvents. A 0.04 M concentration of water was found to be present in high-grade commercial formamide.[56] A measure of water content was made by weighing 5.6 g of the amide into a small beaker with a free surface of about 9 cm². The beaker was exposed to atmosphere under ambient conditions, and the gain in weight noted at various time intervals. A typical result was a gain of 0.2% in 1 h, 1% in 5.5 h, and 10% in 141 h.[56]

Water is totally miscible with molten acetamide. Solid acetamide, on exposure, absorbs considerable amounts of moisture, causing lumping. Hygroscopicity decreases with increasing temperature and is negligible in the range of temperatures 80–140°C. Wallace and Bruins[30] have studied the effects of relative humidity and temperature on the hygroscopicity of acetamide in different moist atmospheres. The residual water content of a carefully prepared melt is lower than that which can be detected with the Karl-Fischer method, and water is probably retained by impurities such as ammonium acetate and acetyl chloride. NMA is highly hygroscopic, whereas DMA is not.[56] The presence of water in amides has been shown to have negligible, positive or negative, effects in many studies. Deliberate additions of small amounts of water (1–5 mM) to purified and dried molten acetamide, containing a supporting electrolyte, were not found to affect the electrochemical behavior of several inorganic and organic depolarizers.[52, 57] Wallace and Bruins[30, 31] described in detail the effect of added water on the melting point of 2.5% KCl–acetamide mixture and found that addition of water was advantageous to the working of the fused acetamide thermal cells. Similarly, introduction of small amounts of water into urea–NH_4NO_3 melt at 63.5°C enabled the anodizing voltage for titanium and aluminum to be raised to the breakdown value of 83 V, with little or no gas formation below 75 V.[58] Water absorption was found to interfere in the chemistry of $LiNO_3$–amide melts. Exposure of the $LiNO_3$–amide melts to the ambient atmosphere for more than a few minutes has been reported to invalidate experimental results. The NH_4NO_3–amide melts are less hygroscopic.[35, 36] It was noted that corrosion reaction rates (and observed current densities or anode potentials) in the nitrate–amide melts are particularly sensitive to moisture. The $LiNO_3$–amide melt is hygroscopic, and much more negative anode potentials were noted when the melt was exposed to ambient atmosphere (10–15% relative humidity) for periods as short as 15 min. Continued investigation of the role of impurity levels of water will be an ongoing area of research, to rationalize these effects.

3.2. Other Impurities

Since two hydrogen atoms are available in the parent amides, these amides are able to form linear chain polymers, dimers, and a number of intermediate forms. Acetamide is inferred to have a high proportion of dimers.[42–44] However, the kinds of problems that these dimers and polymeric impurities can present (or how they may interfere with chemical or electrochemical reactions) are as yet unknown. Appreciable changes are believed to occur in the hydrogen bonding of NMA when small amounts of impurities are present.

DMA contains two impurities.[70] AnalaR NMA contains methylamine and acetic acid, apart from water. Formamide is easily hydrolyzed by acids and bases. It also reacts with peroxides, acid halides, acid anhydrides, esters, and alcohols, while strong dehydrating agents convert it to a nitrile. It is very hygroscopic. Commercial material often contains acids and ammonium formate.

Urea decomposes at temperatures below its melting point at atmospheric pressure. It contains mainly ionic impurities. Solid urea is stable at room temperatures, and biuret formation is negligible. In the presence of water, urea slowly hydrolyzes to ammonium carbamate, and eventually decomposes to NH_3 and CO_2. The possible impurities in commercial-grade urea are ammonia, biuret, cyanuric acid, and triuret.

TMU contains water as its chief impurity. Dimethylammonium chloride, a by-product formed in the preparation of TMU from dimethylamine and phosgene, is found to be troublesome. However, a different route of preparation may be employed to avoid formation of dimethylammonium chloride.[56]

All grades of acetamide are found to contain acetic acid as an impurity, which was inferred to be present from linear sweep voltammetric curves at a platinum indicator electrode[57] (Fig. 1). Most manufacturers mention the presence of acetic acid, up to 1%, in their specifications. Ammonium acetate is the other impurity present in commercial products.

3.3. Criteria of Purity

The criterion used for the detection of ionic impurities is the specific conductance and that for electroactive materials is the residual current in polarogra-

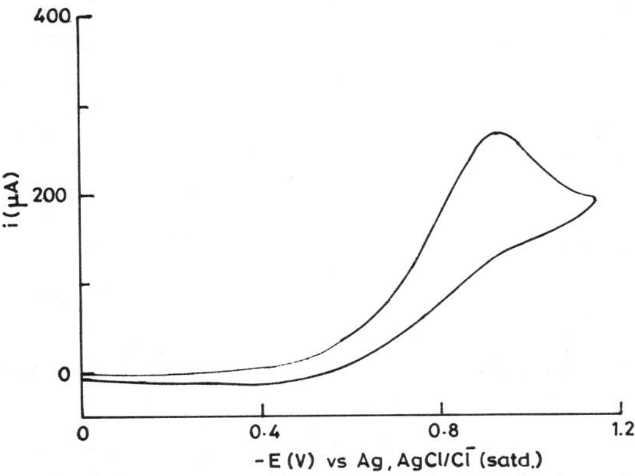

FIGURE 1. Cyclic voltammogram in molten acetamide (supporting electrolyte, $NaNO_3$); $v = 0.06$ V/s, platinum ($A = 0.098$ cm²).

phy or in linear-sweep voltammetry. In the absence of accurate electrocapillary or capacitance data, no criteria can be recommended for the presence of surface-active agents.

FA: Rosin[66] gives specifications and test methods for reagent-grade formamide. The requirements are listed below:

Specific gravity (25°C)	1.132–1.134
Freezing temperature	2°C
Miscibility with water	To pass test
Ammonia, amines, etc.	To pass test
Evaporation residue (max.)	0.1%
Water (max.)	0.3%

Notley and Spiro[59] found a specific conductance of less than 2×10^{-7} Ω^{-1} cm^{-1} with a freezing point of 2.3°C for a material specially prepared for electrochemical investigations.

NMA: The purity criteria recommended[56, 59] are (1) the polarographic range, determined with $0.1\ M\ (C_2H_5)_4NClO_4$ supporting electrolyte ($+0.35$ to -2.75 V), (2) water content of less than 0.01%, and (3) specific conductance of less than $2 \times 10^{-7}\ \Omega^{-1}$ cm^{-1}. Pure NMA has no detectable odor.

DMA: The purity criteria recommended[59,60] are:

Water content	$< 0.01\%$
Evaporation residue	$< 0.001\%$

For anhydrous DMA, the criteria are:

Water content	$< 0.005\%$
Evaporation residue	$< 0.001\%$

AA: The residual current is 0.4μA with the usual dropping mercury electrode at negative potentials prior to the solvent decomposition; with a glassy carbon electrode (area $= 0.05$ cm^2), it is less than 1μA, at a sweep rate of 50 mV/s.[52,57]

4. Purification and Preparation of Solvents

4.1. Acetamide

Weighted amounts of the solid can be placed in a vacuum desiccator for prolonged periods (4 h), transferred to the electrolytic cell with minimum exposure to the atmosphere, and melted in an inert gas atmosphere. The molten solvent is continuously purged with the inert gas for 30–40 min prior to electrochemical studies, since the solubility of oxygen has been found to be unusually high (3–4 mM).[5] Results are found to be reproducible if mercury and glassy carbon electrodes are used as indicator electrodes. However, from experiments with platinum electrodes, an electrochemically active species in appreciable quantities can be inferred to be still present (Fig. 1). The species is thought to be acetic acid. Crystallization from various solvents or distillation under atmospheric or under reduced pressure did not result in the removal of this acid. Prolonged pre-electrolysis with a large platinum electrode and a counter anode at a potential of -1.0 V with respect to the Ag/AgCl, Cl$^-$(satd.) electrode resulted in the current falling from a few milliamperes to 10 μA. Addition of small amounts of acetic acid to this purified melt resulted in linear-sweep voltammetry in an increased current at -1.0 V at a micro platinum electrode. Apparently, this reduction occurs at far negative potentials with mercury and glassy carbon electrodes, and it has not been reported elsewhere.

Walker and Johnson[61] recrystallized acetamide twice from alcohol to remove ammonium acetate and water. The acetamide specific conductance was then reported to be 4.3×10^{-5} Ω^{-1} cm^{-1} at 100°C. Kumler and Porter[62] recommend distillation followed by recrystallization, once from ethyl acetate and three times from chloroform. The sample is stored at 70°C in a desiccator. Belladen[63] recommends repeated distillations at 220 torr and shaking the middle fraction with anhydrous ether, followed by vacuum drying. The specific conductance is reported to be 1.5×10^{-5} Ω^{-1} cm^{-1} at 90°C. Hirano[64] recommends the use of benzene for recrystallization of distilled acetamide and drying in an Aderhelden drier for 5 h ($K = 88 \times 10^{-7}$ Ω^{-1} cm^{-1} at 83°C). Ethyl acetate and ethanol are the other solvents used for recrystallization prior to zone refining. Tsveniashvili[65] suggested the use of benzo-2,1,3-selenodiazolecobalt for electrochemical studies. This was added, in the form of admixtures with known drying agents, to the organic solvents to increase the degree of drying at the final stage of purification (e.g., distillation). The above compound is used especially in electrochemistry for determining trace amounts of H_2O in different solutions.

Berchiesi *et al.*[20] observed that acetamide–$Ca(NO_3)_2$ supercooled and remained as liquid for several days in a glass or porcelain container but crystallized when put in contact with a metallic surface. In their rheological studies, they observed also differences in the real part of the mechanical impedance for measurements performed in glass and metallic containers. All-glass containers may conveniently be used for electrochemical investigations.

4.2. Formamide

4.2.1. Removal of Water

A procedure for the removal of water using molecular sieves is given by Rosin.[6] A 5-cm column is packed with 1 kg of 3-Å molecular sieves in $\frac{1}{16}$ inch pellets. The column is heated electrically to about 60°C to increase the rate of water removal. Formamide is passed through at a rate of 0.2–0.25 liter/h. The water content was found to be about 0.008 to 0.01 M. The column can dry about 20 liters of the liquid. The sieves were washed with water, dried at 100°C, and heated at 360°C in a column, through which a stream of nitrogen was flowing.

Skold and Suurkuusk[156] purified analytical-grade FA for heat capacity studies by fractional freezing followed by drying over 4-Å molecular sieves and fractionally distilling under N_2 at reduced pressure.

4.2.2. Removal of Ions

Ions were removed by a mixed-bed resin that had been specially treated.[67] A 250-g portion of Amberlite IR-120 cation-exchange resin was well washed with water and with 2 M HCl and NaOH. It was then washed with ethanol until the effluent was colorless. After a second cycle of this treatment, the resin was soaked in water. Amberlite IRA-400 anion-exchange resin was treated similarly. The IR-120 resin was drained of water, put in a column, and washed with FA ($\sim$ 7 bed volumes) until the effluent contained less than 0.05 M water. One bed volume of FA containing 2 kg of sulfuric acid was allowed to stand 1 day in the resin. Two liters of 2 M H_2SO_4 in FA were passed through the resin. The anion resin was treated similarly with 2 kg of sodium formamide. The two resins were washed with formamide until the respective effluents were only slightly acidic or slightly basic. The resins were mixed, and about 6 liters of FA were passed through. Notley and Spiro[59] developed the above methods to produce formamide of low water and ion content in large laboratory amounts for electrochemical investigations. Chemically pure grade FA was found by Letaw and Gropp[68] to be polarographically pure in the range they studied, after being held at a pressure of 1 torr, or less, for 2 h.

4.3. N-Methylacetamide

N-Methylacetamide (mp 30°C, bp 70–71°C/2.5–3 mm) is available commercially. It is not suitable for use as received if a high degree of purity is required. To purify, NMA is fractionally distilled under vacuum, then fractionally crystallized twice from its melt.*

Polarographically pure NMA can be prepared[56] by two different methods, both of which are tedious and require extreme care. One of these methods is as follows. Crude solvent contaminated with acetic acid and methylamine is fractionally distilled through a short column under nitrogen at atmospheric pressure. The fraction boiling between 204 and 206°C is collected. This fraction contains acetic acid, methylamine, and an unknown impurity, which is removed by extraction with petroleum ether. At this point, water must be added to bring about a

---

* *Editor's note:* Considerable controversy concerning the purity (as determined from its melting point) of NMA was evident in the early literature. L. A. Knecht, in Ref. 155, has reviewed and rationalized those reports, indicating that very pure samples show time-dependent decomposition.

phase separation. To remove most of the water and petroleum ether which has dissolved in the solvent, the crude product is distilled at normal pressure, keeping the temperature below 130°C. The methylamine is removed as a nonvolatile sulfate by adding 10 ml of concentrated sulfuric acid per liter of solvent, and the crude product is vacuum distilled. The distillate is allowed to stand over freshly ignited calcium oxide for several hours to remove the acetic acid and water. Final purification is accomplished by vacuum distillation in the presence of small amounts of calcium oxide. The purified solvent must be preserved under an atmosphere of dry nitrogen. AnalaR NMA was purified before use in IR spectroscopic studies[69] by passing dry nitrogen through liquid at 130°C, to remove methylamine and acetic acid, and dried over molecular sieves at approximately 30°C for several hours.

4.4. N,N-Dimethylacetamide

N,N-Dimethylacetamide is thermally stable at its atmospheric boiling point and in the presence of acidic/alkaline substances. It decomposes above 350°C to dimethylamine and acetic acid. It is not very hygroscopic. It acts as a dehydrohalogenating agent with halogen compounds. Schmulbach and Drago[70] found that Eastman White Label DMA contained two impurities after shaking with BaO and fractional distillation through a 2.0 × 4.0 cm vacuum-jacketed Podbielmiak column at reduced pressure. The middle fraction boiled at 58.0–58.5°C at 11.4 torr and was retained. Chromatography indicated that the impurities were below 0.01%. DMA can be purified further by shaking with BaO for several days, refluxing with BaO for 1 h, and fractionally distilling under reduced pressure.

4.5. Urea

Urea is recrystallized twice from conductivity water using centrifugal drainage and keeping the temperature below 60°C. The crystals are dried under vacuum at 55°C for 6 h. Levy and Magoulas[157] prepared a 9 *M* solution in conductivity water (keeping the temperature below 25°C), filtered it through a medium-porosity glass filter, and added an equal volume of ethanol. The mixture was kept at −27°C for 2–3 days and filtered. The precipitate was washed and dried in air. Ionic impurities are removed from solution by ion exchangers.

4.6. Tetramethylurea

Pure TMU is best obtained by vacuum distillation through a packed column of some suitable drying agent such as granular BaO. The solvent contains less than 0.01% water. TMU has a convenient boiling point and can be made practically anhydrous by simple distillation. TMU was purified by Foerster and

Miller[71] (see also Ref. 54) for studies of its properties as a solvent for inorganic substances by drying over porous BaO and then distilling in an atmosphere of nitrogen.

5. Stability

All the amide solvents are fairly stable, the acetamides being more stable than the formamides. Decomposition of formamides generally involves hydrolysis by traces of water and is acid–base-catalyzed.[1] In contrast to NMF, the usefulness of which is limited to a certain extent to its instability, NMA is very stable when pure and can be stored for extended periods under nitrogen.

Acetamide is a white solid of hexagonal, short-prismaticine habit which melts at 81.1°C. A metastable form exists at lower temperatures. In the presence of HCl, the compound is hydrolyzed about twice as fast as other amides.

N-Methylacetamide is a white crystalline solid at room temperature which undergoes a phase change at about 2°C with an estimated heat of transition of 100 cal/mol.[72] The fact that NMA is highly hygroscopic necessitates transferring the solvent quickly and, if possible, under an inert atmosphere of nitrogen. For ultrapure solvent, a nitrogen atmosphere is a necessity. NMA solutions of trichloroacetic acid are unstable at 40°C, decomposing to chloroform and CO_2.

Urea tends to decompose at its melting point.

6. Safety

Except for DMA, the amide solvents reviewed here have negligible toxic effects.* Repeated exposure to relatively low levels of DMA was reported to result in liver damage[73] (see also Ref. 53). Du Pont cautioned that DMA should be handled with care and that contact of the liquid with the skin and the eyes should be avoided. Spills on the skin should be immediately and thoroughly flushed with water. Breathing of DMA vapors should be avoided.

Based on analysis of the curves for thermal decomposition of urea–NH_4NO_3 solutions, Rubtsov *et al.*[142] observed no danger of violent reactions, as long as the components remain in solution. It is recommended that a temperature of 120°C not be exceeded in any stage of the fertilizer manufacturing process.

No injuries have been reported in the use of FA, the minimum ignition temperature being 601°C.[143] DMA acts as a dehydrohalogenating agent with halogen compounds. The heat of reaction with highly halogenated substances is

---

* *Editor's note:* Many sources (e.g., Ref. 154) have listed acetamide as having experimental neoplastic carcinogenic properties in laboratory animals, although recent unsubstantiated evidence suggests that these may not be exhibited in *Homo sapiens*.

large, and the reaction may become violent, particularly in the presence of iron.[56]

7. Acid–Base Properties

Gruttner[74] was probably the first to associate the conductivity of pure molten acetamide with self-ionization or autoprotonolysis, as observed with water. The reaction is

$$2CH_3CONH_2 \rightleftharpoons CH_3CONH_3{}^+ + CH_3CONH^-$$

if the proton is solvated by one molecule of acetamide. This ionization lends an amphiprotic acid–base character to acetamide. The acetamide anion is also probably solvated. A detailed investigation of this acid–base behavior was undertaken by Guiot and Tremillon.[24] A hydrogen reference electrode was used in a potentiometric and linear-sweep voltammetric study.

Hydrochloric and nitric acids are fully dissociated in acetamide. Crystalline compounds of the formulas $HNO_3 \cdot CH_3CONH_2$ and $HCl \cdot 2CH_3CONH_2$ are prepared easily.[21, 22] These compounds are formed because of the slightly basic character of acetamide. Potassium acetamide, another easily prepared crystalline solid, is a strong base in acetamide. Acetamide serves as a differentiating solvent for acids, with $HClO_4$ being the strongest acid. The order of increasing acid strength in acetamide is as follows[20–22]:

$$p\text{-toluenesulfonic acid} < \text{picric acid} < HCl < HBr < HNO_3 <$$
$$HClO_4.$$

A hydrogen electrode at 98°C and 1 atm pressure of hydrogen gas is found to obey the relation, $E \simeq$ constant - 0.074 pH, for concentrations of the strong acid in the range of 3×10^{-3} M to 3×10^{-2} M, with the ionic strength kept constant using 1 M tetramethylammonium chloride.[24,25] A normal hydrogen electrode can be set up using these results. The ionic product of molten acetamide at 98°C was determined by measuring the potential of the hydrogen electrode, in the presence of various concentrations of the strong base and various ionic strengths, and was found to be $10^{-14.6\pm0.1}$ at 94°C. Jander and Winkler[21, 22] had earlier reported a pK value of 10.9 at 94°C for ionization. A molybdenum electrode was used in their potentiometric study. The response of this electrode to varying hydrogen ion concentrations was reported to be 90 mV per decade, not 74 mV. Further, the mechanism of establishment of equilibrium at this electrode is not known exactly.

Guiot and Tremillon[24, 25] also have attempted a tentative correlation between the pH scales in molten acetamide and in water. Of the various methods proposed, they adopted a method based on the potential of the ferrocene–ferricenium ion couple and using the acidity scale $R(H)$ proposed by Strehlow.[75] Redox potentials of the couple in aqueous media have been measured in the temperature range 50–80°C and extrapolated to 98°C. The couple behaves reversibly in acetamide at 98°C. Ferrocene is apparently soluble to $1.2 \times 10^{-2} M$ in molten acetamide ($10^{-4} M$ in water). An acidity function is defined by $R°(H) = -FE_s/2.3RT$, with E_s the potential measured in the solvent S; $E_s = -\log (r_H)_s$ [$(r_H)_s$ is an activity coefficient and measures the differences between the activities of a proton in the solvent S and in water.] Measurements in water were of limited accuracy because of the low solubility and the instability of potentials over periods exceeding 15 min. The ionic strength had to be kept at 1. The redox potentials at a pH of 0 in acetamide and a pH of 0 in water are 0.641 and 0.280 V, respectively. Therefore, the relative acidity is 4.7, and the authors concluded that hydrogen ions are more strongly solvated by acetamide than by water and that acetamide is more basic than water at 98°C. They also have reported the dissociation constants of 15 weak acids, based on a potentiometric titration with potassium acetamide. A constant ionic strength was maintained using Tetramethyl ammonium perchlorate.[76, 77] The values obtained were compared with the corresponding dissociation constants in water at 98°C, calculated using the equation of Harned and Robinson. It was concluded that acids of the type HB^+ are stronger in acetamide than in water, while acids of the type HA^- are slightly weaker. Since the dielectric constants of the two solvents are approximately the same at 98°C, differences in solvation play an important role.

Pournaghi *et al.*[76, 77] studied the apparent acidity of metal ions in acetamide at 98°C. Electrodes of different glass materials were used in these potentiometric procedures, with the best results obtained with lithium glass. Hg(II) is a diacid, with pK_1 being much higher than pK_2 ($pK_2 = 4.8$). Bi(III) is a triacid, with the third dissociation corresponding to a very weak acid, while Cu(II) is a weak diacid. The pK_1 and pK_2 are reported to be 5.3 ± 0.15 and 11.3 ± 0.15, respectively.

8. Acetamide Eutectics

There has been, in the recent past, a spurt in the literature on acetamide mixtures with organic and inorganic species, the mixtures having melting points as low as $-5°C$.[26–29, 35, 36] The mixtures are advocated as good solvents. Several of these mixtures tend to supercool. The supercooling phenomena depend on the presence of various cations and anions. Supercooling is observed with LiCNS/NaCNS/KCNS but not with HCOOK/CH$_3$COOK and is better manifested in the presence of Na^+ ion.[35, 36] No supercooling is observed with tetraalkylammonium salts in the melt. Mixtures with very low melting points may also afford solvent systems with a large range of stability in the liquid state and help to study various parameters as a function of temperature (refer also to Chapter 2 in this volume). These mixtures tend to be more ionic than the usual organic solvents and also react less with air and water. In addition, these eutectics may be expected to function well in thermal battery systems.

The advantage of a low melting temperature can be offset by a lower capacity to dissolve inorganic salts. Variations of ionic strength or acid–base nature also may be not possible in the presence of a large concentration of an inorganic salt. The systems containing urea become unstable in the presence of even small quantities of NaOH or CH_3COONa (1.0 M). Ammonia is rapidly evolved.[78] The electrochemical window becomes narrower due to redox processes involving NO_3^- and NH_4^+ ions.[78]

Bleshchinskaya *et al.*[26, 28, 29] obtained the phase diagrams for binary and ternary mixtures containing acetamide. Eutectic compositions and temperatures are shown in Table III. Another series of eutectics has been investigated by the Italian group of workers (Table III). Some of these room temperature eutectics are the outcome of a search by Soviet agricultural scientists for a liquid fertilizer. A few have been studied for application to thermal batteries,[30, 31, 37, 39] while recently the electrochemical behavior of several systems has been reported by McManis *et al.*[36–39] The potential utility of amide melts also was referred to by Lovering and Gale.[40] Photoelectrochemical studies also are feasible in the acetamides, as indicated by work in our laboratories.[78] The acetamide–KCNS eutectic at room temperature is highly suitable for the coordination chemist, with S, N, and O being available as coordination sites. Kerridge[145] at Southampton University has carried out considerable work in this direction. Addition of alkali metal nitrates lowers the melting point of the urea–acetamide eutectics. The true melting points decrease in the order $Li^+ < Na^+ < K^+$. Melts containing Li^+/Na^+ show glass transition near $-40°C$ in differential scanning calorimetry. No spontaneous nucleation can be observed in spite of repeated transitions across the glass transition temperatures, indicating the remarkable stability of the liquid phase. The eutectic containing 0.2 mole fraction of $NaNO_3$ can be stored quiescent for several days at $-20°C$, while the eutectic containing $LiNO_3$ cannot be crystallized at $-20°C$ even after storage for months.

Poberezhyenk *et al.*[144] evaluated the use of KNO_3–$NaNO_3$–$NaNO_2$ melts containing acetamide or urea as heat exchange materials. Melting point–composition diagrams are available. Urea eutectics with $LiNO_3/KNO_3$ and $LiNO_3/NaNO_3$ are stated to be stable to $160–170°C$. Decomposition occurs rapidly beyond this temperature.

Two classes of binary systems of acetamide and electrolytes have been distinguished depending on the types of ions present:

1. The anion is ClO_4^-, NO_3^-, or CNS^- (derived from strong acids) and the cation is from an alkali metal or an alkaline earth metal.
2. The anion is CH_3COO^- or $HCOO^-$ and the cation is from an alkali metal.

The eutectic temperatures of the first class are generally lower than those of the second class, and the stability is higher for the second than for the first class. A wide range of concentrations of sodium or calcium salts is possible without crystallization occurring from the liquid and a stable system can be produced.

TABLE III. Properties of Amide + Inorganic Salt Eutectic Mixtures

Eutectic system	mp (°C)	Viscosity (cP)	Conductivity (r^{-1} cm^{-1})	Density (g/cm^3)
Acetamide (65%) + urea (35%)	54–56	—	—	—
Acetamide[a]	79–81	2.01 (85°C)	2.6×10^{-6} (98°C)	0.9980 (85°C)
Acetamide (60%) + NH$_4$NO$_3$ (40%)	38.0	—	10^{-3}	—
Acetamide (37.4%) + urea (28%) + NH$_4$NO$_3$ (34.6%)[b]	7.5	42.7 (27°C)	10^{-3}	1.2 (27°C)
Acetamide (73.7%) + KCNS (26.3%)	26.0	118.4	(2.747–3.15) $\times 10^{-3}$	1.2105
Acetamide (70.15%) + tetraethylammonium bromide (29.85%)	105	—	—	—
Acetamide (79.6%) + LiNO$_3$	16.5	656.2	4.45×10^{-4}	—
Acetamide (87.6%) + Ca(NO$_3$)$_2$	41.3	191.8	9.42×10^{-4}	—
Ammonium nitrate (53.5%) + urea	~46.0	—	4.45×10^{-4}	—
Formamide (78.5%) + KCNS (21.5%)	−20.5[c]	—	—	—
N-Methylacetamide + NH$_4$NO$_3$	< 0	—	—	—

[a] See Table II.
[b] Viscosity determined by the falling ball method and by a capillary viscometer. The viscosity apparently decreases by seven times when the temperature is changed to 85°C from 25°C. The free energy of activation for diffusion of oxygen and anthraquinone is found to be 7 kcal M^{-1} from diffusion coefficient measurements [78] Viscosity of the ternary melt acetamide, urea and lithium nitrate is quoted as 195 ± 6 cp at 25°C [39].
[c] Incongruent melting point.

The presence of Li$^+$ ion results in a complex behavior during crystallization. Crystallization is regular with the second class of eutectics.

9. Electrochemistry in Amides

9.1. Supporting Electrolytes

Up to 0.6 *M* KNO$_3$, 1 *M* NaClO$_4$, 1 *M* tetramethylammonium nitrate, or 2 *M* CH$_3$COONa can be used conveniently in molten acetamide. Sodium acetate makes the solution basic. This is inferred from the total removal of dissolved oxygen[57] by pyrogallol in solutions containing sodium acetate and from the progressive shift of the $E_{1/2}$ reduction potentials of nitro and other compounds which involve protons in the reduction process.[52] The basic character of the melt can be increased by the addition of NaOH (up to 0.1 *M*), and the acid character by *p*-toluenesufonic acid (PTSA). The negative limit of the potential window is then reduced considerably in the presence of 0.5 *M* PTSA, due to hydrogen evolution. It has been reported that acetamide will hydrolyze in the presence of NaOH and produce ammonia, but we have not observed any ammonia. However, ammonia is seen to evolve even at room temperature from the acetamide–urea eutectic with addition of a very small amount of NaOH or sodium acetate. KNO_3, PTSA, and CH_3COONa can be used at concentrations up to 0.5 M in formamide.

9.2. Working Electrodes

The dropping mercury electrode is found to function satisfactorily in molten acetamide at 85°C at ordinary pressure for polarographic and pulse polarographic studies. The hanging mercury drop (Metrohm) is highly satisfactory for chronopotentiometric and cyclic voltammetric studies. Reproducibility of results with glassy carbon electrodes is generally high except when strong adsorption of reactants/products is operative. When protons are involved in the sequence of an electrochemical reaction, pretreatment of the glassy carbon can play a part in the course of the reaction. Examination of the carbon surface by ESCA revealed the presence of graphitic, carboxyl, and phenolic (quininoid) groups while new carbonyl groups appeared on anodic treatment of the electrode and these carbonyl groups catalyze electrochemical reduction of oxygen[160]. The pretreatment generally given is hand polishing using different grades of emery paper or β-alumina powder and degreasing with acetone, followed by repeated washing with water and then drying. The carbon rod is snug-fitted into a Teflon cylinder with the surface flush with the cylinder edge. A Teflon dispersion can be used to seal any leaks. The Teflon cylinder can then be sealed into a glass tubing using Araldite. Alternatively, the glassy carbon cylinder is inserted into a glass tubing with 3 to 4 mm protruding. The sealing is then done with a thick coating of Epoxy-Patch/clear, with the surface of the carbon electrode left exposed. External electrical contact is made through a metal (copper, platinum, nickel, steel) wire dipping into a few drops of mercury placed on top of the carbon rod. These electrodes behaved reproducibly in molten solvent over a few months. The geometrical area of a typical electrode is 0.075 cm^2, while the area inferred from linear-sweep voltammetric studies of the reduction of Cd^{2+} ion is 0.1 cm^2. Reproducibility of electrochemical behavior has been checked with carbon electrodes of different areas.

The use of platinum electrodes to study reduction of depolarizers is limited, due to interference by acetic acid, which is invariably present in the melt. Further, definition and reproducibility are very poor even with systems which give highly reproducible results with mercury and glassy carbon electrodes (e.g., Cd^{2+}, benzophenones, nitrobenzenes). The pretreatment, when reproducible results can be obtained, (e.g., O_2, quinones) is dipping in a 1:1 nitric acid solution for 15 min, washing, drying, and flame polishing to dull redness.

9.3. Reference Electrodes

Several redox systems have been employed as reference electrodes in acetamide-based electrolytes. (Table IV). Quasi-reference electrodes consisting of a piece of a precious metal such as Ag or Pt in contact with the metal are also

TABLE IV. Reference Electrode Systems Employed in Acetamide-Based Electrolytes

System	Medium	Temperature (°C)	Potential measured (V)	Remarks	Ref.
Hydrogen electrode	Acetamide	98	$+0.02 \pm 0.004$[a]	Stable; unstable if temperature exceeds 100°C	24
Molybdenum electrode	Acetamide	98	—	Reasonable pH response; Nernst slope, 90 mV	21–23
Aluminum electrode	Fused acetamide–KCl salt ($+0.1\ M$ AlCl$_3$)	90	1.150 (open circuit potential) vs. Ag,AgCl electrode	Responds to Al^{3-} ions	31
Silver–silver chloride electrode	Acetamide	85	—[b]	Stable for long periods	57
	Acetamide + urea + LiNO$_3$ or NH$_4$NO$_3$ or KCNS	25–80		Stable for long periods	52,78
Silver wire quasi-reference electrode	Acetamide + urea + NH$_4$NO$_3$ (or LiNO$_3$)	23	-0.10 vs. Cp$_2$Fe/Cp$_2$Fe$^+$ couple	Used only for nickel studies	38
Pb/PbO quasi-reference electrode	Acetamide + urea + NH$_4$NO$_3$ or LiNO$_3$	23	-0.750 vs. reversible Cp$_2$Fe/Cp$_2$Fe$^+$ couple	Reasonable stability for period of several hours in NH$_4$NO$_3$ melts	35

Pb-wire quasi-reference electrode	$Pb(NO_3)_2$ in $LiNO_3$ + acetamide + urea	23	-0.70 vs. Cp_2Fe/Cp_2Fe^+ on Pt	Stable	39
Cu-wire quasi-reference electrode	Acetamide + urea + $LiNO_3$ (or NH_4NO_3)	23	0.30 vs. Cp_2Fe/Cp_2Fe^+ couple	Suited only for copper deposition and dissolution studies	38
Cp_2/Cp_2Fe^+ reversible couple	Acetamide + urea + NH_4NO_3 (or $LiNO_3$) + $Pb(NO_3)_2$ 0.001 mol)	23	0.75 vs. positive of the Pb-wire quasi-reference electrode	Stable	38
	Acetamide	98–130	$0.641 \pm .08$ vs. hydrogen electrode see pp. 19, 28	Stable	24, 25
	Acetamide + KSCN	25–30	—	Satisfactory	78
Hg-Pool electrode (Hg covered with satd. $NaClO_4$) via nonaqueous salt bridge	$NaClO_4$ (satd.) in dimethyl-acetamide, D or F	25	Not reported	Fairly stable; uncertainly in junction potential	80, 81, 84
Calomel (nonaqueous) electrode[c]	Dimethylacetamide	25	—	—[d]	81

[a]Arbitrary scale; extrapolated value.
[b]Polarographic $E_{1/2}$ values measured vs. this electrode are similar to those measured in aqueous solutions against SCE.
[c]Preparation similar to that of the acetone calomel electrode; see (a) P. Arthur and H. Lyons, *Anal. Chem.* **24,** 1422 (1952); (b) T. A. Pinfold and F. Sebba, *J. Am. Chem. Soc.* **78,** 2095 (1956).
[d]No detailed investigation.

used and function in the same way as a pool of mercury—a stable and reproducible potential is produced.[80, 81] An aqueous calomel electrode can be used, connected through an appropriate salt bridge to the nonaqueous system. This may result in an ill-defined aqueous/nonaqueous junction, an indeterminate liquid junction potential, and possible contamination of the nonaqueous system by traces of water. It is more difficult to use such a combination at elevated temperatures.

A hydrogen electrode[22, 23, 82] has been used as a primary reference electrode in several investigations. It is similar in construction to the hydrogen electrode used in aqueous systems, and, apparently, it has a reproducible potential in the amides. However, it is also reported that the electrode is easily poisoned and can take part in the acid–base-catalyzed decomposition of the solvent.

A Hg/Hg(II) electrode has been successfully used in DMF.[83, 84] However, a freshly prepared electrode would seem necessary for reproducibility within ± 1 mV, since the reference electrode potential changes owing to the reduction of Hg(II) ions taking part in the oxidation of the solvent. The order of stability of this electrode in other amide solvents seems to be the same as in DMF. A Cd/CdCl$_2$ electrode has been reported to be most successful in formamide by Pavlopoulos and Strehlow,[82] who consider both calomel and Ag/AgCl electrodes unsatisfactory. Considering the reproducibility reported in formamide, this electrode may be suitable in other formamides. Wallace and Bruins[31] reported the use of an aluminum electrode, which responds to Al^{3+} ions, in acetamide.

An Ag/AgCl electrode,[82, 85−90] either introduced directly into the amide or connected through a salt bridge, has been reported by several workers. While one set of authors suggested that Ag$^+$ ions take part in the oxidation of the solvent, many others have found reproducibility with this reference electrode. The $E°$ value for the cell H$_2$/HCl/AgCl, Ag has been reported by Mandel and Decroly[85, 86] and over a range of temperatures by Agarwal and Nayak.[87, 88] Both sets of authors reported that an extrapolation to zero time becomes necessary owing to the catalytic decomposition of formamide by HCl, with consequent potential drifts. Both Ag/AgBr and Ag/AgCl electrodes have been found suitable in NMF.[89] In DMF, Ag/AgCl functioned satisfactorily at HCl concentrations not exceeding 0.03 M.[90] Time-dependent variation of the potential can be associated with the increased solubility of the silver salt in these solvents, particularly in the presence of high halide concentration.[91] This problem of solubility also is present in formamide. Wallace and Bruins[30] reported the successful use of the Ag/AgCl electrode in acetamide, but Jander and Winkler[21−23] reported that decomposition occurred without a stable potential being obtained. Guiot and Tremillon[24] found this electrode satisfactory in acetamide over a period of several months. The reference electrode compartment was separated from the working electrode compartment.

We have found the Ag/AgCl, Cl$^-$ (saturated in acetamide) electrode to perform satisfactorily in acetamide at 85°C. The reference electrode was confined to a thin-walled glass bulb (1 ml), and contact with the main compartment was made through a wetted asbestos fiber (Fig. 2). The internal resistance was low enough to allow the use of ordinary potentiostats, and the reference electrode potential was stable over 48 h, as indicated by the reproducible potentials of decomposition of the solvent and the $E_{1/2}$ and E_{peak} potentials of depolarizers.

The ferrocene–ferricenium ion redox system, as indicated earlier, has been found to be highly reversible at a platinum electrode in acetamide at 98°C and in aqueous systems over a wide range of temperatures. This system can be tried in other amides, also. A potential difference of 0.641 ± 0.08 V has been reported for the ferrocene electrode and the hydrogen electrode in molten acetamide, and cathodic limits of -1.0 and -0.8 V with respect to Ag/AgCl and Ag/AgCl, NMe$_4$Cl reference electrodes, respectively, have been reported. The use of the ferrocene references electrode for nonaqueous media has been advocated by Bard and Faulkner.[92]

The quasi-reference electrode using molybdenum was reported by Jander and Winkler[21–23] to give reasonable response to changes in pH in acetamide, but Guiot and Tremillon[24] have subsequently pointed out the inaccuracy of the results obtained with this electrode.

A silver electrode was employed by McManis *et al.*[38] as a quasi-reference electrode in the ternary urea–acetamide–NH$_4$NO$_3$ system. The potential of this electrode (internal reference) was reported to be reproducible at 0.10 V. This study also investigated the deposition and dissolution of nickel. A copper wire electrode was similarly used in the study of copper deposition and was reported[38] to measure a potential of -0.30 V versus the ferrocene electrode. A Pb/PbO electrode[37–39] has been used by isolating a lead wire in a melt-filled

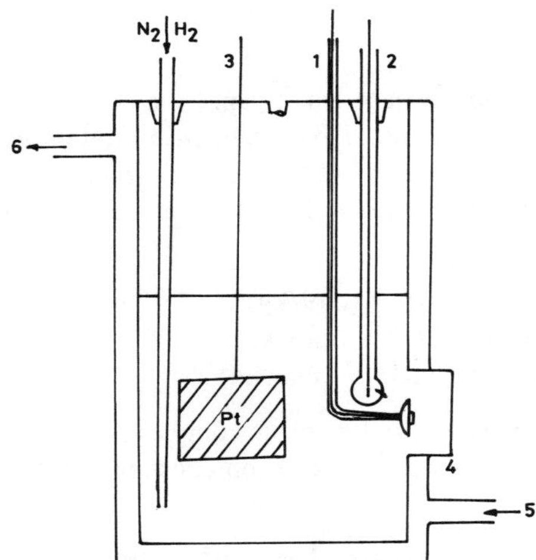

FIGURE 2. Double-walled cell for photoelectrochemical studies. 1, Working electrode (semiconductor electrode); 2, reference electrode [Ag/AgCl, Cl⁻(satd.)]; 3, counter electrode (platinum foil); 4, quartz window; 5, thermostating liquid inlet; 6, thermostating liquid outlet.

capillary tubing, and it was reported to work satisfactorily in the nitrate–amide eutectic. The potential is -0.70 V versus the ferrocene electrode, but this electrode was not found to be satisfactory in melts containing $CuCl_2$ or $NiCl_2$. We have found the bulb type Ag/AgCl, Cl^- (saturated in amide) electrode to function well in the ternary eutectic over a wide range of temperatures.[78] The reference electrode is more positive by 0.040 V than the Ag/AgCl, Cl^- electrode used in molten acetamide, at the same temperature of 85°C.

In summary, the Ag/AgCl, Cl^- couple perhaps is the most satisfactory electrode when used in a separate compartment and suitably connected to the main compartment.

9.4. Double-Layer Studies

Reports of measurements to understand the double layer (electrocapillary and capacitance measurements) are very few in amides. Even these are mostly in formamides.[93–97] Enigmatically, a hump in the capacitance–potential plots is seen in FA and NMF but not in DMF. If the hump is the consequence of preferred solvent orientation, it may be proposed that FA and NMF are adsorbed with the negative ends of their dipoles pointing inwards. Payne's results indicate the opposite orientation,[95, 96] since a positive shift of the potential of zero charge is observed upon addition of FA and DMF to aqueous solutions. Anions seem to be more strongly adsorbed from solutions in NMA than from solutions in NMF, the order of specific adsorption being $I^- > Br^- > Cl^- > PF_6^- > ClO_4^- > NO_3^- > BF_4^-$. The surface charge of a germanium electrode in formamide has been measured by a new technique named the "rapid charging curve technique."[98] There are hardly any measurements of double-layer properties in molten acetamide.

9.5. Complex Formation

Bartocci's group[6–8] in Italy has published a series of papers on the formation of silver halide complexes in molten acetamide. Potentiometric titration of Ag^+ ion with potassium halides was carried out at 87°C. Acetamide nitrate (0.05 mol/kg) was added to the solvent to prevent precipitation of silver acetamide. The solubility products obtained were

$$K_{sp}(AgCl) = (1.9 \pm 0.5) \times 10^{-8} \ (mol/kg)^2$$
$$K_{sp}(AgBr) = (8.5 \pm 0.1) \times 10^{-10} \ (mol/kg)^2$$
$$K_{sp}(AgI) = (2.2 \pm 0.1) \times 10^{-12} \ (mol/kg)^2$$

Complex formation also has been investigated; polynuclear complexes are not seen to be formed. A typical set of overall complete formation constants for the Ag–I system is

$$\beta_1 = 1.1 \times 10^9 \ kg/mol, \quad \beta_2 = 4.98 \times 10^{10} \ kg^2/mol^2 \quad \beta_3 = 5.5 \times 10^{11} \ kg^3/mol^3$$

The stability constant of the first complex is unusually high. $AgCl$, $AgCl_2^-$, $AgBr$, $AgBr_2^-$, and $AgBr_3^{2-}$ are the predominant species found with the other halogens. The solubility product of $AgSCN$ is $(2.5 \pm 0.06) \times 10^{-18}$ mol^2/kg^2. $AgSCN$, $Ag(SCN)_2^-$, and $Ag(SCN)_3^{2-}$ species are found with excess of SCN^- and are less stable than the corresponding chlorocomplexes.[7, 8] It is concluded that Ag^+ ion is more strongly solvated by acetamide than by water. A study of mixed-complex formation, also based on a potentiometric procedure, has been made.[8] $Ag(SCN)Cl^-$ appears to be the predominant complex.

Polarography and cyclic voltammetry have been used to study complex formation in molten acetamides.[57] Halides form a series of complexes with Cd^{2+} ions, $CdBr_3^-$ and CdI_2 being the highest complexes formed. Thiourea forms a weak complex with Cd^{2+} ions. Acetate ions shift the $E_{1/2}$ potential of Cd^{2+} ion to more negative values, the $E_{1/2}$ - log [X] variation being linear. Thiocyanate ions shift the $E_{1/2}$ potential to more positive values, unlike the observations in aqueous media. The limiting currents are considerably reduced. Stability constants of the thiocyanate complexes could not be evaluated using the methods suggested in polarographic literature. Ni^{2+} ion reduction at the mercury electrode is a more reversible process than in water. Acetate ions shift the $E_{1/2}$ potentials to more negative values and decrease the reversibility of the process. Thiourea, thiocyanate ions, and bromide ions shift the $E_{1/2}$ potential to more positive values, with the thiocyanate ions increasing the reversibility of reduction.

9.6. Voltammetric Studies

Several metal ions,[57, 72, 78, 95, 96] oxygen,[100–102] iodine,[103] sulfur,[104] and organic species[52, 105–107] have been investigated in molten acetamide using conventional techniques of polarography, chronopotentiometry, and cyclic voltammetry. Most inorganic salts are found to have appreciable solubility, except for a few, such as Mn^{2+} salts and ferrocyanides.

The electrochemical windows for various electrode materials in acetamide containing different supporting electrolytes are shown in Table V.[52] The potentials are with respect to the $Ag/AgCl,Cl^-$ (satd.) reference electrode in molten acetamide at 85°C.

The negative limit with platinum in the presence of supporting electrolytes such as KNO_3 and CH_3COONa could be more negative if acetic acid is effectively removed and prevented from forming in the acetamide. The hydrogen evolution reaction (h.e.r.) sets the negative limit in acidic melts.[35, 52] Gas bubbles are seen to collect at the electrode surface. The negative limit is associated with the reduction of the solvent, a process similar to the hydrogen evolution reaction from water, while the oxidation process at the positive limiting potential appears to be considerably more complicated. Evolution of ammonia and polymerization of the amides are reported. The results of the electrochemical reduction of

TABLE V. Electrochemical Windows Available in Molten Acetamide at 85°C[a]

Supporting electrolyte	Electrode material		
	Hg	Pt	GCE
1.0 M CH$_3$COONa	-0.1 to -1.6	$+0.3$ to $+0.8$	$+0.7$ to -1.9
0.1 M NaOH	-0.5 to -1.7	0.0 to -1.6	$+0.3$ to -1.8
0.5 M PTSA	$+0.4$ to -0.8	$+0.6$ to -0.3	$+0.8$ to -0.9

[a]References 57, 58, and 96.

various depolarizers bring out the striking similarity in the behavior of water and acetamide.

Oxygen is highly soluble in molten acetamide (4 mM) at 85°C and undergoes a two-step, irreversible reduction at the mercury electrodes.[5] Two electrons are transferred in the first step. The current magnitudes of the second step, $i_{d,2}$ in polarography and $1_{peak,2}$ in linear sweep voltammetry are smaller than those of the first step, and this has been shown to be the result of the dissociation of the peroxide formed in the first step, regenerating oxygen at the interface. Oxygen undergoes irreversible reduction in one step at the platinum electrode, and kinetic control is inferred from the $i_p-v^{1/2}$ variations. The second-order dissociation of peroxide formed in the electrochemical reduction has been considered as a disproportionation process, and the kinetics of disproportionation has been evaluated from the linear-sweep voltammetric[100] and chronopotentiometric[101] results [$K_d = 2 \pm 1 \times 10^3$ M^{-1} s^{-1} at 85°C]. Observations similar to those with platinum are made with glassy carbon, n–MoSe$_2$, n–MoS$_2$, n–WSe$_2$, n–FeS$_2$, and p–MoSe$_2$ electrodes.[78]

Iodine is soluble in molten acetamide but not stable. Complete bleaching of 10^{-4} M solutions occurs within 48 h. An excess of I$^-$ ion is found to retard the bleaching process and provides stability for over 24 h.[78] The I$_2$ molecule is thought to dissociate into a positive I$^+$ and a negative I$_3^-$ species. The formal potential of the I$_3^-$/I$^-$ couple is reported as 0.269 V vs. the Ag/AgCl,Cl$^-$ (satd.) electrode.[103] The anodic oxidation of I$^-$ ion to iodine at the platinum electrode is postulated to occur in two steps through a stable I$_3^-$ species. The current–voltage variations depend on whether they are recorded from the negative or from the positive side, and the difference is attributed to the presence or absence of adsorbed iodide/iodine on the platinum surface.

Iodine is found to be rapidly decolorized in formamide containing PTSA, even at room temperature.[78] The solubility and stability of iodine in formamide containing KNO$_3$ or CH$_3$COONa can be increased by addition of KI.

Sulfur is soluble in molten acetamide, but the reduction of sulfur in molten acetamide is a complicated process.[104] The sulfur species present in the melt appear to vary both with the acid–base nature of the system and the oxygen content. Time-dependent variation of the reduction current and the color of the solution is observed even in deaerated solutions. Sulfur is apparently present as an S_8 species in strongly acidic solutions and undergoes a two-electron reduction process at the platinum electrode. The scheme suggested for the reduction of sulfur in other solvents such as acetonitrile seems to be applicable, but further studies are necessary to establish the mechanism. No defined reduction steps can be observed with the glassy carbon electrode whereas a strong surface interaction with the mercury electrode is indicated.

The behavior of several metal ions (Tl^+, Cd^{2+}, Ni^{2+}, Pb^{2+}, In^{3+}, Cu^{2+}, Ag^+, U^{6+}) has been investigated[57,108] using various electrochemical techniques. We are in agreement with the conclusion of Reid and Vincent,[1] in 1968, that there seem to be few reasons for undertaking analytical polarography in amides. Only a few more ions such as Sr^{2+}, Ca^{2+}, Ba^{2+}, and Li^+ perhaps can be analyzed in DMF. The previous reviewers had quoted that maxima are not observed in FA, DMF, and NMF with reasonable concentrations of depolarizers, while the observation of maxima is a common feature in DMA. Unusual maxima or minima are observed in polarographic studies in acetamide.[57] Gelatin is insoluble in amide solvents while the other usual maximum suppressors, such as methyl cellulose and congo red, do not suppress the maxima. Poly(vinyl chloride) (PVC) in the concentration range of 0.001% to 0.01% proved successful in suppressing the maxima in DMA.

The reduction of Tl^+ ion in molten acetamide is found to be a reversible process.[57] Voltammetric studies of deposition and anodic stripping at the mercury and the glassy carbon electrode indicate a degree of quasireversibility in systems containing Cd^{2+}, In^{3+}, and Cu^{2+}. U^{6+} ion is found to undergo reduction to U^{5+} at the mercury electrode in a medium of low acidity and to U^{4+} in a solution of higher acidity.[99] U^{5+} ion can be further irreversibly reduced to U^{3+} ion. U^{4+} ion can be oxidized at a platinum electrode. U^{5+} ion also undergoes disproportionation.

An irreproducible maximum and minimum are observed in the polarographic reduction of Cu^{2+} ion in molten acetamide[57] containing 0.1 M KNO_3. The corresponding i–E variations in linear-sweep voltammetry are complicated. The usual maximum suppressors have very little effect.

An unusual observation in the polarographic reduction of In^{3+} ion from molten acetamide,[57] with a startling similarity to observations in aqueous media, is the occurrence of a minimum over a range of potentials beyond the limiting current (Fig. 3). This minimum is produced with the addition of KSCN exceeding 0.9 M. The drop times of the dropping mercury electrode also are found to undergo unusual lowering over a small range of potentials in the limiting current regions. No clear explanation has been given for a similar observation in aqueous medium. The only difference between the observations in aqueous and acetamide media is the amount of KSCN required to produce the effect. It may be added that the i–E variations in linear-sweep voltammetry and cyclic voltammetry at the hanging mercury drop electrode are normal in the presence of

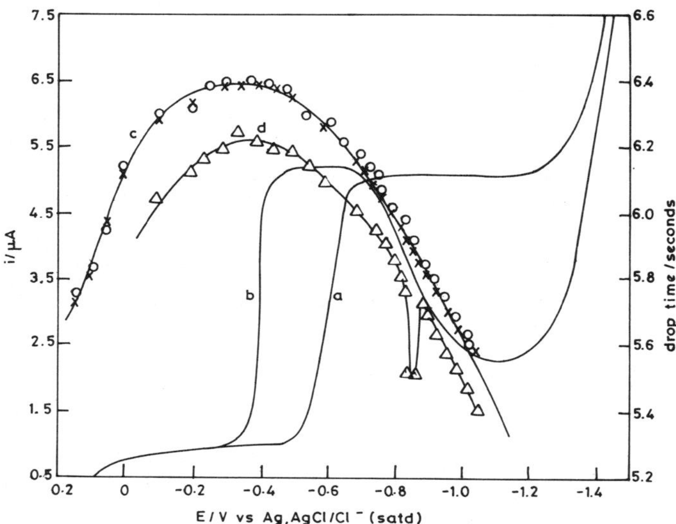

FIGURE 3. Polarograms and drop time curves in molten acetamide (85°C) in presence of KSCN: (a) $0.6\,M$ KNO_3 + 0.89 mM In^{3+}; (b) $0.6\,M$ KNO_3 + $2.0\,M$ KSCN + 0.89 mM In^{3+}; (c) drop time versus E curve, $0.6\,M$ KNO_3; (d) drop time versus E curve, solution as in (b); mt = 6.5 mg; h = 62 cm.

thiocyanate ions. The diffusion coefficients of many metal cations are estimated to be about 2.5×10^{-6}cm^2/s, except for In^{3+} (1.7×10^{-5}cm^2/s).[57]

The investigation of a series of organic compounds—quinones, benzophenones, and nitrobenzene—once again brings out the similarity between acetamide and water as solvents.[52, 57, 78] Benzoquinone, naphthoquinone, phenanthraquinone, and anthraquinone undergo two-electron reversible reductions from molten acetamide containing acetate, KNO_3, or PTSA at the mercury, platinum, and glassy carbon[78] electrodes. The degree of reversibility increases in the order stated above, and the $E_{1/2}$ and $E_{p,c}$ (reduction peak potentials in l.s.v) potentials can be rationalized on the basis of the resonance structures. The $E_{1/2}$ potentials vary linearly with the delocalization energy differences[159] of the respective quinones and the slopes are nearly the same as reported for aqueous systems. Adsorption of hydroquinone at the solid electrodes also is indicated in the cyclic voltammetric studies. A series of substituted benzophenones were investigated recently.[106] All undergo a two-electron, irreversible reduction process involving the —C=O bond. The $E_{1/2}$ shift with the substituent parameter (σ of Hammett), or E_{LUMO}, is linear,[158] as observed in aqueous alcoholic solvents.

The NO_2 group in nitrobenzene undergoes reduction in acidic and slightly basic melts in the same way as in aqueous alcoholic solvents. 4-Methyl-, 2-methyl-, 3-amino-, 3-hydroxy-, and 4-chloronitrobenzene undergo a $4e^- - 4H^+$ reduction process to form phenylhydroxylamine, and an azoxybenzene is formed by further chemical reactions.[52] Nitrosobenzene is formed as a short-lived intermediate. 2-Hydroxy-, 4-amino-, and 2-aminonitrobenzene undergo a $6e^- - 6H^+$ reduction process to form the corresponding anilines. The most interesting observation is that nitrobenzaldehyde is found to exist in two forms in equilibrium, brought about by solvation of the aldehyde group (see Section 2). The equilibrium depends on the acid–base nature of the system, is time dependent, and is dependent on the position of the aldehyde grouping.

In highly alkaline medium ($0.1\ M$ NaOH), the first step in the reduction of nitrobenzenes, except 2-hydroxynitrobenzene, is always a one-electron charge transfer to form a radical anion. This anion then undergoes a three- or five-electron reduction process to yield the final product. The stability of the nitrobenzene radical anion was found to decrease in the order 4-OH > 3-OH > 4-COOH > 4-Cl > 4-CH$_3$ > H > 2-OH.[52, 158] The radical anion of the 2-OH derivative is the least stable since this compound was found to undergo a one-step, six-electron reduction process, even in highly basic medium. The radical anion of the 3-OH derivative is reported to have greater stability than that of the 4-OH derivative in alkaline aqueous solutions.[107] The $E_{1/2}$–σ plot for the one-electron addition step is linear with a better correlation factor than observed in the acidic or the neutral medium.[158] The reaction parameter (ρ) from this plot is different from those obtained in aqueous alcoholic and acetonitrile solutions. Further investigations of this type for a series of amides will prove useful for an understanding of the influence of the solvent on electron transfer process. The p values are 0.76 and 0.29 in molten acetamide containing 0.1M NaOH and 1.0M Na–acetate respectively[158]. The anthraquinone radical anions and dianion are also found to be stable in molten acetamide containing 0.1M NaOH. Other quinones decompose gradually in such alkaline melts.[78]

It may also be pointed out that although a range of potentials free from faradaic processes are available, no well-defined steps are observed for the reduction of any of the nitrobenzenes at the platinum electrode.[52]

A few reports[108–112, 150] on electrochemical studies in NMA, DMA, and TMU have been made in the past, and these are mainly polarographic investigations. The conductivity of TMU solutions is reported to be very low, in spite of appreciable solubilities of salts in TMU, and this may be a deterrent for electrochemical studies. There is one report on the reduction of oxygen from urea solutions.

9.7. Electrochemistry in Amide Mixtures

The acetamide–urea–NH$_4$NO$_3$ eutectic (mp 7.8°C) has a conductivity of $10^{-3}\Omega^{-1}$cm^{-1} at 230°C. Cyclic voltammetric studies have been carried out with a bright platinum electrode. A prominent cathodic reduction is

$$NH_4{}^+ + e^- \rightarrow NH_3 + \frac{1}{2}H_2$$

Analysis of the melt after prolonged electrolysis showed the presence of a small quantity of NO_2. Adsorbed hydrogen atoms on the platinum can be detected, as in studies with platinum in sulfuric acid. NO_2 is formed at the anode, on prolonged electrolysis, by the reaction

$$NO_3{}^- \rightarrow NO_2 + \tfrac{1}{2}O_2 + e$$

Urea has been reported to catalyze reduction of the base melt of $K/Na/LiNO_3$ and results in an additional wave at more positive potentials.[113-115] This new wave is thought to be associated with the reduction of ammonium isocyanate, formed from urea. The catalytic effect is believed to be due to the removal of surface alkali metal oxide by urea. Silver ion reduction was found to be a diffusion-controlled process in this melt, and the diffusion coefficient was determined to be $10^{-7}\,cm^2/s$.[39] The results also indicate the absence of any significant quantities of the silver acetamide adduct, proposed by other researchers. Small shoulders are present in the forward and reverse scans in the cyclic voltammogram.

The anodic corrosion of nickel and copper in the ammonium nitrate eutectic at 120°C has been reported.[38] Experiments were carried out using 22-gauge pure nickel wires and with very fine copper powder. The major corrosion process proposed is

$$M + 2NH_4NO_3 \rightarrow M^{2+} + 2NH_3 + 2NO_3{}^- + H_2$$

The Ni^{2+} ion is incorporated into an amide complex, $[Ni(acetamide)_6]^{2+}$,[116] which is present as a deep blue complex, while a blue-green complex of copper is produced. The Ni^{2+} ion is reduced in a single two-electron step, which is evidenced on long-time electrolysis by the visible production of metallic nickel deposits. The metal ion reduction is inferred to occur at potentials close to that of the solvent reduction. The reduction of Ni^{2+} ion has been shown to be an irreversible process in water, formamide, acetamide, and urea, probably the result of a strong coordination through the oxygen atom. A redox couple is observed at $+0.1$ V (vs. the ferrocene–ferricenium ion couple), depending on the surface treatment applied to the electrode.[38] A similar couple is observed with $ZnCl_2$ in solution.[38] We have observed such couples in air-saturated, ternary eutectic without any depolarizer but with a hanging mercury drop electrode.[78] The shape of the peaks is indicative of surface processes. Cu^{2+} ion undergoes reduction in two steps, of one electron each, and copper metal undergoes a two-step oxidation.[37] Spectroscopic studies provide evidence for these mechanisms.[116]

Investigations in our laboratories[78] have shown that several inorganic and organic depolarizers can be studied in this melt over a temperature range of 30–100°C. Oxygen undergoes reduction at mercury electrodes, apparently in two steps at 30°C, but the first step is complicated by the presence of a small sharp peak in both the cathodic and anodic scans ($E_{peak} = -0.25$ V vs. $Ag/AgCl,Cl^-$

satd.). This peak is very small, or absent, in deaerated solutions and in air-saturated solutions at higher temperatures ($> 50°C$). The solubility of oxygen at room temperature is about 0.195 mM at $30°C$. In a deaerated solution, a potential window of 0 to -1.0 V is available. The potential window is $+0.8$ to -0.6 V with a glassy carbon electrode and the redox peak observed with the hanging mercury drop electrode is not seen. Oxygen undergoes one-step reduction at the glassy carbon electrode. Peroxide decomposes to regenerate oxygen when temperature exceeds $50°C$. Anthraquinone undergoes a one-step, two-electron reversible reduction,[78] the reversibility increasing with the temperature. The other quinones which are stable in molten acetamide decompose gradually with time in the ternary melt. Addition of urea or ammonium nitrate to molten acetamide containing the quinones (except anthraquinone) is observed to cause the decomposition even at $25°C$[78].

10. Applied Aspects

10.1. Anodizing

A summary of anodizing studies conducted in molten salts can be found in Ref. 148. Anodizing using molten nitrates in the temperature range $45-160°C$ has been reported.[117–121] Anodic oxide films of Al, Ti, W, Nb, and Mo have been characterized. Low temperatures are favorable for the production of thicker and more adherent films. The adoption of the low-melting amide systems has been advocated by Lovering[120] and Turner and Lovering.[121] Russian patents report that addition of acetamide (20–80 wt %) or urea to a $NaNO_3$–KNO_3 eutectic mixture used for anodizing steel-welding materials decreased the temperature of operation, while increasing the rate of anodization.[119]

10.2. Electrodeposition

Zn, Cd, Pb, Sn, Co, and Ni can be deposited from solutions in formamide and acetamide.[110, 122] The properties of the deposit were found to be influenced by the presence of dissolved oxygen, and good deposition required closed electrolysis cells. The presence of water in the acetamide did not affect the deposit properties. In fact, the presence of a small amount of water in the melt containing $Zn(CN)_2$ led to excellent deposits of zinc.[110] Anodic dissolution of zinc was apparently more efficient than the corresponding cathodic deposition. Electrodeposition of nickel from primary alkylamides including FA, AA, and urea has been demonstrated.[109–112] The deposition is an irreversible process. Generally, metals lower than Zn in the conventional electromotive force (emf) series are not easily electrodeposited from amide solutions. The technology of titanium coatings is apparently simplified, with a wide range of current densities for operation, by using a molten mixture of urea (36.4–54.6%), acetamide (36.4–54.6%), and $TiCl_4$ (4–7 wt %) at 105°C.[123] In–Sn alloys with a tin composition of 12–18% were electrodeposited from fused acetamide at 110–160°C.[124] Iron can be plated onto a platinum surface from a molten bath of $FeCl_3$ (0.01), NH_4NO_3 (0.19), CH_3CONH_2 (0.48), and urea (0.32 mole fraction), using a current of $10mA/cm^2$.[125–129]

10.3. Batteries

10.3.1. High-Temperature Thermal Batteries

Thermal batteries are highly reliable and have long shelf life. The methods of activation are often complex. An organic solvent of low melting point with a capacity to dissolve ionic solids is useful in this context. Wallace[32] reported the functioning of cells of the type metal/metallic chloride in acetamide/AgCl/Ag (Table VI). The metal is Al, Mg, Zn, or Cd. The effective temperature range is 77–180°C. $ZnCl_2$ is highly stable in the melt, and zinc cells also could be recharged. No gassing was observed even under high-current-discharge conditions in the temperature range 90–130°C. The emf decreases linearly with temperature. The maximum current output is limited by the internal resistance of the molten eutectic. The $Cd/CdCl_2$ electrode also was found to be stable when discharged in the temperature range 70–136°C. Aluminum cells were found to deteriorate rapidly with hydrogen evolution at the aluminum.[31] This is possibly due to the high acidity of the melt containing $AlCl_3$. There is a general tendency for metal anodes to react with the acetamide as the temperature is increased. Addition of small quantities of water (2%) permitted operation in the temperature range 70–80°C, and the performance was similar to that of the anhydrous cells at 100°C.[30,31]

TABLE VI. Characteristics of Fused Acetamide–Salt Cells[a]

Cell	Temperature (°C)	Open-circuit potential (V)	Conductivity (r^{-1} cm^{-1})
Mg/0.10 *M* $MgCl_2$–acetamide/AgCl/Ag	90	1.835	—
	125[b]	1.860	
Al/0.10 *M* $AlCl_3$–acetamide/AgCl/Ag	90[b]	1.150	—
Zn/0.10 M $ZnCl_2$–acetamide/AgCl/Ag	90	1.058	0.5×10^{-3} (90°C)
			0.9×10^{-3} (130°C)
Cd/0.10 M $CdCl_2$–acetamide/AgCl/Ag	90	0.739	0.3×10^{-3} (90°C)
			0.7×10^{-3} (130°C)

[a]References 30 and 31.
[b]Hydrogen evolution starts at this temperature.

10.3.2. Ambient Temperature Thermal Batteries

The amide eutectics are considered attractive candidates for ambient temperature battery electrolytes because of their remarkable stability in the supercooled liquid phase, their low toxicity, and their relatively high conductivities. After activation, the molten electrolyte may cool and even supercool but the battery is still capable of delivering electrical energy. Such a characteristic meets the demands of military usage, which requires a wide temperature range of operation (-40 to $70°C$). The acetamide eutectic with $LiNO_3/NH_4NO_3$ has the necessary characteristics for such use. McManis *et al.*[128, 129] have done extensive studies on the discharging of lithium and calcium anodes in these melts and proved the viability of a battery system. These studies show that very many silver salts[39] can be used as cathode materials. Silver chromate, oxide, or orthophosphate cathodes, in a slightly acidic melt, provide attractive cell voltages (~ 2.4 V), current densities, and discharge characteristics (constant voltage over 200 s at $10mA/cm^2$). However, cells with silver halides are not recommended as they show wide behavioral variations between experiments.

Calcium reacts rapidly with acetamide at $100°C$ but is stable in the eutectic.[35–37] In NH_4NO_3-rich melts, the anode reaction is

$$4Ca + 10NH_4NO_3 \rightarrow 4Ca(NO_3)_2 + N_2O + 10NH_3 + 5H_2O$$

No passive films are formed.

In $LiNO_3$-rich melts, the processes are more complicated. The nitrate ion and the acetamide are reduced, and hydrogen can be formed. High current flow results in passive oxides being formed ($2Li^+ + O^{2-} \rightarrow Li_2O$).

A gray-black film of CaO is formed on calcium metal in contact with the melt containing $LiNO_3$, and gas evolution thus is very low. Galvanostatic discharge rates are high for calcium anodes in NH_4NO_3 melts (lack of passive film), while they are significantly lower in $LiNO_3$ melts. The corrosion rates and anode potentials are found to vary with the presence of water, particularly in $LiNO_3$ melts, which are highly hygroscopic.[37] Cells based on calcium anodes and AgCl cathodes in eutectic amides have been shown to have good discharge characteristics (1.1 V for over 30 min at 1-mA/cm^2 discharge rate, using a 0.32-cm^2 anode).

10.3.3. Lithium Organic Battery

A recent report on the use of dimethylacetamide (DMA) as a solvent in a nonaqueous lithium battery is that of Tobishima Shin-Ichi *et al.*[130] The use of DMF and other amides (DMA and DMAA [N-N dimethylacetoacetamide] was prompted by the observation that in the primary and secondary batteries, not only conductivity, but also the cycling efficiencies are closely related to the solvation state of Li$^+$ ion. DMF is found to react with lithium, while DMA and DMAA are expected to have a lower reactivity because the electron density on the oxygen atom in DMA and DMAA is higher than in DMF. LiClO$_4$ is soluble to 1 *M* in DMA and in mixed solvents containing more than 90 vol % of DMA. The conductivity of LiClO$_4$ is found to increase with the proportion of DMA, up to 50 vol%, and is twice that in propylene carbonate (PC) alone. The viscosity and the dielectric constant of the mixture of DMA and PC are lower than those of PC, and the lower viscosity is responsible for the higher conductivity. The lithium cycling efficiency follows the order: PC (67%) $\approx$ PC/DMA (65.8%) > PC/DMAA (60.9%) > PC/DMF (51.2%), with a small decrease in the efficiency with increasing DMA concentration. This small variation, in spite of an increased conductivity, is attributed to the preferential solvation of Li$^+$ ion by DMA. It was recently reported that DMA, as a cosolvent with propylene carbonate in ambient temperature lithium batteries, inhibits dendrite formation.[131]

11. Photoelectrochemical Studies

The reported corrosion of semiconductors in aqueous media prompted us to undertake photoelectrochemical studies in molten acetamides and in formamide. Very similar results to those in water are obtained with the layered dichalcogenides MoS$_2$, MoSe$_2$, and WSe$_2$ in molten acetamide with such redox systems as quinone–hydroquinone and I$_2$/I$^-$. The *i–E* variations have been obtained by cyclic voltammetry, flat-band potentials from the Mott–Schottky plots, and power characteristics by discharging across standard resistances.

One side of the semiconductor is brushed with an indium–gallium amalgam layer and attached to a brass holder with silver epoxy and cured at 400°C for an hour. The brass holder is then embedded into a Teflon cylinder. All metal areas, except the semiconductor surface, are coated heavily with Epoxy-Patch-0.5/clear, from the Dexter Corporation, USA. A leak-proof bonding, stable for 36–40 h in the melt, is obtained. It is observed that other bonds, such as those of Araldite or even Epoxy-Patch/white, peel away quickly in the melt. External connection is made through the brass holder.

The design of the glass electrolytic cell depends on the shape of the brass holder (Fig. 2). It is convenient to use an electrode in the shape of a hockey stick so that illumination of the semiconductor surface is simple and straightforward. If the electrode holder is straight and vertical, illumination from the underside of the glass cell with attendant mirror arrangements becomes necessary. Fusing a large, optically flat window to the bottom of a double-walled glass cell poses difficulties. Double-walled glass cells are necessary to circulate liquids (water or oil) at the required temperature to keep the solvent in the molten state. The disadvantage of the hockey stick arrangement is that the optical window has to be on the side of the cell. Since cylindrical glass cells are used, it becomes necessary to introduce a projection on the side and then fuse the optical window. If the projection is more than 4–5 mm, then the solid trapped on the projection may not be melted completely by the circulating liquid.

Experiments have been carried out successfully at 85°C with both the vertical and hockey stick types of holders. The former was necessary for experiments with rotating electrodes under constant illumination. The power characteristics are generally poor. The flat-band potentials with respect to an $Ag/AgCl,Cl^-$ (satd.) electrode are shown in Table VII. The shifts of the flat-band potentials with the concentration of the depolarizers paralleled the observations in aqueous solutions. The power characteristics are low in acetamide (Table VIII) and a little higher in formamide.

The capacitances are obtained using a lock-in amplifier (PAR-128A). A 5-mV alternating voltage at 1000 or 3000 Hz is superimposed over the polarizing potential (Fig. 4). The 90°C out-of-phase output signals are recorded at various potentials and the capacitances calculated with the help of a calibrated chart using standard capacitances.

A few experiments also have been carried out in the ternary eutectics, and further investigations are in progress. The potential window with n-$MoSe_2$, for example, is -0.3 to $+0.7$ V. Photocorrosion of $n-MOSe_2$ sets in at $+0.5V$. The power characteristics are only a little better than those observed in acetamide alone and decrease with increase in temperature. However, this investigation has yielded values of various parameters over a wide temperature range and thus should be of help for a better understanding of semiconductor physics and chemistry.

Anthraquinone is stable in acetamide, or in the ternary eutectic, but phenanthraquinone, naphthaquinone, and benzoquinone decompose rapidly in the

TABLE VII. Flat-Band Potentials with Reference to $Ag/AgCl,Cl^-$ (satd.) Electrode in Molten Acetamide at 85°C ($f = 1$ kHz)[78]

	Flat band potential, V		
Electrolytes	n-$MoSe_2$	n-WSe_2	n-MoS_2
0.5 M KNO_3	$+0.025$	-0.165	$+0.160$
0.5 M KNO_3 + $10^{-1}M$ I^-	-0.050	-0.195	$+0.090$
0.5 M KNO_3 + $10^{-1}M$ I^- + $10^{-3}M$ I_2	-0.100	-0.215	$+0.045$
0.5 M KNO_3 + $10^{-1}M$ hydroquinone	-0.005	-0.195	$+0.120$
0.5 M KNO_3 + $10^{-1}M$ hydroquinone + $10^{-3}M$ quinone	-0.015	-0.205	$+0.115$
Urea + acetamide + NH_4NO_3 (25°C)*	$+0.025$	-0.165	$+0.170$

*V_{fb} becomes more positive by 20 mV at 85°C $Ag/AgCl,Cl^-$ (satd) in ternary amide. Electrode is more positive than the reference electrode in molten acetemide by 40mV at 85°C

TABLE VIII. Power Characteristics of Photoelectrochemical Cells in Molten
Acetamide at 85°C

Cell			E_{photo} (mV)	I_{photo}, I_{sc} (mA/cm²)	Fill factor	η (%)
n-MoSe$_2$	0.5 M KNO$_3$ $+10^{-1}M$ I$^-$ $+$ 10^{-3} I$_2$	Pt	120	0.85	0.15	1.0
n-WSe$_2$	0.5 M KNO$_3$ $+10^{-1}M$ H$_2$Q $+$ 10^3 Q	Pt	290	1.45	0.29	2.3

ternary eutectic. Similar problems may arise with other organic redox systems/depolarizers. The rate of decomposition varies with the quinone, and it is the presence of urea that is responsible for the decomposition. Addition of urea to an acetamide solution of phenanthraquinone results in the decomposition of phenanthraquinone, which is otherwise stable in acetamide.

A somewhat different photoelectrochemistry has been reported by Pucciarelli and co-workers[132, 133] in the binary acetamide–KSCN mixture. The 25 mol % KSCN mixture melts at 25°C. In spite of a very high coefficient of viscosity, this mixture is a good solvent, particularly for coordination chemistry[145] and electrochemistry studies. Voltammetric studies show a stable range of 1.6 V with a platinum electrode. The electroreduction process is as follows:

$$2CH_3CONH_2 + 2e^- \rightarrow H_2 + 2CH_3CONH^-$$

oxidation is:

$$SCN^- \rightarrow SCN^{\cdot} + e^-$$

The formation of parathiocyanogen, (SCN)$_x$, and/or of polytrithiocyanogen, [(SCN)$_3$]$_x$, can occur, as it does in the electrolysis of thiocyanate mixtures.[134–142] Low current densities permit the formation of soluble (SCN)$^{2-}$ or (SCN)$^{3-}$ ion species, while high current densities lead to the formation of orange-yellow deposits on the anode. The polymerization of SCN radicals to produce appreciable amounts of the solid product is apparently low at 80°C. However, the growth

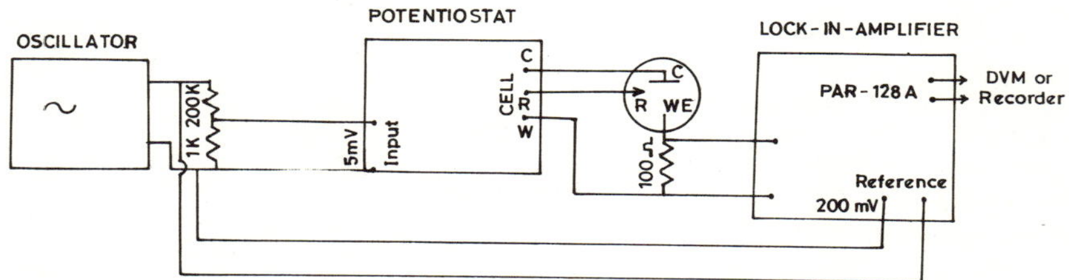

FIGURE 4. Block diagram of semiconductor electrode capacitance measurements.

of and coverage by solid products is rapid at 150°C. The solid product is found to be photoactive. On irradiation with a xenon lamp, a positive photopotential lasting for a few minutes is followed by a drift to a more negative photopotential, reaching a limit of about 0.2 V. The maximum amount of deposit consequent to a complete oxidation of $(SCN)^{3-}$ species corresponds to a total charge of 50 mC/cm^2. Similar behavior is observed with aluminum and carbon substrates.

12. Future Perspectives

The amides, with a variety of dielectric constants, structural characteristics, and solubilizing and ionizing capacities, form a group of solvents of great promise for both basic and applied electrochemistry. Redox potentials, a pH scale, electrical double layer and electrode kinetic parameters, and complex formation are some of the aspects that require further attention. The study of Hammett plots will help in understanding the role of the solvent in electron transfer processes. Accurate determination of diffusion coefficients in various solvents will help to correlate ion solvation with viscosity.

The low-temperature binary and ternary eutectics are a viable alternative to the haloaluminate systems. These materials are simpler to handle, and they are also nonexplosive. No toxicity has been reported. A limitation perhaps is their instability in the presence of strong base. These amides may prove useful in electroanalytical chemistry for detection of metal ions at the micro and submicro level. Water purification for such analyses is a time-consuming process and can be avoided. Further, trace metal analysis in organic polymers and microanalysis of organic substances may be possible in these organic solvents.

It will be interesting to study the effects of the polymeric structures of the amides on electrodeposition characteristics. The good solubilizing and ionizing properties can be taken advantage of in electrochemical energy conversion. The higher solubility of oxygen in acetamide can be made use of in the study of fuel cells. As a cosolvent in organic batteries, the amides may help in increasing the cycling efficiencies. Their use in ambient temperature thermal batteries will undoubtedly be of great consequence. Their wide temperature range of liquid stability can be exploited in photoelectrochemical studies. The variation of dielectric constant and its effects, carrier density, and surface states with temperature can be investigated.

Electroorganic synthesis has been explored so far to a negligible extent in these solvents, and this is yet another promising field. The electrogeneration of radical anions of high stability in alkaline melts not containing urea can be taken advantage of in this field.

Similarly, electro-initiated polymerization and electropolymerization, a field of current interest, may prove to be successful in the amide melts.

References

1. R. S. Reid and C. A. Vincent, *J. Electroanal. Chem.* **18**, 427 (1968).
2. J. W. Vaughn, in: *The Chemistry of Non-aqueous Solvents* (J. J. Lagowski, ed.), Vol. 2, pp. 191–264, Academic Presss, New York (1967).
3. O. Popovych and R. P. T. Tomkins, *Non-aqueous Solution Chemistry*, Wiley-Interscience, New York (1981).
4. Z. Borkowska, *J. Electroanal. Chem.* **244**, 1 (1988).
5. K. L. N. Phani and R. Narayan, *J. Electroanal. Chem.* **134**, 291 (1982).
6. V. Bartocci, M. Gusteri, R. Marassi, F. Pucciarelli, and P. Cescon, *J. Electroanal. Chem.* **94**, 153 (1978).
7. M. Gusteri, V. Bartocci, and F. Pucciarelli, *J. Electroanal. Chem.* **102**, 199 (1979).
8. M. Gusteri, V. Bartocci, F. Castellani, and G. Scarponi, *J. Inorg. Nucl. Chem.* **43**, 3400 (1981).
9. F. Castellani, G. Berchiesi, F. Pucciarelli, and V. Bartocci, *J. Chem. Eng. Data* **26**, 150 (1981).
10. V. Bartocci, P. Passamonti, and F. Pucciarelli, *Gazz. Chim. Ital.* **117**, 375 (1987).
11. G. Berchiesi, G. Vitali, and A. Amico, *J. Chem. Eng. Data* **30**, 208 (1985).
12. F. Castellani, G. Berchiesi, F. Pucciarelli, and V. Bartocci, *J. Chem. Eng. Data* **27**, 45 (1982).
13. A. Amico, G. Berchiesi, C. Cametti, and A. Di Biasio, *J. Chem. Soc., Faraday Trans. 2* **83**, 619 (1987).
14. G. Berchiesi, G. Vitali, P. Passamonti, and R. Plowiec, *J. Chem. Soc., Faraday Trans. 2* **83**, 1257 (1987).
15. R. Plowiec, A. Amico, and G. Berchiesi, *J. Chem. Soc., Faraday Trans. 2* **81**, 217, (1985).
16. G. Berchiesi, G. Vitali, and A. Amico, *J. Mol. Liq.* **32**, 99 (1986).
17. G. Berchiesi, G. G. Lobbia, V. Bartocci, and G. Vitali, *Thermochim. Acta* **70**, 317 (1983).
18. G. Berchiesi, G. G. Lobbia, M. A. Berchiesi, and G. Vitali, *J. Therm. Anal.* **29**, 729 (1984).
19. G. G. Lobbia and G. Berchiesi, *Thermochim. Acta* **74**, 251 (1984).
20. G. Berchiesi, private communication (1988).
21. G. Jander and G. Winkler, *J. Inorg. Nucl. Chem.* **9**, 24 (1959).
22. G. Jander and G. Winkler, *J. Inorg. Nucl. Chem.* **9**, 32 (1959).
23. G. Jander and G. Winkler, *J. Inorg. Nucl. Chem.* 9, 39 (1959).
24. S. Guiot and B. Tremillon, *J. Electroanal. Chem.* **18**, 261 (1968).
25. S. Guiot, *Ann. Chim. (Paris)* **4**, 235 (1969).
26. L. S. Bleshchinskaya, K. S. Sulaimankulou, and M. D. Davranov, *Russ. J. Inorg. Chem.* **28**, 607 (1983).
27. S. A. Kudrya, D. A. Tkalenko, and L. P. Anthropov, *Zh. Khim.* **1974**(11), 84; *CA* **83**, 33575.
28. L. S. Bleshchinskaya, K. S. Sulaimankulou, and M. D. Davranov, *Izv. Akad. Nauk Kirg. SSR* **6**, 35-b (1982); *CA* **98**, 96530v.
29. L. S. Bleshchinskaya, K. S. Sulaimankulou, and M. D. Davranov, *Izv. Akad. Nauk Kirg. SSR(1)* 50–51 (1983); *CA* **98**, 205177.
30. R. A. Wallace and P. F. Bruins, *J. Electrochem. Soc.* **114**, 209 (1967).
31. R. A. Wallace and P. F. Bruins, *J. Electrochem. Soc.* **114**, 212 (1967).
32. R. A. Wallace, *Inorg. Nucl. Chem. Lett.* **9**, 601 (1973).
33. R. A. Wallace, *J. Inorg. Nucl. Chem.* **35**, 3641 (1973).
34. R. A. Wallace, *Inorg. Chem.* **11**, 414 (1972).
35. G. E. McManis, A. N. Fletcher, D. E. Bliss, and M. H. Miles, *J. Electroanal. Chem.* **190**, 171 (1985).
36. G. E. McManis, A. N. Fletcher, D. E. Bliss, and M. H. Miles, *Electrochim. Acta* **31**, 1271 (1986).

37. G. E. McManis, A. N. Fletcher, D. E. Bliss, and M. H. Miles, *J. Appl. Electrochem.* **16,** 101 (1986).

38. G. E. McManis, A. N. Fletcher, D. E. Bliss, and M. H. Miles, *J. Appl. Electrochem.* **16,** 229 (1986).

39. G. E. McManis, A. N. Fletcher, D. E. Bliss, and M. H. Miles, *J. Appl. Electrochem.* **16,** 920 (1986).

40. D. G. Lovering and R. J. Gale (eds.), *Molten Salt Techniques,* Vol. 1, Plenum Press, New York (1983), p. 5.

41. P. Walden, *Z. Phys. Chem.* **46,** 175 (1903).

42. G. R. Leader and J. F. Gormley, *J. Am. Chem. Soc.* **73,** 5731 (1951).

43. S. Mizushima, T. Simanouti, S. Nagakura, K. Kwatani, M. Tsuboi, H. Baloa, and O. Fujioka, *J. Am. Chem. Soc.* **72,** 3490 (1950).

44. L. R. Dawson, J. E. Berger, J. W. Vaughn, and H. C. Eckstrom, *J. Phys. Chem.* **67,** 281 (1963).

45. A. J. Parker, *Q. Rev. London* **16,** 163 (1962).

46. J. Miller and A. J. Parker, *J. Am. Chem. Soc.* **83,** 117 (1961).

47. L. Weeda and G. Somsen, *Recl. Trav. Chim. Pays-Bas* **86,** 263 (1967).

48. G. Somsen and J. Coops, *Recl. Trav. Chim. Pays-Bas* **84,** 985 (1965).

49. L. Weeda and G. Somsen, *Recl. Trav. Chim. Pays-Bas* **85,** 159 (1966).

50. G. Somsen, *Recl. Trav. Chim. Pays-Bas* **85,** 517 (1966).

51. G. Somsen, *Recl. Trav. Chim. Pays-Bas* **85,** 526 (1966).

52. R. Saraswathi, Ph.D. thesis, Indian Instituture of Technology, Madras, 1988.

53. M. M. Baizer, *Organic Electrochemistry,* Marcel Dekker, New York (1983).

54. R. P. Bell, in: *Advances in Physical Organic Chemistry* V. Gold, ed.), Vol. 4, Academic Press, New York (1966).

55. P. Zuman, *J. Electroanal. Chem.* **75,** 523 (1977).

56. J. A. Riddick, W. B. Bunger, and T. K. Sakano, *Organic Solvents—Physical Properties, Preparation and Purification Methods,* 4th ed., *Techniques of Chemistry* (A. Weissberger, series ed.), Vol. 2, Wiley-Interscience, New York (1986).

57. K. L. N. Phani, Ph.D thesis, Indian Institute of Technology, Madras (1984).

58. M. Y. Abdullahi, K. P. D. Clark, and D. G. Lovering, Extended Abstracts, No. 81-2, Denver Meeting of the Electrochemical Society (1981).

59. J. M. Notley and M. Spiro, *J. Chem. Soc. B* **61,** 362, (1966).

60. Aldrich Fine Chemicals Catalogue, Aldrich Chemical Co., Inc., Milwaukee, Wisconsin (1986).

61. J. W. Walker and F. M. G. Johnson, *J. Chem. Soc.* **87,** 1597 (1905).

62. W. D. Kumler and C. W. Porter, *J. Am. Chem. Soc.* **56,** 2549 (1934).

63. L. Belladen, *Gazz. Chim. Ital.* **57,** 407 (1927).

64. K. Hirano, *Bull. Chem. Soc. Jpn.* **38,** 842 (1965).

65. V. S. Tsveniashvili, *Elektrokhimiya* **21,** 1142 (1985); *CA* **103,** 168723.

66. J. Rosin, *Reagent Chemicals and Standards,* 5th ed., Van Nostrand, Princeton, New Jersey (1967).

67. G. R. Leader and J. F. Gormley, *J. Am. Chem. Soc.* **73,** 5731 (1965).

68. H. Letaw and A. H. Gropp, *J. Phys. Chem.* **57,** 964, (1953).

69. M. C. R. Symond, T. A. Shippey, and P. P. Rastagi, *J. Chem. Soc., Faraday Trans.* **76,** (4), 2251 (1980).

70. C. D. Schmulbach and R. S. Drago, *J. Am. Chem. Soc.* **82,** 4484 (1960).

71. D. R. Foerster and R. Miller, 10th Annual Report on Petroleum Research Fund, American Chemical Society, Washington, D. C., p. 15 (1966).

72. J. W. Vaughn, in: *The Chemistry of Non-aqueous Solvents* (J. J. Lagowski, ed.), Vol. 2, p. 225, Academic Press, New York (1967).

73. H. J. Horn, *Toxicol. Appl. Pharmacol.* **3**, 12 (1961).

74. B. Gruttner, *Z. Anorg. Allg. Chem.* **270**, 223 (1952).

75. H. Strehlow, Electrode Potentials, in: *The Chemistry of Non-aqueous Solvents* (J. J. Lagowski, ed.), Academic Press, New York (1966).

76. M. Pournaghi, J. Devynck, and B. Tremillon, *Anal. Chim. Acta* **89**, 321 (1977).

77. M. Pournaghi, J. Devynck, and B. Tremillon, *Anal. Chim. Acta* **97**, 365 (1978).

78. S. Sampath and R. Narayan, unpublished results.

79. P. H. Given and M. E. Peover, *Nature* **182**, 1226 (1958).

80. P. H. Given and M. E. Peover, *J. Chem. Soc.* **1959**, 1602.

81. M. Musha, T. Wasa, and T. Kiyoo, in: *Modern Aspects of Polarography* (T. Kambara, ed.), pp. 169–175, Plenum Press, New York (1966).

82. T. Pavlopoulos and Strehlow, *Z. Phys. Chem. (Frankfurt)* **2**, 89 (1954).

83. G. Demange-Guerin and J. Badoz-Lambling, *Bull. Soc. Chim. Fr.* **1964**, 3277.

84. M. Breant and N. Van Kiet, *Bull. Soc. Chim. Fr.* **1965**, 3638.

85. M. Mandel and P. Decroly, *Nature* **182**, 794 (1958).

86. M. Mandel and P. Decroly, *Trans. Faraday Soc.* **56**, 29 (1960).

87. R. K. Agarwal and B. Nayak, *J. Phys. Chem.* **70**, 2568 (1966).

88. R. K. Agarwal and B. Nayak, *J. Phys. Chem.* **71**, 2062 (1967).

89. E. Luska and C. M. Criss, *J. Phys. Chem.* **70**, 1496 (1966).

90. E. C. Eckstrom, D.-G. Oei, and L. R. Dawson, *J. Phys. Chem.* **67**, 281 (1953).

91. Yu. M. Povarov, V. E. Kazarinov, Yu. M. Kessler, and I. V. Safonova, *Dokl. Akad. Nauk SSSR* **155**, 1411 (1964); *CA* **60**, 15203.

92. A. J. Bard and L. R. Faulkner, *Electrochemical Methods: Fundamentals and Applications,* pp. 701–702, Wiley, New York (1980).

93. R. Payne, in: *Advances in Electrochemistry and Electrochemical Engineering* (P. Delahay, ed.), Vol. 7, pp. 1–76, Wiley-Interscience, New York (1970).

94. R. Payne, in: *Physical Chemistry of Organic Solvent Systems* (A. K. Covington and T. Dickinson, eds.), Plenum Press, New York (1970).

95. R. Payne, *J. Phys. Chem.* **73**, 3598 (1969).

96. R. Payne, *J. Phys. Chem.* **71**, 1548 (1967).

97. S. Minc, J. J. Jastrzebska, and M. Brzostowska, *J. Electrochem. Soc.* **108**, 1160 (1961).

98. C. M. Slansky, *J. Am. Chem. Soc.* **62**, 2430 (1940).

99. N. Petit, *J. Electroanal. Chem.* **31**, 375 (1971).

100. K. L. N. Phani and R. Narayan, *J. Electroanal. Chem.* **189**, 135 (1985).

101. K. L. N. Phani and R. Narayan, *J. Electroanal. Chem.* **193**, 283 (1985).

102. R. Narayan and K. L. N. Phani, Proceedings of the First International Symposium on Molten Salt Chemistry and Technology, Kyoto, Japan, April 20–22, 1983, pp. 157–162.

103. V. Bartocci, F. Pucciarelli, M. Gusteri, and F. Castellani, *Ann. Chim. (Rome)* **74**, 239 (1984).

104. R. Saraswathi, K. L. N. Phani, and R. Narayan, unpublished observations.

105. K. L. N. Phani and R. Narayan, *J. Electroanal. Chem.* **189**, 155 (1985).

106. R. Saraswathi and R. Narayan, *Proc. Indian Acad. Sci., Chem. Sci.* **97**(3–4), 403 (1986); *Bull. Electrochem.* **4**(2), 157 (1988).

107. C. Corvaja, G. Farnia, and E. Vianelo, *Electrochim. Acta* **11**, 919 (1966).

108. A. J. Arvia and D. Posadas, in: *Encyclopedia of Electrochemistry of the Elements* (A. J. Bard, ed.), Vol. III, pp. 212–421, Marcel Dekker, New York (1975).

109. F. Yntema and L. F. Audrieth, *J. Am. Chem. Soc.* **52**, 2693 (1930).

110. S. Sultan and P. K. Tikoo, *Surf. Technol.* **21**, 233 (1984).

111. L. F. Audrieth, and H. W. Nelson, *Chem. Reviews* **8**, 335–52 (1931).

112. K. K. Bansal and M. L. Anand, *J. Ind. Chem. Soc.* **50**, 33 (1983).

113. V. D. Prisyazhnyi, D. A. Tkalenko, N. A. Chmilenko, and S. A. Kudrya, *CA* **93**, 122382.

114. L. I. Antropov, D. A. Tkalenko, and S. A. Kudrya, *Sov. Electrochem.* **11**, 1701 (1975).

115. D. A. Tkalenko, S. A. Kudrya, and A. A. Rudnitskaya, *Sov. Electrochem.* **14**, 117 (1978).

116. M. E. Stone and K. E. Johnson, *Can. J. Chem.* **49,** 3836 (1971).

117. D. Inman and D. G. Lovering, in: *Comprehensive Treatise of Electrochemistry* (J. O'M. Bockris, B. E. Conway, and S. U. M. Khan, eds.), Vol. 7, pp. 625–627, Plenum Press, New York (1983).

118. D. G. Lovering and K. P. D. Clark, in: *Proceedings of the Fourth International Symposium on Molten Salts* (M. Blander, D. S. Newman, G. Mammantov, M. L. Saboungi, and K. E. Johnson, eds.), Vol. 84-2, Electrochemical Society, New Jersey pp. 603–610 (1984).

119. L. I. Antropov, D. A. Tkalenko, S. A. Kudrya, E. I. Kozlov, A. A. Rudnitskaya, N. M. Vorpai, and N. A. Chimilenko, *Tovarnze Znaki* **1981** (4), 89; USSR Patent 800,249 (1981); *CA* **94,** 147516.

120. D. G. Lovering, *Trans. IMF* **61,** 113 (1983).

121. A. K. Turner and D. G. Lovering, *Trans. IMF* **58,** 109 (1980).

122. A. Brenner, in: *Advances in Electrochemistry and Electrochemical Engineering* (P. Delahay and C. W. Tobias, eds.), Vol. 5, pp. 205–248, Interscience, New York (1967).

123. N. Kh. Tumanova, N. M. Sarnavskii, M. U. Prikhod'ko, A. V. Chatverikov, L. V. Bogdanovich, and I. M. Mukha, Russian Patent 876,800 (1981); *CA* **96,** 767655 (1982).

124. E. M. Golubchik, *Zashch. Met.* **20**(2), 286 (1984); *CA* **100,** 217864.

125. G. E. McManis, A. N. Fletcher, and D. E. Bliss, U.S. Patent 4,624,755; *CA* **106,** 92543.

126. G. E. McManis, A. N. Fletcher, and D. E. Bliss, U.S. Patent 4,624,753; *CA* **106,** 75010.

127. G. E. McManis, A. N. Fletcher, and D. E. Bliss, U.S. Patent 4,624,755; *CA* **106,** 92544.

128. G. E. McManis, M. H. Miles, and A. N. Fletcher, *J. Power Sources* **15,** 141 (1985).

129. G. E. McManis, M. H. Miles, and A. N. Fletcher, *J. Power Sources* **16,** 243 (1985).

130. S.-I. Tobishima, M. Arakawa, and J.-I. Yamaki, *Electrochim. Acta* **33,** 239 (1988).

131. J. O. Besenhard, J. Guertler, P. Komenda, and M. Josowicz, Proceedings of the Symposium on Primary and Secondary Ambient Temperature Lithium Batteries, 1988, Proceedings of the Electrochemical Society Meeting, Pennington, New Jersey, 1988, Vol. 88-6, pp. 618–626; *CA* **108,** 207654.

132. F. Pucciarelli, V. Bartocci, F. Castellani, M. Gusteri, P. Cescon, and M. Bragadin, *Ann. Chim. (Rome)* **73,** 697 (1983).

133. M. Bragadin, G. Scarponi, G. Capodaglio, F. Ossola, V. Bartocci, and F. Pucciarelli, *Mol. Cryst. Liq. Cryst.* **121,** 345 (1985).

134. F. Seel and E. Muller, *Chem. Ber.* **88,** 1747 (1955).

135. F. Seel and D. Wesemann, *Chem. Ber.* **86,** 1107 (1953).

136. R. E. Panzer and M. J. Schaer, *J. Electrochem. Soc.* **112,** 1136 (1965).

137. F. Pucciarelli, P. Cescon, and H. Heyrovsky, *J. Electrochem. Soc.* **126,** 972 (1979).

138. A. J. Calandra, M. E. Martins, and A. J. Arvia, *Electrochim. Acta* **16,** 2057 (1971).

139. K. E. Dennig and K. E. Johnson, *Electrochim. Acta* **12,** 1391 (1967).

140. H. Kerstein and R. Hoffman, *Chem. Ber.* **57,** 491 (1924).

141. B. Cleaver, A. J. Davies, and D. J. Schiffrin, *Electrochim. Acta* **18,** 747 (1973).

142. Yu. I. Rubtsov, I. I. Strizhevski, E. B. Moshkovich, A. I. Kazakov, and L. P. Andrienko, *Khim. Prom-st. (Moscow)* **2,** 93 (1988).

143. N. I. Sax, *Dangerous Properties of Industrial Materials,* Reinhold, New York (1963).

144. M. M. Poberezhyenk, S. A. Kudrya, and T. G. Minchenko, *Geliotechnika* **3,** 22 (1984).

145. D. H. Kerridge, private communication; *J. Chem. Soc., Dalton Trans.* **11,** 2701 (1988).

146. J. J. Kingsley and K. C. Patil, *Mater. Lett.* **6,** 477 (1988).

147. W. Rubin and D. G. Lovering, in: *Molten Salt Technology* (D. G. Lovering, ed.), pp. 185–222, Plenum Press, New York (1982), p. 198.

148. D. H. Kerridge and D. G. Lovering, in: *Molten Salt Technology* (D. G. Lovering, ed.), pp. 123–152, Plenum Press, New York (1982).

149. Y. Marcus, in: *Molten Salt Technology* (D. G. Lovering, ed.), pp. 457–498, Plenum Press, New York (1982).

150. J. J. Lagowski (ed.), *The Chemistry of Nonaqueous Solvents,* Vol. III, Academic Press, New York (1978).
151. S. L. Culp and J. A. Caruso, *Anal. Chem.* **41,** 1329 (1969).
152. T. Sezuki, M. M. Miyake, T. Hatsushika, and Y. Hayakawa, Proceedings of the First International Symposium on Molten Salt Chemistry and Technology, Kyoto, Japan, 1983, pp. 20–22.
153. P. F. Schmidt, U.S. Patent (1959), 2909470.
154. N. I. Sax, *Dangerous Properties of Industrial Materials,* 6th ed., Van Nostrand Reinhold, New York (1984).
155. L. A. Knecht, *Pure Appl. Chem.* **27,** 281 (1971).
156. R. Skold and J. Suurkuusk, *J. Chem. Thermodynamics,* **8,** 1075 (1976).
157. M. Levy and J. P. Magoulas, *J. A. C. S.,* **84,** 1315 (1962).
158. R. Saraswathy and R. Narayan, *J. Elec. Chem. Soc. of India,* **38** 250–254, 255–262 (1989).
159. A. Streetwieser, Jr., *Molecular Orbital Theory for Organic Chemists,* J. Wiley, New York (1961), p. 254.
160. K. M. Sandberg, L. Atanasoka, R. Atanasoki, and W. H. Smyrl, *J. Electroanal. Chem.,* **220,** 161–168 (1987).

2

Physicochemical Properties of Liquid Organic Salts Using Chromatographic Techniques

Colin F. Poole, Kenneth G. Furton, Rena M. Pomaville, Salwa K. Poole, and Brian R. Kersten

1. Introduction

In spite of several technological advances where it would be useful to explore new solvents of higher polarity and specific selectivity than those associated with common nonionic solvents, the liquid organic salts have received little attention. One reason for this is the widely held belief that the melting points and the onset of thermal decomposition or boiling points of the liquid organic salts are too close together to provide a useful operating temperature range. That this is not always the case has been illustrated in several recent studies, with the identification of liquid organic salts having liquid temperature ranges exceeding 100°C, and in a few cases, 200°C. These favorable properties have led to a renaissance of interest in the solvent properties of the liquid organic salts. Further, the increasing interest in low-temperature molten salts and their chemistry and solvent properties necessitates new techniques for their evaluation and characterization, from the standpoints of phase separation, physical properties, solubilities, and so on.

Gas chromatography is a widely used technique for determining thermodynamic properties of pure substances or solvent properties of binary mixtures in which one component, the sample, is generally present at infinite dilution and the other, the stationary phase, is most commonly an involatile liquid.[1-5] For pure substances, the latent heat of vaporization, the boiling point, the vapor

Colin F. Poole, Kenneth G. Furton, Rena M. Pomaville, Salwa K. Poole, and Brian R. Kersten • Department of Chemistry, Wayne State University, Detroit, Michigan 48202.

pressure, and molecular weight information are readily obtained. From studies of band broadening, liquid- and gas-phase diffusion coefficients and reaction rate data (for on-column reactions) can be obtained. The free energy, enthalpy, and entropy of mixing or solution and the infinite-dilution solute activity coefficients can be determined from retention measurements. Virial coefficients, a measure of gas-phase imperfections for gas-solute vapors, can be measured by high-pressure gas chromatography.

The popularity of the gas chromatographic method for physicochemical measurements is due to its simplicity when compared to static methods, small sample size requirements, the ability to measure properties of impure samples, and the ease with which temperature can be varied. Gas chromatographic methods provide a less ambiguous interpretation of physicochemical properties than measurements made by liquid chromatography.[1,6] A lack of knowledge of the precise composition of the stationary phase and the absence of quantitative models for the accurate description of retention in liquid chromatography are the principal reasons for this. As a consequence, our discussion will necessarily focus on the more accessible data obtainable from gas chromatography, with a brief mention of the possibilities of using liquid chromatography for the study of solvent properties of liquid organic salts.

2. Retention Model for Gas Chromatography

Retention in gas chromatography is frequently a complex process involving partition and adsorption with different types of liquid films and interfaces existing simultaneously. A comprehensive model for retention in gas–liquid chromatography (GLC) can be written in the form of Eq. (1), in which retention arises from gas–liquid partitioning with the bulk liquid phase and with a structured liquid layer close to the support surface, by gas–liquid adsorption at the bulk liquid and structured liquid interfaces, and by adsorption at the gas–solid support interface.[7]

$$V_N{}^* = V_L K_L + \delta\, V_L K_{SL} + (1 - \delta)\, A_S K_{ASL} + A_L K_A + A_S K_S \qquad (1)$$

where $V_N{}^*$ is the net retention volume per gram of packing, V_L the volume of liquid phase per gram of packing, K_L the gas–liquid partition coefficient, δ a constant equal to 1 when $d_f < d_s$ and 0 when $d_f \geq d_s$, with d_f the film thickness and d_s the thickness of the structured liquid-phase layer, K_{SL} the gas–liquid partition coefficient for the structured liquid phase layer, A_S the gas–solid interfacial area per gram of packing, A_L the liquid surface area per gram of packing, K_{ASL} the adsorption coefficient at the gas-structured liquid-phase interface, K_A the coefficient for adsorption at the bulk gas–liquid interface, and K_S the coefficient for adsorption at the support interface. Equation (1) is a linear equation with five unknowns (K_L, K_{SL}, K_{ASL}, K_A, and K_S). To solve this equation, the phase characteristics V_L, A_L, and A_S and $V_N{}^*$ must be known for a minimum of five column packings prepared with different liquid loadings. A value for δ must also be found by a combination of intuition and iteration. This approach is far too unwieldy for general use, and some simplifications are needed.

First of all, we have no realistic way of experimentally determining the position of the interface between the surface-modified and bulk liquid phase layers. In most cases, we are interested in the partition and sorption properties of the bulk liquid, and the presence of a structured liquid film is considered no more than a nuisance. By making measurements at comparatively high phase loadings, we can assume that the properties of the bulk liquid dominate the retention equation and, to a first approximation, Eq. (1) can be simplified as indicated by Eq. (2).

$$V_N{}^* = V_L K_L + A_L K_A + A_S K_S \tag{2}$$

For diatomaceous earth supports with surface areas from 1 to 3 m^2/g, a 0.01 to 2% (w/w) liquid-phase loading should be adequate to cover the support with a monolayer, depending on the orientation of the adsorbed molecules. At higher phase loadings, the liquid exists in deep pools formed in the pores. For solvents with good wetting characteristics, it is likely that the solvent first spreads to form a monolayer and then accumulates in the smallest pores under the influence of capillary forces, filling the larger pores and cavities later at the same time as the adsorbed layer thickens.[8] For liquids that do not wet the support well, the liquid will tend to be distributed discontinuously in isolated pools and microdroplets, which may persist on the surface up to relatively high phase loadings. These considerations make general statements of the optimum phase loadings for determining solution properties difficult. There is a general consensus that at phase loadings less than 5% (w/w) on diatomaceous supports, Eq. (2) may not be applicable, while phase loadings in excess of 10% (w/w) seem to be a reasonable compromise. Final proof that Eq. (2) is a realistic model of solute retention at high phase loadings comes from the fact that the thermodynamic constants derived from its use are in excellent agreement with those for the bulk liquid determined by nonchromatographic techniques.[1-3]

3. Quantitative Validation of the Retention Model

There are many situations in gas–liquid chromatography in which solute retention can be adequately described by a partition model in which Eq. (2) can be simplified to Eq. (3).

$$V_N = K_L V_L \tag{3}$$

In fact, Martin's suggestion that interfacial liquid adsorption might contribute substantially to retention in GLC was initially met with much skepticism.[9-11]

Although this is now well established, it is by no means dominant or significant for all solute/solvent combinations. It seems reasonable to anticipate that in those cases in which the liquid phase and the solute differ substantially in solvent strength, the solubility of one in the other will be low and the contribution of interfacial adsorption to retention may be substantial. This is amply borne out by the retention of hydrocarbons on liquid organic salts, which generally show significant retention by interfacial adsorption and a variable contribution from partitioning. A convenient test of Eq. (3), that is, a qualitative indication of the absence of interfacial adsorption, is to construct a plot of V_N^* against phase loading for several columns containing different phase loadings. V_N^* (the net retention volume per gram of packing) is normally corrected to a common temperature, typically 25°C. Usually four phase loadings are adequate to test the applicability of Eq. (3). In the case of tetra-*n*-butylammonium tetrafluoroborate (Fig. 1), retention of all solutes occurs by a gas–liquid partitioning mechanism as indicated by the regular increase in retention volume with similar incremental increases in the phase loading combined with a zero intercept for each line when extrapolated to zero phase loading.[12] We can apply Eq. (3) to these data and evaluate the gas–liquid partition coefficient from the slope of the line by convert-

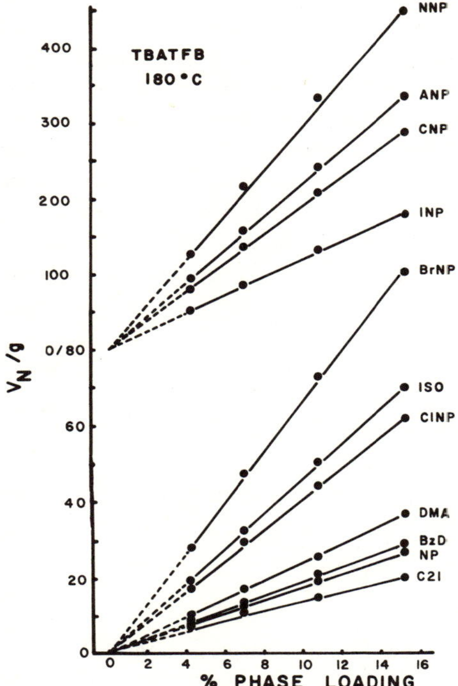

FIGURE 1. Plots of V_N^* (V_N/g) corrected to 25°C as a function of the percent phase loading of tetra-*n*-butylammonium tetrafluoroborate at 180°C on Chromosorb W-AW as support. Identification of test solutes: NNP, nitronaphthalene; ANP, acetonaphthone; CNP, cyanonaphthalene; INP, iodonaphthalene; BrNP, bromonaphthalene; ISO, isoquinoline; CINP, chloronaphthalene; DMA, 2,6-dimethylaniline; BzD, benzodioxan; NP, naphthalene; C21, *n*-heneicosane. (Reproduced with permission from Ref. 12. Copyright Elsevier Scientific Publishing Co.)

ing the *x*-axis to volume units. On the other hand, we see that interfacial adsorption cannot be ignored as a retention mechanism for many solutes on triethanolammonium thiocyanate (Fig. 2).[13] Several solutes, for example, nonanone, tetradecane, octanol, and hexadecane show only a slight change in retention (which in some cases is negative) as the phase loading is increased and at zero phase loading show a substantial intercept value. This behavior is typical of retention by interfacial adsorption. In this case, the nature of the experiment does not indicate whether adsorption occurs at the gas–liquid or the gas–solid interface, or at both interfaces simultaneously. However, Eq. (3) is clearly not applicable in these cases, and resort to Eq. (2) for interpretation is required. On the same phase, we see that naphthalene and nitrobenzene are retained largely by partitioning with some contribution from interfacial adsorption while benzodioxan, acetophenone, 2,6-dimethylphenol, and 2,6-dimethylaniline are retained solely by a partitioning mechanism. It is the individual physical properties of the solute and liquid phase as well as the column temperature which establish the retention mechanism in gas–liquid chromatography.

Plots of V_N^* against phase loading also are one means of studying the coating properties of liquid organic salts on diatomaceous supports.[14] In nearly all cases studied, the liquid organic salts show good support wetting charac-

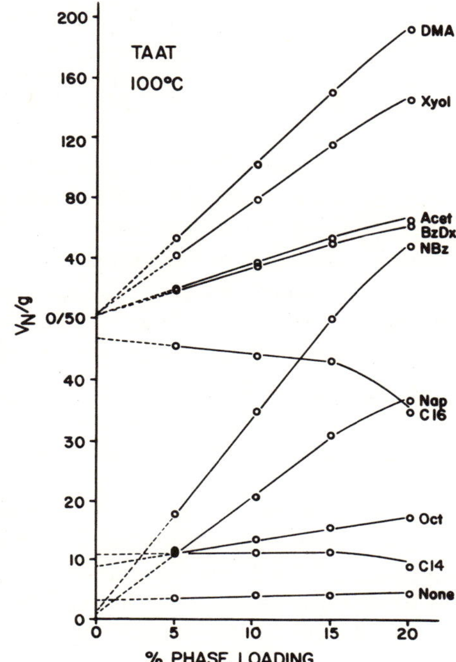

FIGURE 2. Plots of V_N^* (V_N/g) corrected to 25°C as a function of the percent liquid-phase loading of triethanolammonium thiocyanate at 100°C on Chromosorb W-AW as support. Identification of test solutes: None, nonanone; C14, tetradecane; Oct, octanol; C16, hexadecane; Nap, naphthalene; NBz, nitrobenzene; BzDx, benzodioxan; Acet, acetophenone; Xyol, 2,6-dimethylphenol; DMA, 2,6-dimethylaniline. (Reproduced with permission from Ref. 13. Copyright Elsevier Scientific Publishing Co.)

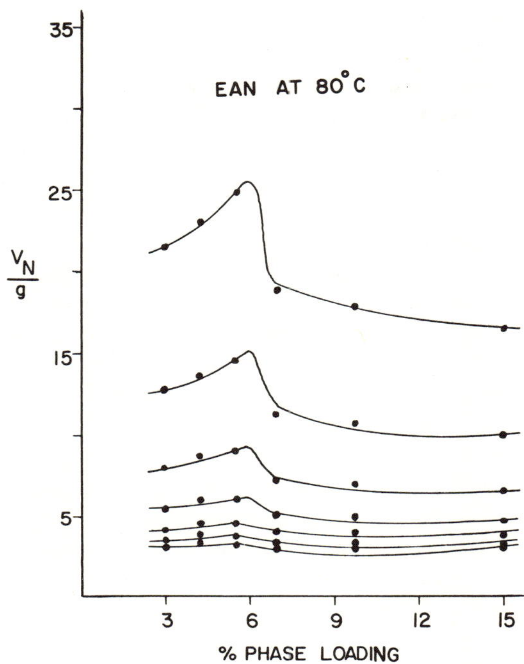

FIGURE 3. Plots of V_N^* (V_N/g) corrected to 25°C as a function of the percent liquid-phase loading of ethylammonium nitrate on Chromosorb W-AW as support at 80°C. Test solutes are *n*-alkanes. (Reproduced with permission from Ref. 14. Copyright Elsevier Scientific Publishing Co.)

teristics, with the notable exceptions of ethylammonium nitrate and propylammonium nitrate. These phases retain hydrocarbons almost entirely by interfacial adsorption (Fig. 3). A very distinct hump in the plots for the hydrocarbons at a phase loading of ca. 5.5% is clearly visible. This hump is most probably caused by a wetting transition due to coalescence of the liquid phase from microdroplets to a continuous film.[15] Thus, for these salts, film formation cannot be considered complete at phase loadings below ca. 8% (w/w) on Chromosorb W.

4. Determination of Distribution Constants from Retention Data

The quantitative evaluation of distribution constants for solutes retained by a mixed retention mechanism can be obtained from Eq. (2) after suitable rearrangement. If we divide both sides of Eq. (2) by V_L, then the gas–liquid partition coefficient is obtained as the intercept on the y-axis corresponding to $1/V_L = 0$ from a plot of V_N^*/V_L against $1/V_L$.

$$\frac{V_N^*}{V_L} = K_L + (A_L K_A + A_S K_S) \cdot \frac{1}{V_L} \tag{4}$$

In a similar manner, if we divide Eq. (2) by the liquid-phase surface area,

$$\frac{V_N^*}{A_L} = K_L \frac{V_L}{A_L} + K_A + \frac{A_S}{A_L} K_S \tag{5}$$

a plot of V_N^*/A_L against $1/A_L$ allows evaluation of the gas–liquid phase adsorption coefficient from the intercept as $1/A_L$ approaches zero. Liquid organic salts are excellent support-deactivating agents, and at high phase loadings it can reasonably be assumed that $A_S K_S \approx 0$. Equations (4) and (5) then permit the direct evaluation of the gas–liquid partition coefficient and gas–liquid adsorption coefficient from the experimentally derived parameters V_N^*, V_L, and A_L for several column packings with different phase loadings at a constant temperature. Values for V_N^* and V_L are easily determined experimentally, and gas–liquid partition coefficients may be determined with reasonable accuracy. Values for the liquid surface area are very difficult to determine with acceptable accuracy, and, therefore, only approximate values for the gas–liquid adsorption coefficients are available generally. We will return to this point later.

Several points in deriving Eqs. (4) and (5) are important. First of all, it is assumed that the contributions from the individual retention mechanism are independent of one another and additive. Second, Eqs. (4) and (5) are strictly applicable at infinite dilution with respect to the partition mechanism and at zero coverage for the adsorption mechanisms, or, alternatively, at constant concentration with respect to the ratio of sample size to amount of liquid phase. Here, we will discuss the infinite dilution/zero coverage method as this is the technique that has been used to determine all partition coefficient data for liquid organic salts to date.[14, 16–19] The constant-concentration experiment should be equally applicable to obtain reliable data and is outlined in detail in Refs. 20–23. The condition of infinite dilution is easier to achieve with respect to partitioning than adsorption, since in general partition isotherms exhibit greater linearity. Practically, the condition of infinite dilution is met when the solute peak is symmetrical, its retention time is unchanged with changes in sample size, and, for normal phase loadings, the amount of sample injected corresponds to about 0.1 to 10 μg or 0.0001 to 0.01 μl of liquid. Detection is not a problem at these levels with the sensitive flame ionization detector. For volatile samples, the injection of 5 to 50 μl of headspace vapors is a convenient technique to ensure that the infinite dilution criterion is met.

A typical plot of V_N^*/V_L versus $1/V_L$ used to determine gas–liquid partition coefficients for a series of solutes is shown in Fig. 4. The absence of curvature in the data indicates good wetting of the support and the probable absence of

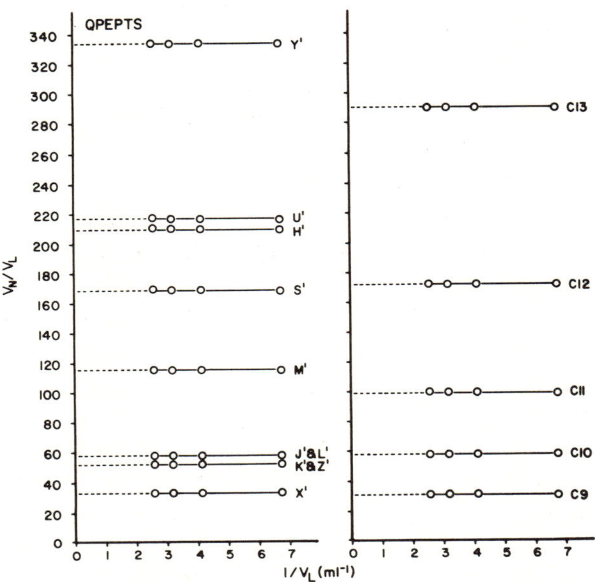

FIGURE 4. Plots of V_N^*/V_L vs. $1/V_L$ for a series of polar test solutes and *n*-alkanes on tetraethylammonium 4-toluenesulfonate on Chromosorb W-AW as support at 120°C. Identification of test solutes: X', benzene; Y', 1-butanol; Z', 2-pentanone; U', nitropropane; H', 2-methyl-2-pentanol; J', iodobutane; K', 2-octyne; L', dioxane; M', *cis*-hydrindan; C9 to C13, *n*-alkanes with 9 to 13 carbon atoms, respectively. (Reproduced with permission from Ref. 16. Copyright American Chemical Society.)

solute–support interactions. The horizontal plots parallel to the *x*-axis are typical of solute–solvent partitioning free of interfacial adsorption. Deviation from this behavior, as shown by the *n*-alkanes, for example, is typical of a mixed retention mechanism. In both cases, extrapolation of the experimental data to $1/V_L = 0$ gives an accurate value for the gas–liquid partition coefficient.

To calculate the gas–liquid adsorption coefficient, the liquid phase surface area must be determined for each column packing. Liquid surface areas have been measured by the Brunauer–Emmett–Teller (BET) method, the continuous flow method, and the thin-film method.[2, 3, 24] None of these methods is free of potential errors because of approximations made in the model used to interpret the experimental data and because of cracking of the film upon abrupt freezing in the BET and continuous flow methods. Differences in areas are also found between measurements using gases and organic vapors as adsorbates in the BET method. The chromatographic/tensiometer values of Martire *et al.*[25], determined for thiodipropionitrile on Chromosorb W and P at different phase loadings, have frequently been adopted as a reasonable approximation for the liquid surface area for any phase of interest. However, the manner in which to interpret liquid surface areas obtained in different ways clearly remains an open question.

5. Experimental Considerations for Determining Gas–Liquid Partition Coefficients

In this section, we will formally define the distribution constants discussed above, indicate all necessary corrections for the measurement of V_N^*, and provide a convenient method of determining the density, and consequently, the liquid-phase volume as a function of temperature for liquid organic salts. Also provided is a compilation of liquid ranges and decomposition temperatures for all organic salts studied by gas chromatography.

5.1. Definition of Distribution Constants

The gas–liquid partition coefficient, K_L, and gas–liquid phase adsorption coefficient, K_A, used in Eqs. (1) to (5) can be defined as follows. The partition coefficient K_L is a measure of the vapor pressure of the solute above a solution, and its value is determined both by the vapor pressure of the pure solute and the extent of its interaction with the solvent.

$$K_L = \frac{\text{Moles of solute per unit volume of bulk liquid}}{\text{Moles of solute per unit volume of gas phase}}$$

It is a dimensionless quantity. In thermodynamic terms, concentration units should be replaced by solute activities. The adsorption coefficient K_A is defined as

$$K_A = \frac{\text{Moles of solute at surface in excess of that in the bulk, per unit area}}{\text{Moles of solute per unit volume of gas phase}}$$

It pertains to a two-dimensional concentration of solute on the liquid surface and has units of distance (mm or cm).

5.2. Calculation of Retention Volumes

The net retention volume, V_N, is the adjusted retention volume corrected for mobile-phase compressibility and is calculated from Eq. (6); for use in Eqs. (1) to (5), it is usually expressed per gram of column packing, V_N^* (Eq. 7).

$$V_N = (t_R - t_M) F_o \left(\frac{T_c}{T_a}\right) \left(1 - \frac{P_W}{P_a}\right) \frac{3}{2} \left(\frac{P^2-1}{P^3-1}\right) \tag{6}$$

$$V_N^* = V_N/W \tag{7}$$

where t_R is the solute retention time, t_M the system dead time, F_o the carrier gas flow rate at the column outlet, T_c the column operating temperature (K), T_a the ambient temperature (K), P_W the vapor pressure of water at T_a (mm Hg), P_a the ambient pressure (mm Hg), P the column pressure drop (P_i/P_a), with P_i the column inlet pressure, and W the weight of column packing. The solute retention time is the time between the instant of sample introduction and when the detector senses the maximum of the retained peak. It is comprised of two parts, the adjusted retention time, or the time the solute spends in the stationary phase, and the system dead time (also known as the column dead time or void volume), which is equivalent to the time required for an unretained solute to reach the detector from the point of injection. All solutes spend the same time in the mobile phase, and this time is determined by the retention time of methane (flame ionization detector) or air (thermal conductivity detector) at the column operating temperature. The solubility of methane, or town gas, in the liquid organic salts at elevated temperatures can reasonably be assumed to be negligible. Note that in Eq. (6) it is the adjusted retention time $(t_R - t_M)$ that appears, so the accuracy with which t_M is determined is not critically important when $t_R \gg t_M$, provided that t_M is measured with reasonable accuracy. It is convenient to determine t_R and t_M, or their difference, in a single injection of headspace vapors of the solute together with a small volume of methane.

The carrier gas flow rate is conveniently measured using a soap film bubble meter at the column outlet, generally the detector exit port, when all sources of auxiliary gas are switched off. For the highest accuracy, the bubble meter should be thermostated to within about 0.5°C and, if needed, interfaced to the column via a spiral of copper tubing thermostated at the same temperature as the bubble meter to ensure equilibration of the carrier gas at ambient temperature. The carrier gas flow is corrected for the difference in temperature between the column and the bubble meter and for the vapor pressure of the soap film, assumed to be the same as that of pure water. Values for the saturation vapor pressure of water as a function of temperature can be found in standard handbooks of physical constants or in Ref. 26. Since gases are compressible, elution volumes are corrected to a mean column pressure by multiplying them by the pressure gradient correction factor. This is defined in terms of the column pressure drop measured as the inlet pressure divided by the outlet pressure. This can be determined directly with a mercury manometer connected to appropriate and convenient positions at either end of the column and isolated from the main column flow by valves when measurements are not being made. However, more usually, the column outlet pressure is assumed to be the same as the ambient atmospheric pressure, and the column inlet pressure is measured by an accurately calibrated pressure gauge or a mercury manometer. The latter is a simple U-tube filled with mercury connected to the column flow line after the flow controller via an on/off valve. This allows isolation of the manometer when not in use. For accurate work, pressures should be read to at least 1.0 mm Hg. It should be noted that the inlet pressure is actually the manometer pressure plus the value for ambient atmospheric pressure.

Other important elution volumes that we will use later are the specific retention volume and the molar retention volume. The specific retention volume, V_g^o, is the net retention volume at 0°C for unit weight of stationary phase.

$$V_g^o = \frac{V_N \, 273.16}{W_L T_c} \tag{8}$$

Where W_L is the weight of liquid phase in the column. The specific retention volume is related to the gas–liquid partition coefficient by Eq. (9).

$$V_g^o = \frac{273.16 K_L}{T_c \rho_c} \tag{9}$$

where ρc is the liquid-phase density at the column temperature.

The molar retention volume is defined as the volume of carrier gas, corrected to 0°C, required to elute half of the solute band from an ideal column containing one mole of salt and across which there is no pressure drop.

$$V_M^o = V_g^o \cdot M \tag{10}$$

where M is the gram molecular weight of the liquid organic salt.

The experimental value of V_g^o based on Eq. (8) is derived assuming a partition model. To obtain accurate values for the solution process in systems in which interfacial adsorption is significant, a value for K_L must first be determined and a new value for V_g^o calculated using Eq. (9).

5.3. Organic Salt Liquid Densities

Liquid organic salt densities as a function of temperature can be determined using a modified Lipkin type bicapillary pycnometer (Fig. 5).[17] Commercially available pycnometers designed for measurements with volatile liquids are not suitable for use with molten salts since the strain generated by the melting and recrystallization of the salt frequently causes shattering of the glass. Pycnometers made from heavy-wall capillary glass tubing are strong enough to solve this problem. A T-piece, made from glass rod, is welded to the capillary arms to increase the stability of the pycnometer and to provide a convenient arm on which to attach a clamp for manipulating the pycnometer at its operating temperature. The side arms are calibrated in 8 scale divisions, each with 10 intermediate scale divisions between major scale divisions. The volume of liquid required to fill the pycnometer to its working range is about 500–550 μl.

The pycnometer is calibrated with water, free from air (boiled and cooled without agitation shortly before use), by taking several readings at various temperatures and fillings. The volumes at the various temperatures are calculated by dividing the weight of water by the known densities. A correction, C, was added to the observed value of the density to take account of all buoyancy errors.[27]

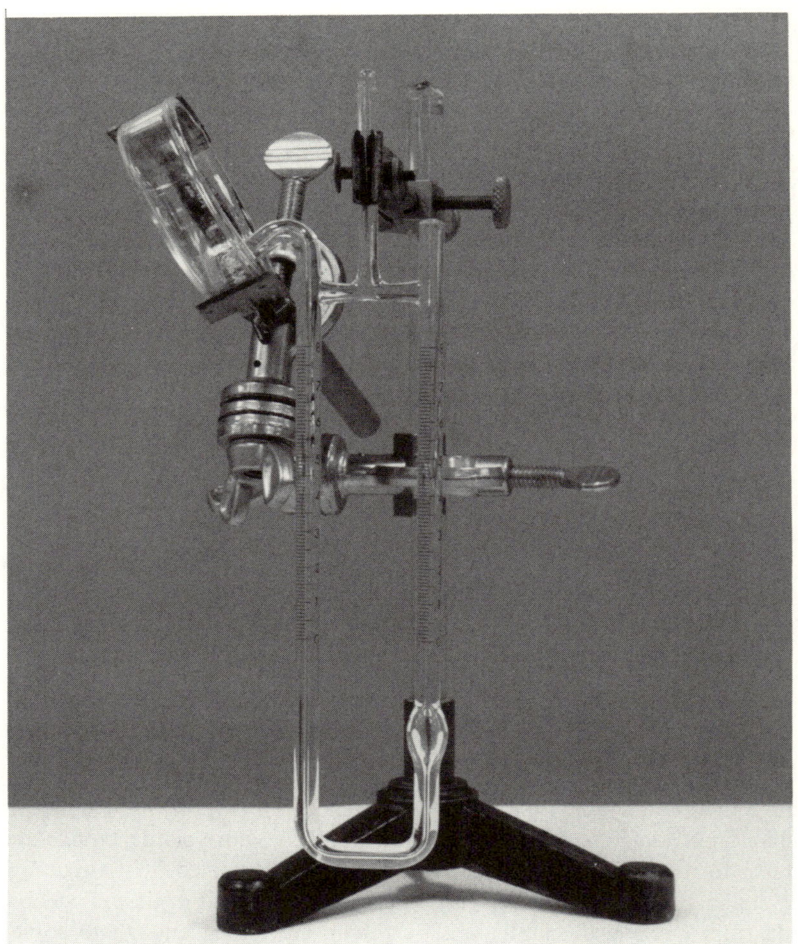

FIGURE 5. Bicapillary pycnometer.

$$C = 0.0012 \, (1 - \rho_A) \tag{11}$$

where ρ_A is the density in air. From these data, a pycnometer volume calibration equation, dependent on the scale reading and, to a lesser extent, the analysis temperature, can be generated; for example,

$$V = (4.8 \times 10^{-6} \, t + 7.43 \times 10^{-3}) \, r + 0.4226 \tag{12}$$

where V is the volume of the pycnometer (ml), t is the analysis temperature (°C), and r is the pycnometer scale reading (0–16).

To determine the density of the liquid salts, the pycnometer is attached to a small-ring stand with a thermometer clamp via the free arm of the T-piece. A small culture dish (60 × 15 mm) also is fastened to the stand and positioned so that the tip of the siphon arm of the pycnometer is in the well of the dish (the dish is clamped with the side toward the pycnometer below the horizontal). A small amount of salt, 450–550 mg, is added to the dish, and the whole apparatus is placed in a convection oven set to a temperature ca. 10°C higher than the melting point of the salt. The liquid salt is then drawn into the pycnometer by capillary action, and the filling process is completed by the siphon action of the bent arm. Once the salt has filled the pycnometer within the scale-reading region, the apparatus is slowly returned to room temperature, the outside of the pycnometer is cleaned with a volatile solvent, and wiped with a damp paper tissue, and the pycnometer is hung from a wire hook in an analytical balance. After allowing sufficient time for the pycnometer to come to the balance temperature, the pycnometer is weighed and then immersed in an oil bath to a height sufficient to include the entire measuring scale. The oil bath temperature is then slowly raised above the melting point of the salt, and several scale readings are recorded at various temperatures. Afterward, the pycnometer is returned to the convection oven, care being taken not to allow the salt to recrystallize, and heated to ca. 50°C above the melting point of the salt, and the salt is rapidly removed by suction. Finally, the pycnometer is thoroughly cleaned with a volatile solvent, such as methanol, dried, and weighed empty, as before.

Liquid organic salt densities as a function of temperature were fitted to the general equation

$$\rho t = A - B(t) \tag{13}$$

where ρt is the liquid organic salt density at temperature t (°C), and A and B are experimental constants fitted by linear regression. The experimentally derived constants A and B and the temperature range of the measurements for all liquid organic salts studied are summarized in Table I.

5.4. Preparation of Column Packings

The column preparation procedure involves several steps commencing with the choice of support, coating solvent, and coating technique. Several methods are available for drying the coated support, filling the column with the column packing material, and determining an accurate value for the liquid-phase loading. A more extensive discussion of column preparation procedures than is given here can be found in Ref. 26.

The common supports for gas chromatography are diatomaceous earths (microamorphous silica with small amounts of alumina and metallic oxide impurities), porous fluorocarbon powders, and glass beads. Neither the fluorocarbon powders nor the glass bead supports are suitable for use with liquid organic salts because of their low liquid loading capacity and the salt's poor wetting characteristics. Several types of diatomaceous supports are available, as indi-

TABLE I. Density as a Function of Temperature for Liquid Organic Salts

Liquid organic salt	Temperature range of measurement (°C)	Coefficients		Standard deviation in the intercept
		A	$B \times 10^4$	
A. *Tetra-n-butylammonium salts*				
Pentacyanopropenide	122–130	1.1640	7.453	0.0009
Picrate	100–158	1.1608	6.854	0.0009
Bromide	122–136	1.1592	8.836	0.0012
Trifluoromethanesulfonate	120–142	1.1332	8.092	0.0006
4-Toluenesulfonate	102–134	1.0898	7.397	0.0004
Tetrabutylborate	124–134	1.0774	7.982	0.0011
Sulfamate	106–128	1.0719	7.669	0.0003
Ethanesulfonate	138–156	1.0580	7.346	0.0004
Methanesulfonate	100–134	1.0421	7.321	0.0003
Nitrite	100–126	1.0098	9.254	0.0005
Nitrate	124–148	1.0086	6.645	0.0004
Thiocyanate	126–144	0.9719	6.188	0.0009
Chloride	72–100	0.9659	6.892	0.0002
Benzenesulfonate	105–145	1.1278	9.466	0.0036
Pentafluorobenzenesulfonate	75– 95	1.2368	8.914	0.0021
Octanesulfonate	80–130	1.0537	8.509	0.0014
Perfluorooctanesulfonate	85–140	1.4813	13.911	0.0060
Tris(hydroxymethyl)methyl- aminopropanesulfonate	120–140	1.2358	10.150	0.0160
3-(Cyclohexylamino)propanesulfonate	95–130	1.0508	7.290	0.0019
3-(*N*-Morpholino)propanesulfonate	70–130	1.0497	6.800	0.0028
N-(2-Acetamido)-2-aminoethanesulfonate	105–130	1.1275	8.741	0.0029
2-(Cyclohexylamino)ethanesulfonate	80–130	1.0612	8.459	0.0061
B. *4-Toluenesulfonates salts*				
Ethylammonium	132–154	1.2346	7.316	0.0008
Diethylammonium	114–140	1.1807	7.855	0.0006
Triethylammonium	102–122	1.1800	8.724	0.0006
Tetraethylammonium	104–124	1.1655	7.756	0.0004
Dipropylammonium	116–138	1.1311	7.148	0.0004
Tetrapropylammonium	110–134	1.1193	7.615	0.0002
Hexylammonium	136–152	1.1273	7.482	0.0007
Diallylammonium	92–112	1.1665	7.425	0.0004
Dimethyl-*n*-butylammonium	110–130	1.1464	7.405	0.0005
Tetrapentylammonium	96–124	1.0508	7.230	0.0007
C. *Miscellaneous*				
Ethylammonium nitrate	25–55	1.1469	4.850	
Propylammonium nitrate	25–55	1.0888	4.200	
Butylammonium thiocyanate	25–55	0.9637	5.550	
sec-Butylammonium thiocyanate	25–55	1.0273	5.700	
Dipropylammonium thiocyanate	25–55	0.9773	5.300	

TABLE II. Physical Properties of Chromosorb Supports

Property	Chromosorb (Johns-Manville)				
	A	G	P	W	750
Color Type	Pink Flux, calcined	White Flux, calcined	Pink calcined	White Flux, calcined	White Flux, calcined
Density (g/ml)	0.48	0.58	0.47	0.24	0.37–0.42
Surface area (m^2/g)	2.7	0.5–0.8	4.0–8.0	1.0	0.5–1.0
Surface area (m^2/ml)	1.3	0.29	1.88	0.29	0.16–0.35
Maximum phase loading (%w/w)	25	5	30	20	12
pH	7.1	8.5	6.5	8.5	8.0

cated in Table II.[28] Any of these might be used to prepare efficient column packings. The white supports have a smaller surface area compared to the pink supports. The use of sodium carbonate flux during calcining imparts a slightly basic character to the white material compared to the natural acidic character of the pink material. Acid washing to remove surface metallic impurities results in a more inert surface and is recommended for analytical and physicochemical studies (acid-washed supports are commercially available). Silanization to reduce the concentration of surface silanol groups is a further common deactivation treatment given to diatomaceous supports. This is not appropriate when the supports are to be coated with liquid organic salts as the latter do not wet such nonpolar surfaces adequately for efficient column preparation.[29] Thus, acid-washed and unsilanized supports are generally used with liquid organic salts. The particle size of the support used depends on the intended application. The column efficiency (as measured by the height equivalent of a theoretical plate) is directly proportional to the particle size, while the pressure drop across the column for a particular column length and mobile-phase velocity is inversely proportional to the square of the particle diameter. Thus, for analytical applications, for which the highest practical efficiency is desired, relatively long columns packed with 80–100 mesh (177–149 μm) or 100–120 mesh (149–125 μm) are selected. For the purpose of making physicochemical measurements, it is desirable to operate columns with a minimum pressure drop so as to minimize the need for additional correction terms due to nonideality of the carrier gas (see below). Thus, for these applications, columns should be as short as possible (commensurate with obtaining sufficient retention at the column temperature to ensure adequate precision in determining t_R), they should have a high phase loading, and they should be packed with a coarse support such as 60–80 mesh (250–177 μm). The pink supports have a much higher loading capacity than the white supports as well as a higher surface area. When it is desired to determine gas–liquid partition coefficients, a support of low surface area is advantageous since this minimizes the contribution from interfacial adsorption to retention. Thus, column packings containing from 10 to 20% (w/w) on Chromosorb W-AW or from 2.5 to 5% (w/w) on Chromosorb G-AW, both 60–80 mesh, are a reasonable choice (Chromosorb G is about $2\frac{1}{2}$ times more dense than Chromosorb W so, in spite of its lower capacity, a given column volume will contain a similar amount of liquid phase). Chromosorb P and specially prepared silicas of higher surface area (Spherosil, Porasil), etc.) are preferred when an accurate estimate of the gas–liquid adsorption coefficient is desired. For Chromosorb P, phase loadings in the range 20–30% (w/w) are generally used in studies involving a mixed retention mechanism. The influence of support surface area as a necessary optimization parameter for the separation of hydrocarbons on a liquid organic salt is discussed elsewhere.[19] On account of the fragile nature of supports, they should always be handled carefully. New batches of support should be sieved to remove fines. It is not unusual for a new bottle of support to contain more than 25% of material smaller than the mesh size indicated. Removal of fines should be accomplished before coating.

Supports are usually coated by one of several evaporation methods.[26] The coating solvent should be a good solvent for the organic salt without reacting with it chemically and should be of sufficient volatility for easy evaporation. Generally, useful solvents for organic salts are methylene chloride, dioxane, methanol or ethanol, and acetonitrile. These solvents also are available in high purity and without additives that may contaminate the prepared packings. Conventionally, column packings are prepared on a weight per weight basis and quoted as a percentage of liquid phase. The support is coated by adding it slowly to a flask containing the organic salt in solution, for example, by letting it trickle into a gently swirled solution through a fluted filter paper with a hole at its apex. Excess solvent can be removed conveniently by rotary evaporation, with care taken (owing to the fragile nature of the supports) to avoid rapid rotation of the flask or violent bumping of the damp solid. After coating, the damp support may be air-dried, oven-dried, or dried in a fluidized bed drier. Hygroscopic packings should be dried in a stream of warm nitrogen without exposure to the atmosphere, such as in a fluidized bed drier. Fluidized bed drying, in general, is fast and may lead to more efficient and permeable column packings because fines developed during the coating procedure are carried away by the gas passing through the packing. The design of a suitable fluidized bed drier, which uses a fine porous glass frit as a disrupter of the nitrogen flow and a hot plate to warm the incoming gas, is shown in Fig. 6.

Copper, aluminum, stainless steel, nickel, or glass tubes bent into various shapes to fit the dimensions of the column oven provide the container for column packings. Glass columns are preferred because of their inertness and because they allow visual inspection of the condition of the packing. Deleterious interactions between polar solutes and metal surfaces are known to occur, which makes

FIGURE 6. Fluidized bed drier for drying column packings.

them less useful for column preparation in spite of their greater mechanical strength.[26, 30] Columns 0.5–3.0 m in length with internal diameters of 2–4 mm can be packed conveniently by the tap-and-fill method, aided by vacuum suction (Fig. 7). One end of the column is terminated with a short glass wool plug and attached via a hose to a water pump aspirator. A small filter funnel is attached to the other end, and the packing material is added in small portions. The column bed is consolidated as it forms, by gentle tapping of the column sides with a rod or with the aid of an electric vibrator. Excessive vibration is detrimental to the

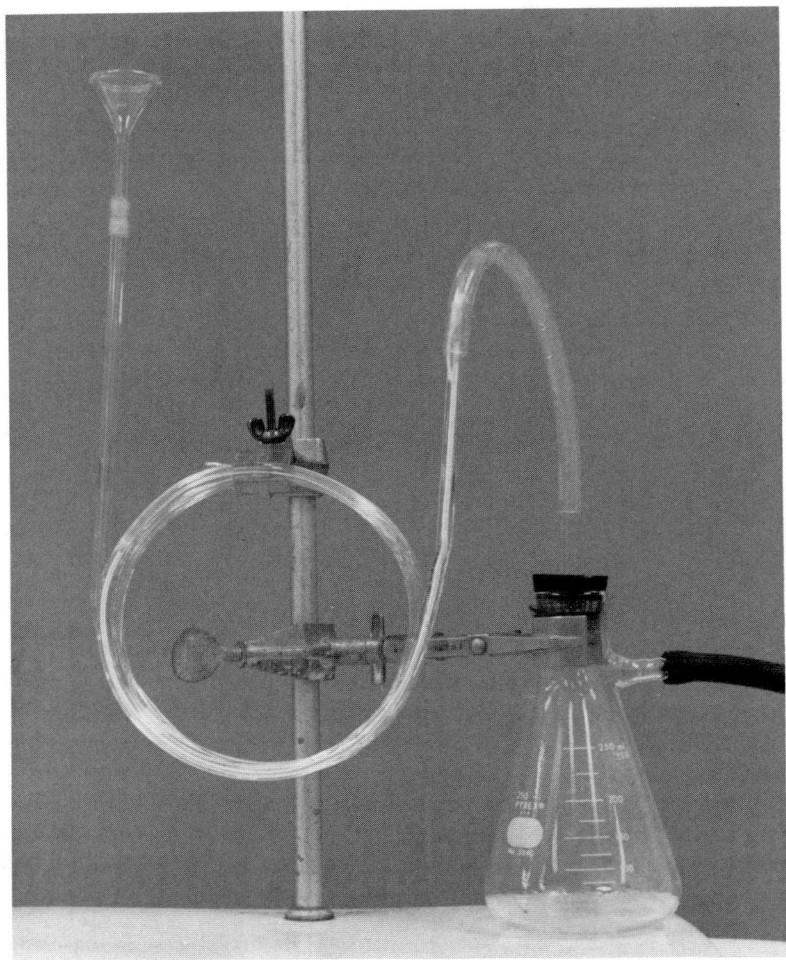

FIGURE 7. Apparatus for column filling by the tap-and-fill method with vacuum suction.

preparation of an efficient column and should be avoided. When sufficient packing has been added, the funnel is removed and a glass wool plug is inserted. The length of the plug is usually dictated by the design of the injector.

The packed column requires a conditioning period before use. The column is installed in the oven with the detector end disconnected and with a low flow of carrier gas passing through it. The conditioning temperature is usually slightly higher, ca. 20°C, than the temperature at which the column will be used, but it should be well below the thermal limit for the salts when physicochemical measurements are to be made. The required conditioning period is usually less than one hour for the organic salts and is complete when a stable detector baseline is obtained.

5.5. Accurate Determination of Phase Loadings

The amount of liquid phase coated on the support is frequently less than the nominal amount weighted out owing to losses of phase to the walls of the coating vessel and from uneven coating. The latter arises when heavily loaded particles adhering to the walls of the coating flask remain behind after the coated support is removed. For physicochemical measurements, the weight of salt in the column must be known accurately. This is usually determined by either combustion,[31] evaporation,[32] or Soxhlet extraction[31, 33, 34] of a sample of the column packing. Soxhlet extraction is generally more convenient for the liquid organic salts. The packing, ca. 0.5–1.0 g, accurately weighted into a fine-porosity glass sintered crucible with a small piece of filter paper on top, is placed in a standard Soxhlet apparatus. Glass beads are used to raise the level of the crucible above that of the siphon arm. The packings are then extracted at a rate of about 10–15 cycles per hour with, generally, the same solvent as used for coating. Extraction of the salt normally requires about 4 h, but overnight extraction is generally used to ensure completion. After extraction, the packing is oven-dried to constant weight, cooled, and accurately weighed. As a check on the procedure, the weight of extracted salt also should be determined. The most likely source of error is carry-over of fine particles from the column packing, if they are present in sufficient amounts. The solvents used for the extraction should be high-quality chromatographic grade, and a clean oven and desiccator are needed for the drying steps. If the evaporation method is used, then blanks should be run with each packing to allow for the weight loss from the support when it is raised to a high temperature.

5.6. Determination of Liquid Ranges

The useful liquid range for an organic salt is defined at its lower end by its melting point and at its upper end by its vapor pressure or intrinsic thermal instability. The definition of the former is obvious, but for the upper temperature limit there is no general consensus and we have adopted the following criteria. The onset of thermal instability was determined as the lowest temperature at which a physical change, such as discoloration or bubbling of a sample of salt contained in a melting point capillary, could be observed by eye. The maximum allowable column operating temperature was determined as the highest temperature at which a column packing prepared with the salt could be maintained for 24 h without change in retention or peak shape in a test chromatogram obtained at a lower temperature before and after the conditioning period. For nonionic phases, a vapor pressure of 0.1–0.01 mm Hg at the maximum operating

TABLE III. Liquid Ranges for Organic Salts Melting Above Room Temperature

Salt	Melting point (°C)	Decomposition temperature[a] (°C)	Column temperature limit (°C)	Useful liquid range (°C)
A. *Tetrabutylammonium salts*[b]				
Perfluorobenzenesulfonate	51		210	159
4-Toluenesulfonate	55	230 (B)	200	145
Benzenesulfonate	78		210	132
Tetrafluoroborate	162	320 (D)	290	128
Trifluoromethanesulfonate	112	270 (D)	240	128
Picrate	90	200 (D)	200	110
Methanesulfonate	79	220 (D)	180	101
Tris(hydroxymethyl)methyl-aminopropanesulfonate	110	230 (D)	210	100
N-(2-Acetamido)-2-amino-ethanesulfonate	84		160	76
Pentacyanopropenide	120	285 (D)	190	70
Chloride	41	200 (D)	110	69
Ethanesulfonate	117	225 (B)	180	65
Nitrate	118	180 (B)	170	52
Sulfamate	110	190 (D)	150	50
Camphorate	138	225 (D)	180	42
Dihydrogenphosphate	154	200 (B)	190	36
Nitrite	81	180 (B)	110	29
Tetraphenylborate	236	270 (D)	260	24
Hydrogensulfate	171	260 (D)	190	19
Hexafluorophosphate	247	270 (D)	260	13
Perchlorate	208	220 (D)	220	12
B. *4-Toluenesulfonate salts*				
Tetrapentylammonium	55	225 (B)	190	135
Tripropylaammonium	75	220 (D)	180	105
Tetraethylammonium	85	250 (D)	190	105
Diethylammonium	105	240 (B)	210	105
Triethylammonium	78	220 (D)	180	102
Ethylammonium	121	250 (B)	220	99
Tributylammonium	82	260 (B)	180	98
Trimethylammonium	93	200 (B)	190	97
Tetrapropylammonium	101	220 (B)	190	89
Diallylammonium	76	170 (B)	165	89
Butylammonium	125	220 (B)	210	85
Dimethylbutylammonium	107	200 (B)	190	83
Dipropylammonium	102	260 (B)	180	78
Dibutylammonium	135	240 (B)	210	75
Hexylammonium	128	250 (B)	180	52
Propylammonium	137	240 (B)	180	43

TABLE III. (Continued)

Salt	Melting point (°C)	Decomposition temperature[a] (°C)	Column temperature limit (°C)	Useful liquid range (°C)
C. Thiocyanate salts				
Triethanolammonium	37		150	113
Isobutylammonium	33	190 (D)	120	87
Ethylammonium	47		125	78
Dibutylammonium	55	210 (D)	130	75
Tributylammonium	60	210 (D)	130	70
Propylammonium	51	195 (D)	120	69
Isopropylammonium	70	210 (D)	130	60
Benzylammonium	82	185 (D)	115	33
Cyclohexylammonium	100	230 (D)	130	30
tert-Octylammonium	93	220 (D)	120	27
tert-Butylammonium	96		120	24
Octylammonium	122	270 (D)	130	8
	gel	210 (B)	130	
D. Miscellaneous				
Tetrahexylammonium tetrafluoroborate	90		230	140
Tetraoctylammonium tetrafluoroborate	123		170	47
Tetrapropylammonium tetrafluoroborate	247		260	13
Tetrapentylammonium 4-toluenesulfonate	55	225 (B)	190	135
1-Methyl-3-ethylimidazolium chloride	80	180 (D)	170	90
Tributylmethylammonium camphorate	65	205 (B)	170	105
Tributylbenzylphosphonium chloride	162		240	78
Ethylpyridinium bromide	114		160	46
Sodium isovalerate	180		280	100

[a]B: bubble formation observed; D: decomposition observed
[b]Bromide, iodide, perrhenate, tetrabutylborate, piperazine-*N,N'*-bis(2-ethanesulfonate), and *N*-2-hydroxyethyl-piperazine-*N'*-2-ethanesulfonate decompose or vaporize in the region of their melting points. Pentacyanopropenide degrades slowly at temperatures above its melting point.

temperature is assumed to be a reasonable limit for general column operation. The column upper temperature limits given in Tables III and IV are in accord with the above definition and provide a qualitative indication of the relative vapor pressures of the different salts. From Tables III and IV, which are comprehensive with respect to salts studied by gas chromatography, it can be seen that the vapor pressure controls the upper column temperature limit more frequently than thermal decomposition, particularly for alkylammonium salts containing weak nucleophilic anions. The room temperature liquid organic salts have in some instances been used as mobile phases in high-pressure liquid chromatogra-

TABLE IV. Thermal Stability of Room Temperature Liquid Organic Salts

Salt	Decomposition temp.[a] (°C)	Column temp. limit (°C)
Ethoquad 18/25[b]	320 (D)	300
Tetrabutylammonium perfluorooctanesulfonate		220
Tetrabutylammonium 3-(*N*-morpholino)propanesulfonate		180
Tetrabutylammonium octanesulfonate		180
Tributylheptylammonium camphorate		170
Tetrabutylammonium 3-(Cyclohexylamino)propanesulfonate		160
Triethylhexylammonium triethylhexylboride[c]	210 (D)	130
Tetraheptylammonium chloride		130
Dipropylammonium thiocyanate	180 (B)	130
Ethylammonium nitrate	170 (D)	120
Butylammonium thiocyanate	190 (D)	120
sec-Butylammonium thiocyanate	200 (D)	120
Tetrahexylammonium benzoate		110
Propylammonium nitrate	120 (D)	110
Tributylammonium nitrate	120 (D)	80

[a]B: bubble formation observed; D: decomposition observed
[b]Stearylmethyldipolyoxyethyl(15)ammonium chloride.
[c]This salt is unstable in air.

phy, as well as providing a convenient link between the chromatographic and spectroscopic methods of determining solvent characteristics.[35, 36]

5.7. Determination of Phase Transitions

Solid–solid and solid–liquid phase transitions are easily determined from plots of log V_g^o versus $1/T$ in gas chromatography. This particular application is well reviewed in Ref. 37. Wasik successfully detected seven phase transitions for sodium stearate in the range 60–290°C by gas chromatography, in good agreement with data determined from calorimetery and dilatometry.[38] Similar techniques have been employed for the liquid organic salts in Tables III and IV. For these low-molecular-weight salts, the only phase transition observed corresponds to the bulk melting point for the salt. The data obtained for ethylpyridinium bromide are fairly typical (Fig. 8).[39] The two test solutes show different behavior since they are retained by different retention mechanisms. A relatively large break at the melting point is observed for naphthalene, which is retained by interfacial adsorption below the melting point and partitioning above it. Butanol is retained largely by interfacial adsorption before and after melting and therefore shows a smaller break at the transition temperature since the change in liquid surface area at the melting point is comparatively small. The melting point transition is frequently accompanied by a fairly large increase in column efficiency (Fig. 9).[40] The efficiency of many of the alkylammonium salts at tem-

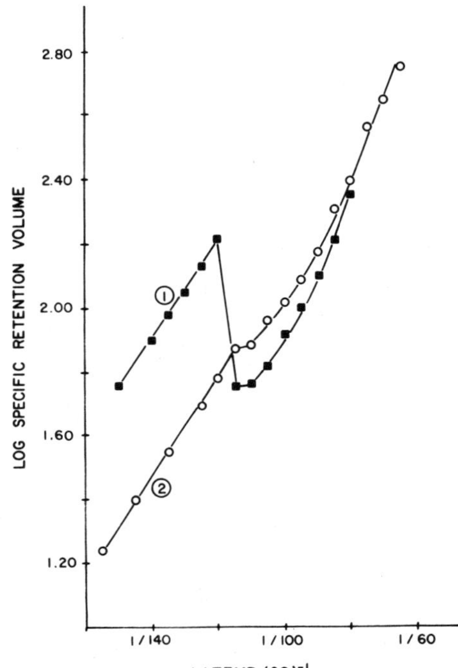

FIGURE 8. Observation of the solid → liquid phase transition temperature for ethylpyridinium bromide. The test solutes are (1) naphthalene and (2) *n*-butanol. (Reproduced with permission from Ref. 39. Copyright Friedr. Vieweg & Sohn.)

peratures below their melting points is too low for their successful use as sorbents in gas–solid chromatography.

6. Correlation of Solute Retention with Ion Type Using Molar Retention Volumes

The influence of ion selectivity on solute retention is conveniently measured from the change in the molar retention volume of a series of test solutes, selected for their ability to express specific intermolecular interactions with a series of liquid organic salt phases possessing a common ion.[16, 40, 41] When only the solvent characteristics are to be compared, the molar retention volume should be corrected for the influence of interfacial adsorption on retention, as discussed previously (Section 5.2). The molar retention volumes for a series of test solutes on 13 tetra-*n*-butylammonium salts are shown in Fig. 10. Dispersive interactions, represented by the retention of toluene, are relatively weak and fairly constant except for the ions of large molecular volume (tetra-*n*-butylborate, pentacyanopropenide, and picrate). Orientation interactions, represented by bromobenzene (μ_D = 1.55), *o*-dichlorobenzene (μ_D = 2.2), benzaldehyde (μ_D =

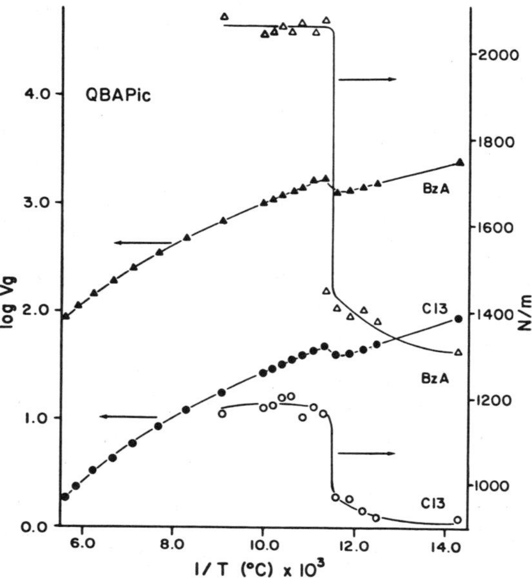

FIGURE 9. Plots of log of the specific retention volume and column efficiency (plates per meter) as a function of column temperature for benzaldehyde (BzA) and *n*-tridecane (C13) on the liquid organic salt tetrabutylammonium picrate. (Reproduced with permission from Ref. 40. Copyright Elsevier Scientific Publishing Co.)

2.96), and nitrobenzene (μ_D = 4.18), are stronger than the dispersive interactions but show a similar variation as a function of the anion type. Thus, it is probable that an increase in dispersive interactions rather than orientation interactions is responsible for the observed increase in retention. The strength of orientation interactions is presumably governed by the magnitude of coulombic forces within the salts, in which case variation in the coulombic forces with changes in the anion must be comparatively small or the coulombic forces are of sufficient magnitude that the orientation interactions have reached saturation. The largest variation in retention is seen for the proton donor solutes aniline and hexanol. Here the nature of the anion, and, in particular, its relative basicity, has a large influence on the retention of these solutes. Unfortunately, no independently derived scale of anion basicities for liquid organic salts is available for comparison with the data in Fig. 10. However, some useful correlations exist with the pK_a values of anions of a similar kind. For example, a good correlation exists between the pK_a values of the sulfonate anions (methanesulfonate, trifluoromethanesulfonate, ethanesulfonate, and 4-toluenesulfonate) and the molar retention volumes of the proton donor solutes butanol (r^2 = 0.995) and 2-methyl-2-pentanol (r^2 = 0.998).[41]

Changes in the molar retention volume are shown in Fig. 11[16] for a series of test solutes similar to those used above on 10 4-toluenesulfonate salts. In this

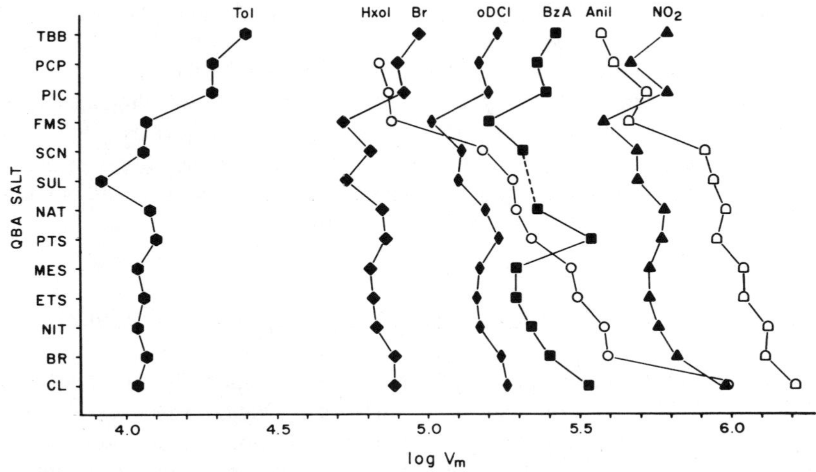

FIGURE 10. The influence of the anion type for a series of tetra-*n*-butylammonium salts on the molar retention volume of toluene (Tol), *n*-hexanol (Hxol), bromobenzene (Br), *o*-dichlorobenzene (oDCI), benzaldehyde (BzA), aniline (Anil), and nitrobenzene (NO$_2$). Identification of anions: TBB, tetra-*n*-butylborate; PCP, pentacyanopropenide; PIC, picrate; FMS, trifluoromethanesulfonate; SCN, thiocyanate; NAT, nitrate; SUL, sulfamate; PTS, 4-toluenesulfonate; MES, methanesulfonate; ETS, ethanesulfonate; NIT; nitrite; BR, bromide; CL, chloride. (Reproduced with permission from Ref. 41. Copyright Elsevier Scientific Publishing Co.)

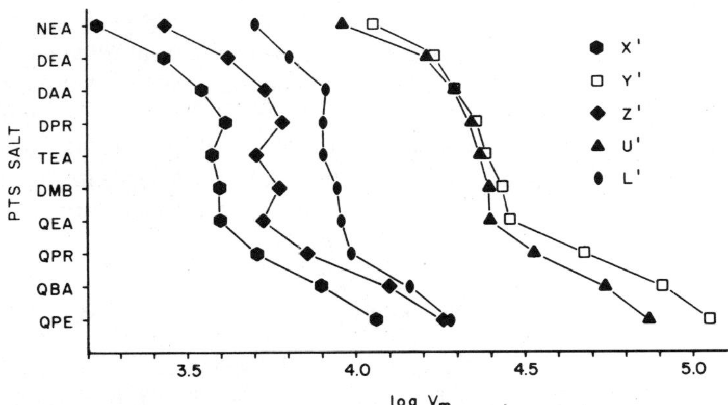

FIGURE 11. The influence of cation type for a series of 4-toluenesulfonate salts on the molar retention volume of X$'$ = benzene, Y$'$ = butanol, Z$'$ = 2–pentanone, U$'$ = nitropropane, L$'$ = dioxane. Identification of cations: NEA, ethylammonium; DEA, diethylammonium; DAA, diallylammonium; DPR, dipropylammonium; TEA, triethylammonium; DMB, dimethylbutylammonium; QEA, tetraethylammonium; QPR, tetrapropylammonium; QBA, tetrabutylammonium; QPE, tetrapentylammonium. (Reproduced with permission from Ref. 16. Copyright American Chemical Society).

case, the similarity of the plots is striking and suggests that the variation in retention with changes in the structure of the alkylammonium cations is largely independent of the various specific interactions expressed by the test solutes. The four tetraalkylammonium salts at the bottom of the figure show a linear increase in retention as a function of the alkyl chain length for all probes. Only their slopes are different. Thus, changing the size and type of the cation primarily changes the contribution made by dispersion to the retention of all solutes, independent of the nature of any specific interactions.

7. Determination of Thermodynamic Properties from Retention Data

The thermodynamic properties of a solution are usually described in terms of their deviations from the properties attributed to an ideal (or perfect) solution. An ideal solution obeys Raoult's law, with a chemical potential for each component that varies linearly with the logarithm of composition throughout the whole concentration range. In gas chromatography, adherence to Raoult's law is uncommon, and the concept of the activity coefficient is introduced to account for the deviations from ideality. In this case, it is assumed that the true active concentration is not the molar concentration but some fraction or multiple of it:

$$P_1 = P_1^\circ Y_1 x_1 \tag{14}$$

where P_1 is the vapor pressure of the component and is equal to its partial pressure in the gas phase, P_1° is the saturation vapor pressure of the pure solute at the given temperature, Y_1 is the solute activity coefficient, and x_1 is the mole fraction of solute in the solution. At low concentration, corresponding to infinite dilution, Henry's law is obeyed and the activity coefficient (also the gas–liquid partition coefficient) can be assumed to have a constant value. At infinite dilution, the solute molecules are not sufficiently close together to exert any mutual attractions, and the environment of each consists entirely of solvent molecules. This is the region of interest for gas chromatographic measurements. In the Henry's law region, the activity coefficient and specific retention volume are related by

$$V_g^\circ = \frac{273\,R}{M_2 Y_1^\infty P_1^\circ} \tag{15}$$

where R is the universal gas constant, Y_1^∞ is the solute activity coefficient at infinite dilution, and M_2 is the molecular weight of the solvent. If P_1° is measured in mm Hg, T in K, and R is taken to be 62,370 cm$^3\cdot$(mm Hg)/(mol$\cdot$K), then Eq. (15) can be conveniently rewritten as

$$Y_1^\infty = \frac{1.704 \times 10^7}{M_2 V_g^\circ P_1^\circ} \tag{16}$$

The specific retention volume and the gas–liquid partition coefficient are related by Eq. (9), and, of course, both must be corrected for the influence of interfacial adsorption. In fact, Y_1^∞ defined by Eq. (16) is an approximation reasonably well suited for the qualitative comparison of solution properties. For some purposes, it is necessary to make a correction for gas-phase imperfections, as indicated by Eq. (17).

$$\ln Y_1^\infty(0) = \ln \frac{273\,R}{P_1^\circ M_2 V_g} - \frac{(B_{11} - v_1^\circ)}{RT}\,P_1^\circ + \frac{(2B_{12} \times v_1^\infty)}{RT}\,P_o\,J_3^4 \tag{17}$$

where $Y_1^\infty(0)$ is the activity coefficient at zero total pressure, B_{11} is the second virial coefficient characterizing interactions between two solute molecules, B_{12} is the second virial coefficient characterizing interactions between solute and carrier gas molecules, v_1° is the partial molar volume of the liquid solute, v_1^∞ is the molar volume of the solute at infinite dilution in the liquid phase, P_o is the column outlet pressure, and $J_3^4 = \frac{3}{4}(P^4-1)/P^3-1)$, with P the column pressure drop. The middle term in the right-hand expression is known as the fugacity, and sometimes activity coefficients are corrected for fugacity only. When accurate data are required, this is rarely justified. Equation (17) requires a knowledge of the solute–solute and solute–gas second virial coefficients, which are unknown for many solutes of interest in gas chromatography. If they are not known, they must be calculated from the critical constants for the pure substances using the McGlashan–Potter equation and appropriate combining rules.[2, 3, 42] The corrections for gas-phase imperfections are small (ca. 0.5–5%) for hydrogen, helium, or nitrogen as carrier gas and column pressure drops less than 5 psi. The magnitude of the correction for gas-phase imperfections increases with the column pressure drop, dictating that this be maintained as small as possible for physicochemical measurements by using moderate flow rates, short columns, and coarse packings, as discussed earlier. Generally speaking, calculation of Y_1^∞ by Eq. (16) yields values that are 1–5% below the fully corrected value obtained using Eq. (17). To determine excess functions, Y_1^∞ should be known to at least 1% of its true value, and thus the fully corrected activity coefficient is always needed for the calculation of these values.

The excess functions of free energy, enthalpy, and entropy are determined from the fully corrected activity coefficient, using standard equations (Eqs. 18–20).

$$\Delta G^E = RT \ln Y_1^\infty \tag{18}$$

$$\Delta H^E = \frac{\partial \ln Y_1^\infty}{\partial(1/T)} = -RT^2 \frac{\partial \ln Y_1^\infty}{\partial T} \tag{19}$$

$$\Delta S^E = \frac{\Delta H^E - \Delta G^E}{T} \tag{20}$$

The excess functions, also referred to as mixing, measure the transfer of a solute from an infinitely dilute ideal solution to the infinitely dilute real solution. As such, it could be argued that the excess functions rather than those of solution provide a more exacting description of solute-solvent interactions. There are, however, certain limitations on the gas chromatographic experiment for determining excess functions. Activity coefficients, in general, show only small variations with temperature, and this variation and the experimental precision with which Y_1^∞ can be conveniently determined may be close. Many authors have concluded that calorimetric methods are usually far more convenient and accurate for determining excess theromodynamic functions, except in those circumstances where the solute is unavailable in a pure form or is only available in limited quantities, inadequate for calorimetry. Since the Gibbs excess free energies are not determined by the temperature variation of the activity coefficients, they should be easier to determine with higher accuracy than the enthalpies and entropies.

The solution process in gas chromatography can be imagined as a two-stage process in which the solute is first condensed from a gas to a liquid followed by mixing of the two liquids. The change is free energy for this process, and analogously that of the enthalpy and entropy, can be represented by

$$\Delta G^{soln} = \Delta G^E + \Delta G^V \tag{21}$$

where ΔG^{soln} is the partial Gibbs free energy of solution, ΔG^E is the partial Gibbs free energy of mixing, and ΔG^V is the partial Gibbs free energy of vaporization. The Gibbs free energy of solution can be calculated directly from the partition coefficient. The variation of the heat of vaporization with temperature is given by the Clausius–Clapeyron equation:

$$\frac{d\ln P_1^\circ}{dT} = \frac{\Delta H^V}{RT^2} \quad \text{or} \quad \ln P_1^\circ = -\frac{\Delta H_V}{RT} + A \tag{22}$$

The heats of vaporization and solution are reasonably constant over a small temperature range (10–30 K), which enables solution enthalpies to be evaluated directly from plots of $\ln V_g^\circ$ versus $1/T$. Curvature in these plots is generally observed over an extended temperature range. Usually, between three and five measurements over a 15–30 K range are used to evaluate ΔH^{soln}.

7.1. Selection of Standard States

The magnitude of thermodynamic properties calculated by gas–liquid chromatography depends on the definition of the selected standard states. A brief description of the standard states and their associated equations for free energy, enthalpy, and entropy are given in Table V. The selection of individual standard states depends on the nature of the problem and the experimental data available. For example, when molarity is selected as the standard concentration, the cal-

TABLE V. Definitions of Standard States and Thermodynamic Constants Used in Gas–Liquid Chromatography

	Molarity	Mole fraction	Molality
Standard states	Solute in gas phase at temperature T, unit molar concentration, and behaving ideally being transferred to solution in the liquid phase at temperature T, unit molar concentration, and having molecular interactions characteristic of the infinite dilute solution	Solute in the gas phase at temperature T, 1 atm pressure, and behaving ideally transferred to solution in the liquid phase at temperature T, solute mole fraction = unity, the solute having molecular interactions characteristic of infinite dilution	Solute in the gas phase at temperature T, 1 atm pressure, and behaving ideally transferred to solution in the liquid phase at temperature T, solute molality = unity, the solute having molecular interactions characteristic of infinite dilution
Gibbs free energy	$\Delta G_k^\circ = -RT \ln K_L$ $= -RT \ln \dfrac{\vartheta \gamma T \rho_2}{273}$	$\Delta G_x^\circ = -RT \ln \dfrac{M_2 Vg}{273R}$	$\Delta G_M^\circ = -RT \ln \dfrac{1000 V_g}{273R}$
Enthalpy	$\Delta H_k^\circ = RT^2\,(d \ln K_L/dT)$ $= RT^2\,(d \ln V_g/dT)$ $+ RT\,(1 - \alpha_2 T)$	$\Delta H_x^\circ = RT^2\,(d \ln V_g/dT)$	$\Delta H_M^\circ = RT^2\,(d \ln V_g/dT)$
Entropy	$\Delta S_k^\circ = (\Delta H_k^\circ - \Delta G_k^\circ)/T$	$\Delta S_x^\circ = (\Delta H_x^\circ - \Delta G_x^\circ)/T$	$\Delta S_M^\circ = (\Delta H_M^\circ - \Delta G_M^\circ)/T$

culation of thermodynamic functions requires a knowledge of the liquid-phase density and thermal expansion coefficient, the latter defined by

$$\alpha_2 = \left(\frac{1}{v_2}\right)\left(\frac{\partial v_2}{\partial T}\right) \tag{23}$$

where α_2 is the liquid-phase thermal expansion coefficient, and v_2 is the molar volume for the liquid phase. Elimination of these terms from the expression for the equilibrium constant greatly simplifies the calculation of the thermodynamic properties of solution.[43, 44] Using the mole fraction standard state, neither density nor thermal expansion data are required for the liquid phase, a ready comparison can be made between solution and condensation thermodynamic properties for a given solute (these differences reflect the difference between solute–solute and solute–solvent interactions), and a comparison of the solution thermodynamics of different solutes in the same solvent is more meaningful since the same starting point for each solute (1 atm ideal gas state) is used. Activity coefficients with molarity as the standard state include the influence of intermolecular interactions in the pure liquid solute.

                                           Colin F. Poole et al.

When gas chromatographic thermodynamic studies were extended to polymeric solvents, certain inadequacies in the molarity and mole fraction standard states were identified.[45] For polymeric solvents, molecular weights generally are not known with certainty and are difficult to specify for polydisperse polymers. Furthermore, molarity and mole fraction activity coefficients tend to infinity as the molecular weight approaches the same limiting value within a family of structurally identical polymers, indicating that the activity coefficient is not a meaningful measure of the interaction of a solute with its surroundings in this case. To overcome this difficulty, the weight fraction and molal activity coefficients were introduced. Both activity coefficients tend to a finite value as the molecular weight increases, remaining almost constant for homologous solvents of molecular weight greater than 10,000.

Experimentally, the different enthalpies and entropies of solution are determined from plots of $\ln V_g^{\circ}$ versus $1/T$ for a short temperature range using the following equations:

$$\ln V_g^{\circ} = -\frac{\Delta H_x^{\circ}}{R}\left(\frac{1}{T}\right) + \frac{\Delta S_x^{\circ}}{R} - \ln\left(\frac{M_2}{273R}\right) \tag{24}$$

$$\ln V_g^{\circ} = -\frac{\Delta H_M^{\circ}}{R}\left(\frac{1}{T}\right) + \frac{\Delta S_M^{\circ}}{R} - \ln\frac{1000}{273R} \tag{25}$$

where ΔH_x° and ΔS_x° are the partial mole fraction enthalpy and entropy of solution, respectively, ΔH_M° and ΔS_M° are the partial molal enthalpy and entropy of solution, respectively, and M_2 refers to the use of the molal standard state and is contained in the definition for ΔH_M° etc. The enthalpies are determined from the slope of the plots and the entropies from a specific value for V_g° by incorporating the value for ΔH°. The enthalpies of solution for the mole fraction and molality standard states are identical. This is a consequence of the use of identical vapor-phase standard states and the use of infinite dilution as the standard state for the liquid phase. The partial molar enthalpy of solution (ΔH_k°) is different:

$$\Delta H_x^{\circ} = \Delta H_M^{\circ} \tag{26}$$

$$\Delta H_k^{\circ} = \Delta H_x^{\circ} + RT - \alpha_2 RT^2 \tag{27}$$

For the molarity standard state, the RT term arises as a consequence of the adoption of different vapor-phase standard states, and the $\alpha_2 RT^2$ term accounts for changes in molarity resulting from thermal expansion. Determination of the partial molar enthalpy of solution using equations analogous to Eqs. (24) and (25) also involves temperature and density terms that make the calculation less straightforward. In this case, ΔS_k° is usually calculated directly from ΔG_K° and ΔH_k°, using the relation $\Delta G_k^{\circ} = \Delta H_k^{\circ} - T\Delta S_k^{\circ}$.

7.2. Instrumental Considerations for Determining Thermodynamic Parameters

We have noted in previous sections how the reproducibility and absolute error in experimental variables influence the reliability of thermodynamic data. It is useful at this point to bring these data into focus and to consider the potential pitfalls of using commercial analytical instruments for thermodynamic measurements.

A typical modern analytical gas chromatograph will most likely provide mass flow control to an accuracy of $\pm 0.3\%$ of a preset value, an indication of the column pressure drop to ± 0.5 psi, an oven temperature set point accuracy of $\pm 3°C$, and constancy of the oven set point temperature to $\pm 0.2°C$. For a 1°C change in oven temperature, the specific retention volume will vary by perhaps 3%. Thus, temperature stability of about 0.2°C is at the upper limit for obtaining reliable thermodynamic data, and a stable temperature of $\pm 0.02°C$ is probably optimum. Similarly, the column pressure drop should be determined to at least a few mm Hg, and a value of about 0.1 mm Hg would be optimum. However, the largest source of experimental error generally will be the accuracy with which the liquid-phase volume can be determined. It is unlikely that this will be known to better than $\pm 0.2\%$, and in many cases a larger error can be anticipated. Additional errors can also be anticipated when corrections for interfacial adsorption are required and estimates for gas-phase inperfections have to be made. Here, absolute errors rising to 5% or so would not be unusual. Such an analytical instrument would not be compromised for determining free energies, enthalpies, and entropies of solution. However, it would be unsuitable for determining excess functions, since an accuracy of at least 1% or better for the activity coefficient is needed. Such precision requires special instruments that have been described in the literature.[2, 3] The determination of experimental errors and their contribution to the accuracy of thermodynamic quantities are discussed in Refs. 31 and 46–48.

7.3. Solution Thermodynamic Properties for Liquid Organic Salts

The selectivity of a solvent is normally expressed in terms of the relative capacity of the solvent to enter into specific intermolecular interactions. To evaluate these interactions, a series of test solutes is needed, such as those indicated in Table VI. The partial free energy of solution for the compared solvents provides an indication of how changes in solvent structure influence selective interactions.[17] For example, Fig. 12 illustrates the influence on solvent selectivity of varying the nature of the anion for a series of tetrabutylammonium salts. Changing the anion only weakly influences the dispersive contribution, with an average value for the partial molar free energy of solution for

TABLE VI. Selection of Test Solutes for Evaluating Specific Intermolecular Interactions by Gas Chromatography

Test solute	Dipole moment (μ_D)	Boiling point (°C)	Interactions expressed
Benzene	0	80	Primarily dispersion with weak proton acceptor and induction properties
Butanol	1.75	118	Orientation properties with both proton donor and proton acceptor capabilities. Retained most strongly by proton acceptor solvents
2-Pentanone	2.82	102	Orientation properties with proton acceptor but not proton donor capabilities
Nitropropane	3.59	131	Primarily orientation properties with weak proton acceptor properties
Pyridine	2.37	115	Orientation properties with strong proton acceptor but not proton donor capabilities
Dioxane	0.45	101	Weak orientation properties with strong proton acceptor properties. Proton donor properties are absent

benzene of about -2700 cal/mol. Similarly, orientation interactions are strong but do not vary significantly as the anion is changed. The average value for the partial molar free energy of solution for nitropropane is ca. -4100 cal/mol. The proton acceptor capabilities of the salts show a tremendous range as a function of the anion type, with the partial molar free energy of solution for butanol varying from -5707 cal/mol for the chloride salt to -3385 cal/mol for the picrate salt. There is no evidence for significant proton donor selectivity, which is not too surprising since the anions studied lack acidic protons.

For the 4-toluenesulfonates, varying the properties of the cation induces different trends in the change of solvent selectivity (Fig. 13). There is a general trend of increasing interaction for all test solutes with increasing cation size for the tetraalkylammonium salts. The underlying reason for this is most probably a general increase in the strength of dispersive interactions. It is also particularly noticeable how closely the free energy of solution for the orientation probe (nitropropane) and for the proton donor/acceptor probe (butanol) track each other as the cation is varied. Since the proton acceptor interactions measured for dioxane show far less variation among these salts, the proton donor interactions are probably not significantly changed by varying the cation. Therefore, the dominant interaction influenced by the cation is dispersion.

Mole fraction and molal partial free energies of solution for the solutes in Table VI are given in Ref. 17. Additional values for some room temperature liquid alkylammonium nitrate and thiocyanate salts[14, 49, 50] and Ethoquad 18/25[18] are given elsewhere.

The partial mole fraction enthalpy and entropy of solution for a number of

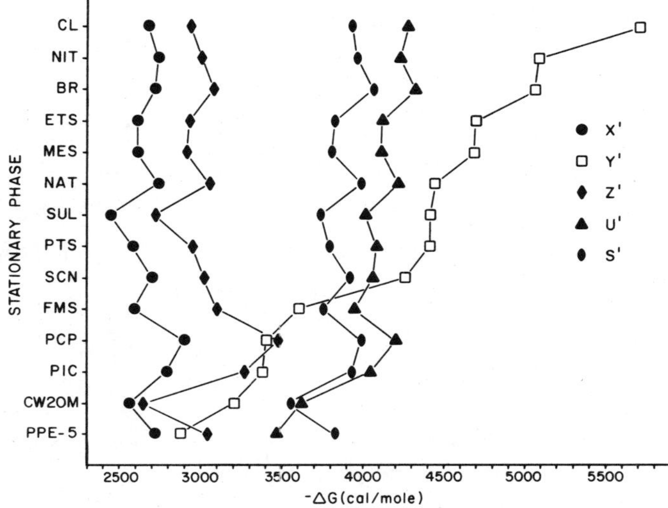

FIGURE 12. Variation in the partial molar Gibbs free energy of solution at 120°C as a function of the anion type for a series of tetrabutylammonium salts. Anions are identified in the caption to Fig. 10; CW20M, Carbowax 20M; PPE-5, 5-ring poly(phenyl ether); X', benzene; Y', butanol; Z', 2-pentanone; U', nitropropane; S', pyridine. (Reproduced with permission from Ref. 17. Copyright Elsevier Scientific Publishing Co.)

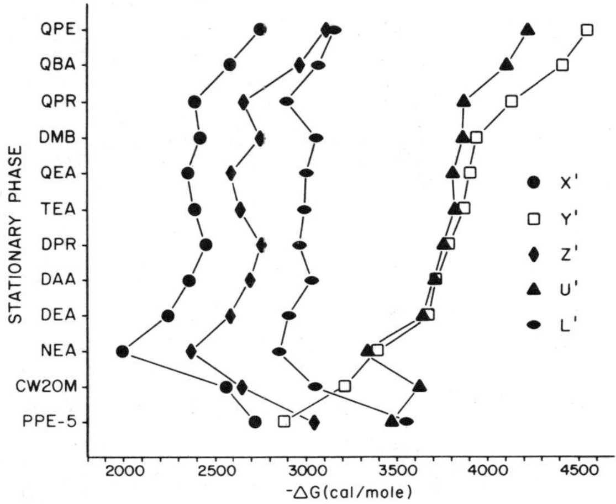

FIGURE 13. Variation in the partial molar Gibbs free energy of solution at 120°C as a function of the cation type for a series of 4-toluenesulfonate salts. The cations are identified in the caption to Fig. 11 and the test solutes in that to Fig. 12; L' = dioxane. (Reproduced with permission from Ref. 17. Copyright Elsevier Scientific Publishing Co.)

TABLE VII. Partial Mole Fraction Enthalpy (kcal/mol) and Entropy [cal/mol·K] of Solution for Aromatic Selectivity Probes on Tetra-*n*-butylammonium Salts

Anion	Naphthalene		Bromobenzene		Benzaldehyde		Benzonitrile		Acetophenone	
	$-\Delta H_x^\circ$	$-\Delta S_x^\circ$	$-\Delta H_x^\circ$	$-\Delta S_x^\circ$	$-\Delta H_x^\circ$	$-\Delta S_x^\circ$	$-\Delta H_x^\circ$	$-\Delta S_x^\circ$	$-\Delta H_x^\circ$	$-\Delta S_x^\circ$
Thiocyanate	13.1	20.1	10.4	17.0	11.7	18.0	12.1	18.4	12.0	17.8
Chloride	12.6	18.7	10.5	16.8	11.8	17.7	12.0	17.4	12.3	18.5
Methanesulfonate	12.1	17.6	9.8	15.3	10.9	16.0	13.9	16.2	11.8	17.4
Ethanesulfonate	12.0	17.8	10.0	16.2	11.0	16.8	13.6	16.6	11.7	17.6
Picrate	11.9	16.3	9.2	13.6	10.3	14.3	10.6	14.2	11.5	15.9
4-Toluenesulfonate	11.9	16.8	8.4	11.4	10.6	14.8	11.2	15.4	11.6	16.5
Sulfamate	11.7	17.1	9.8	15.7	—	—	12.1	18.4	11.6	17.3
Trifluoromethanesulfonate	11.6	16.4	9.0	13.7	10.4	15.0	10.5	14.6	11.2	15.8
Carbowax 20M	12.1	10.1	10.0	8.4	11.0	9.4	11.5	9.8	11.9	10.6
OV-275	12.1	14.5	9.5	11.7	10.7	12.1	10.9	12.0	11.8	13.9

Anion	Nitrobenzene		Benzodioxan		2,6-Dimethylaniline		Octanol	
	$-\Delta H_x^\circ$	$-\Delta S_x^\circ$	$-\Delta H_x^\circ$	$-\Delta S_x^\circ$	$-\Delta H_x^\circ$	$-\Delta S_x^\circ$	$-\Delta H_x^\circ$	$-\Delta S_x^\circ$
Thiocyanate	12.6	18.5	13.1	20.1	14.5	22.4	15.2	25.7
Chloride	12.7	18.1	13.0	19.3	—	—	—	—
Methanesulfonate	12.1	16.9	12.7	18.5	14.1	20.7	15.3	24.1
Ethaanesulfonate	12.0	17.3	12.6	18.8	14.2	21.4	15.6	25.3
Picrate	11.3	15.0	12.0	16.7	13.1	18.6	12.7	20.5
4-Toluenesulfonate	12.1	16.6	12.6	18.1	14.0	20.6	15.1	24.0
Sulfamate	11.9	16.6	12.5	18.2	13.9	20.6	14.6	23.4
Trifluoromethanesulfonate	11.1	15.1	11.8	16.8	12.7	18.5	12.8	20.7
Carbowax 20M	12.3	10.6	13.0	11.8	14.0	14.0	12.5	13.0
OV-275	11.6	12.8	12.7	15.4	13.3	16.2	12.2	17.3

aromatic solutes are summarized in Table VII. The data were calculated using the temperature variation of the specific retention volume and Eq. (24). The strengths of solute–salt interactions are reflected in the ΔH_x° values. In all cases, these are negative—that is, heat evolved—and of similar magnitude for the salts and the polar nonionic phases Carbowax 20M [a poly(ethylene glycol)] and OV-275 [a poly(dicyanoethylsilicone)], except for octanol. The ΔH_x values for octanol are generally much larger for the organic salts than for the nonionic phases, again testifying to the unusual strength of proton acceptor interactions in the liquid organic salts.

The entropy differences in Table VII are more difficult to interpret. The values for the salts are in all cases greater than for the nonionic phases. Since the entropy has a negative sign, the entropy term contributes in a negative sense to the free energy of solution. Dissolution of the test solutes in the liquid organic salts is a less random process than for the nonionic phases. This may reflect restricted maneuverability of the test solutes in the liquid organic salts or structural changes due, in part, to the large differences in size between the two types of solvents.

TABLE VIII. Activity Coefficients at Infinite Dilution (Uncorrected for Gas-Phase Imperfections) for Test Solutes in Some Liquid Tetrabutylammonium Salts at 120°C

Anion	Benzene	Butanol	2-Pentanone	Nitropropane	Pyridine
Chloride	1.07	0.06	1.41	0.60	0.58
Nitrite	1.01	0.14	1.26	0.63	0.55
Bromide	1.51	0.15	1.21	0.58	0.50
Tetrabutylborate	0.55	0.17	0.33	0.57	0.44
Ethanesulfonate	1.06	0.20	1.23	0.64	0.58
Methanesulfonate	1.09	0.21	1.30	0.66	0.61
4-Toluenesulfonate	0.96	0.25	1.07	0.58	0.53
Sulfamate	1.37	0.30	1.71	0.77	0.68
Nitrate	1.00	0.30	1.16	0.62	0.52
Thiocyanate	1.02	0.37	1.19	0.74	0.56
Trifluoromethanesulfonate	1.04	0.77	0.95	0.76	0.60
Picrate	0.69	0.88	0.66	0.58	0.43
Pentacyanopropenide	0.70	0.98	0.59	0.54	0.44
4-Morpholinepropane- sulfonate	0.68	0.19	1.00	0.54	—
N-(2-Acetamido)2- aminoethanesulfonate	1.62	0.43	2.35	1.08	—
3-Cyclohexylamino-1- propanesulfonate	0.65	0.18	1.02	0.59	—
Octanesulfonate	0.70	—	0.95	0.65	—
Perfluorooctanesulfonate	0.84	0.78	0.84	0.81	—
Benzenesulfonate	0.93	0.30	0.79	0.64	—
Perfluorobenzenesulfonate	0.59	0.43	0.69	0.67	—

TABLE IX. Activity Coefficients at Infinite Dilution (Uncorrected for Gas-Phase Imperfections) for Test Solutes in Some Liquid 4-Toluenesulfonate Salts at 120°C

Cation	Benzene	Butanol	2-Pentanone	Nitropropane	Dioxane
Ethylammonium	4.72	1.78	4.89	3.33	2.56
Diethylammonium	2.73	1.19	3.10	1.89	2.03
Triethylammonium	2.01	0.82	2.58	1.34	1.61
Tetraethylammonium	1.90	0.72	2.49	1.24	1.44
Tetrapropylammonium	1.47	0.43	1.84	0.92	1.34
Tetrabutylammonium	0.96	0.25	1.05	0.58	0.90
Tetrapentylammonium	0.66	0.18	0.74	0.42	0.68
Diallylammonium	2.13	1.02	2.44	1.58	1.55
Dipropylammonium	1.80	0.90	2.17	1.41	1.62
Dimethylbutylammonium	1.89	0.74	2.21	1.25	1.46

Values of the infinite-dilution coefficients for the test solutes of Table VI in all liquid organic salts studied to date are summarized in Tables VIII–X. Typical infinite-dilution activity coefficients for nonionic solvents used in gas chromatography have values in the range 0.3 to 70. Positive deviations from Raoult's law, $\gamma_1^\infty > 1$, are common for the high-molecular-weight solvents widely used in gas chromatography. For the liquid organic salts, the range of values observed is not unusual and indicates that the solvation of organic solutes in these media does not occur by any extraordinary means. Activity coefficients that are less than one indicate strong intermolecular interactions and are observed primarily for alcohols in the tetrabutylammonium salts.

A convenient ranking of solvents for comparative purposes is in terms of their solvent strength (or polarity). The solvent strength is best defined as the capacity of a solvent for various intermolecular interactions. Since solvent strength is not a unique property of a molecule but a composite expression for different selective interactions, there is no single substance that can be defined as polar. One solution to this problem is to equate the polarity of a stationary phase

TABLE X. Activity Coefficients at Infinite Dilution (Uncorrected for Gas-Phase Imperfections) for Test Solutes in Some Room Temperature Liquid Organic Salts at 30°C

Salt	Benzene	Butanol	2-Pentanone	Nitropropane	Dioxane
Ethylammonium nitrate	18.94	4.55	9.78	9.27	2.85
Propylammonium nitrate	12.94	2.95	6.90	7.08	2.60
sec-Butylammonium thiocyanate	5.17	1.16	1.68	2.99	0.80
Butylammonium thiocyanate	5.26	1.36	1.99	3.48	1.02
Dipropylammonium thiocyanate	2.07	1.29	1.92	1.86	1.44

with its reluctance to retain a nonpolar solute, such as a hydrocarbon. In thermodynamic terms, this reluctance to retain a hydrocarbon solute can be expressed as the partial molar free energy of solution for a methylene group and is calculated from the following equation[17,49,50]:

$$\Delta G^{CH_2} = -2.3RTb \qquad (28)$$

TABLE XI. Dispersive Polarity Ranking of Liquid Organic Salts

Salt	$-\Delta G^{CH_2}$(cal/mol)
	120°C
Tetrabutylammonium Tetrabutylborate	440
Tetrapentylammonium 4-Toluenesulfonate	426
Tetrabutylammonium octanesulfonate	420
Tetrabutylammonium 3-cyclohexylamino-1-propanesulfonate	419
Tetrabutylammonium pentafluorobenzenesulfonate	407
Tetrabutylammonium pentacyanopropenide	404
Tetrabutylammonium picrate	399
Tetrabutylammonium 4-morpholinepropanesulfonate	389
Tetrabutylammonium perfluorooctanesulfonate	385
Tetrabutylammonium ethanesulfonate	383
Tetrabutylammonium 4-toluenesulfonate	383
Tetrabutylammonium chloride	379
Tetrabutylammonium benzenesulfonate	378
Tetrabutylammonium methanesulfonate	374
Tetrabutylammonium trifluoromethanesulfonate	372
Dipropylammonium 4-toluenesulfonate	367
Tetrabutylammonium bromide	367
Tetrabutylammonium nitrite	365
Tetrabutylammonium nitrate	361
Tetrabutylammonium thiocyanate	358
Tetrabutylammonium sulfamate	356
Dimethylbutylammonium 4-toluenesulfonate	347
Diallylammonium 4-toluenesulfonate	347
Tetrapropylammonium 4-toluenesulfonate	345
Tetrabutylammonium *N*-(2-acetamido)-2-aminoethanesulfonate	323
Diethylammonium 4-toluenesulfonate	311
Triethylammonium 4-toluenesulfonate	302
Tetrabutylammonium 3-[tris(hydroxymethyl)methylamino]-1-propanesulfonate	285
Ethylammonium 4-toluenesulfonate	264
Tetraethylammonium 4-toluenesulfonate	237
	80°C
Dipropylammonium thiocyanate	386
Butylammonium thiocyanate	347
sec-Butylammonium thiocyanate	300
Propylammonium nitrate	267
Ethylammonium nitrate	141

where ΔG^{CH_2} is the partial molar Gibbs free energy of solution for a methylene group, and b is the slope of a plot of log K_L versus carbon number for a homologous series of n-alkanes or 2-alkanones. On some salts, the retention of n-alkanes occurs almost entirely by interfacial adsorption, and more soluble test solutes such as the 2-alkanones must be sought in order to make meaningful solution measurements.[49, 50] Similar values for ΔG^{CH_2} are usually obtained for any homologous series differing only in chain length, provided that the chosen homologous series produces reasonable gas–liquid partition coefficients. Values of ΔG^{CH_2} for all the liquid organic salts studied to date are summarized in Table XI. The salts are ordered by increasing solvent strength going from tetrabutylammonium tetrabutylborate (least polar) to tetraethylammonium 4-toluenesulfonate (most polar) on the scale measured at 120°C. Although equivalent free energy values at 120°C are unavailable for the nitrate salts at the bottom of Table XI, it is clearly obvious that they are significantly more polar than the other salts. For the 4-toluenesulfonate salts, the polarity decreases with increasing size of the cations in the order tetrapentylammonium < tetrabutylammonium < tetrapropylammonium < tetraethylammonium. Except for the lowest member of the series, there is a linear correlation between the value for ΔG^{CH_2} and the number of carbon atoms attached to the cation.[16, 17] The range of ΔG^{CH_2} values for the 20 tetrabutylammonium salts is not very great, varying from -323 to -440 cal/mol on a scale that goes from -237 to -520 cal/mol when the nonionic phases are included. Ignoring extreme values, 14 salts have values between -389 and -356 cal/mol. This is in keeping with general observations on the retention of polar, non-proton-donor solutes discussed earlier. The salts in Table XI cover the entire polarity range of values for nonionic solvents except for the aliphatic and aromatic hydrocarbons, and the range of ΔG^{CH_2} values extends considerably beyond the smallest value of ΔG^{CH_2} measured for a nonionic phase.[17, 49, 50]

8. Solute Interactions in Inorganic Melts

Inorganic molten salts are more numerous and better characterized than their organic analogues and have been the subject of several investigations by gas–liquid chromatography.[51–53] Only hydrocarbon and aromatic solutes could be chromatographed on these salts owing to the poor peak shapes observed for polar solutes. Even for nonpolar solutes, the recovery of injected mass for small sample sizes was incomplete. All the available evidence points to interfacial adsorption as the dominant retention mechanism. Indeed, in the most recent studies of Butler and Sibbald[54] and Koser et al.,[55] it was concluded that most separations reported could be explained by assuming that the inorganic salts behave simply as support-deactivating agents, with solute–support interactions dominating the retention mechanism. In support of this conclusion, it was shown by contact angle measurements that inorganic salt melts, even at very high concentrations, are unable to completely wet silica-based surfaces in common use for column preparation in gas chromatography. Similar behavior was observed in a study of inorganic crystal hydrate melts.[29] In passing, it should be noted that inorganic melts have proven useful for the separation of volatile metal halides that were either too involatile or too reactive for analysis using conventional column packings. The separation mechanism is probably based on the formation of reversible halide complexes, which could be characterized thermodynamically using gas chromatography.[56, 57] Barber *et al.*[58] investigated the retention behavior of a wide range of organic solutes on the stearate salts of the bivalent metals manganese, cobalt, nickel, copper, and zinc. The selective retention of amines by the liquid salts was noted, and thermodynamic data (free energies, heats of solution, activity coefficients) were determined for several solutes. Cartoni *et al.*[59] made similar studies using the low-melting-point chelates of beryllium, aluminum, nickel, zinc, and copper n-nonyl-β-diketones as the stationary phase. Strong interactions were noted between alcohols and the metal complexes, whereas the retention of alkanes, aromatic hydrocarbons, and ketones was very low. In the above two studies, the solvent selectivity was due to the formation of a reversible complex by solute–metal interactions. Gas–liquid chromatography is well suited to studying reactions of this kind over a wide range of experimental conditions. The necessary theoretical models are well developed although the technique has not been widely employed by inorganic chemists.

9. Conclusions

The liquid organic salts are a new class of polar selective solvents ready for technological development. This development is slowed at the moment only by a lack of fundamental knowledge of their chemistry and physical properties. Gas chromatography is a useful tool for establishing their solvent characteristics in a manner that permits a comparison with the properties of nonionic solvents. Early studies have already demonstrated that the solvation of organic solutes in liquid organic salts is not an unfavored or unusual process, the principal difference between the ionic and nonionic solvents being the magnitude of the interactions involved. In this respect, the liquid organic salts can be considered as a unique class of solvents with properties not available with any of the common nonionic solvents, either singularly or in mixtures.

References

1. D. C. Locke, *Adv. Chromatogr.* **14**, 87 (1976).
2. R. J. Laub and R. L. Pecsok, *Physiochemical Applications of Gas Chromatography*, Wiley, New York (1979).
3. J. R. Conder and C. L. Young, *Physicochemical Measurements by Gas Chromatography*, Wiley, New York (1979).

4. K. L. Mallik, *An Introduction to Nonanalytical Applications of Gas Chromatography*, Peacock Press, New Delhi (1976).
5. R. J. Laub, in: *Inorganic Chromatographic Analysis* (J. C. Macdonald, ed.), pp. 13–186, Wiley, New York (1985).
6. T. L. Hafkenscheid and E. Tomlinson, *Adv. Chromatogr.* **25**, 1 (1986).
7. R. N. Nikolov, *J. Chromatogr.* **241**, 237 (1982).
8. V. G. Berezkin, *J. Chromatogr.* **159**, 359 (1978).
9. R. L. Martin, *Anal. Chem.* **33**, 347 (1961).
10. R. L. Martin, *Anal. Chem.* **35**, 116 (1963).
11. D. E. Martire, in: *Progress in Gas Chromatography* (J. H. Purnell, ed.), pp. 93–120, Wiley, New York (1968).
12. S. C. Dhanesar, M. E. Coddens, and C. F. Poole, *J. Chromatogr.* **349**, 249 (1985).
13. M.E. Coddens, K. G. Furton, and C. F. Poole, *J. Chromatogr.* **356**, 59 (1986).
14. B. R. Kersten and C. F. Poole, *J. Chromatogr.* **399**, 1 (1987).
15. J. R. Conder, N. K. Ibrahim, G. J. Rees, and G. A. Oweimreen, *J. Phys. Chem.* **89**, 2571 (1985).
16. K. G. Furton and C. F. Poole, *Anal. Chem.* **59**, 1170 (1987).
17. K. G. Furton and C. F. Poole, *J. Chromatogr.* **399**, 47 (1987).
18. K. G. Furton, S. K. Poole, and C. F. Poole, *Anal. Chim. Acta* **192**, 49 (1987).
19. S. K. Poole, K. G. Furton, and C. F. Poole, *J. Chromatogr. Sci.* **26**, 67 (1988).
20. J. R. Conder, *J. Chromatogr.* **39**, 273 (1969).
21. J. R. Conder, D. C. Locke, and J. H. Purnell, *J. Phys. Chem.* **73**, 700 (1969).
22. H.-L. Liao and D. E. Martire, *Anal. Chem.* **44**, 498 (1972).
23. C. Eon, A. J. Chatterjee, and B. L. Karger, *Chromatographia* **5**, 28 (1972).
24. J. Serpinet, *J. Chromatogr. Sci.* **12**, 832 (1974).
25. D. E. Martire, R. L. Pecsok, and J. H. Purnell, *Trans. Faraday Soc.* **61**, 2496 (1965).
26. C. F. Poole and S. A. Schuette, *Contemporary Practice of Chromatography*, Elsevier, Amsterdam (1984).
27. Annual Book of ASTM Standards, Vol. 5.01, D941, pp. 420–425, American Society for Testing and Materials, Philadelphia, Pennsylvania (1986).
28. Diatomite Supports for Gas-Liquid Chromatography, product brochure MD-37, Johns-Manville, Denver, Colorado.
29. C. F. Poole, H. T. Butler, M. E. Coddens, S. C. Dhanesar, and F. Pacholec, *J. Chromatogr.* **289**, 299 (1984).
30. D. C. Fenimore, J. H. Whitford, C. M. Davis, and A. Zlatkis, *J. Chromatogr.* **140**, 9 (1977).
31. R. J. Laub, J. H. Purnell, P. S. Williams, M. W. P. Harbison, and D. E. Martire, *J. Chromatogr.* **155**, 233 (1978).
32. N. D. Petsev, V. H. Petkov, and C. Dimitrov, *J. Chromatogr.* **114**, 204 (1975).
33. L. Mathiasson, *J. Chromatogr.* **174**, 201 (1979).
34. E. F. Sanchez, J. A. G. Dominguez, J. G. Munoz, and M. J. Molera. *J. Chromatogr.* **299**, 151 (1984).
35. C. F. Poole, B. R. Kersten, S. S. J. Ho, M. E. Coddens, and K. G. Furton. *J. Chromatogr.* **352**, 407 (1986).
36. P. H. Shetty, P. J. Youngberg, B. R. Kersten, and C. F. Poole, *J. Chromatogr.* **411**, 61 (1987).
37. P. F. McCrea, in: *Progress in Gas Chromatography* (J. H. Purnell, ed.), pp. 87–136, Wiley, New York (1968).
38. S. P. Wasik, *J. Chromatogr. Sci.* **14**, 516 (1976).
39. F. Pacholec and C. F. Poole, *Chromatographia* **17**, 370 (1983).
40. K. G. Furton and C. F. Poole, *J. Chromatogr.* **349**, 235 (1985).
41. K. G. Furton, C. F. Poole, and B. R. Kersten, *Anal. Chim. Acta* **192**, 255 (1987).
42. R. J. Laub, *Anal. Chem.* **56**, 2110 (1984).

43. E. F. Meyer, *J. Chem. Educ.* **50**, 191 (1973).
44. R. C. Castells, *J. Chromatogr.* **350**, 339 (1985).
45. D. E. Martire, *Anal. Chem.* **46**, 626 (1974).
46. J. E. Oberholtzer and L. B. Rogers, *Anal. Chem.* **41**, 123 (1969).
47. M. Goedert and G. Guiochon, *Anal. Chem.* **45**, 1188 (1973).
48. O. Wicarova, J. Novak, and J. Janak, *J. Chromatogr.* **65**, 241 (1972).
49. B. R. Kersten, C. F. Poole, and K. G. Furton, *J. Chromatogr.* **411**, 43 (1987).
50. B. R. Kersten and C. F. Poole, *J. Chromatogr.* **452**, 191 (1988).
51. W. W. Hanneman, C. F. Spenser, and J. F. Johnson, *Anal. Chem.* **32**, 1386 (1960).
52. L. R. Snowdon and E. Peake, *Anal. Chem.* **50**, 379 (1978).
53. G. Dahlmann, H. J. K. Koser, and H. H. Oclert, *J. Chromatogr.* **171**, 398 (1979).
54. A. C. Butler and R. R. Sibbald, *J. Chromatogr. Sci.* **23**, 352 (1985).
55. H. J. K. Koser, H. H. Oclert, and J. Stremmler, *J. Chromatogr.* **285**, 289 (1984).
56. R. S. Juvet, V. R. Shaw, and M. A. Hahn, *J. Am. Chem Soc.* **91**, 3788 (1969).
57. F. M. Zado and R. S. Juvet, in: *Gas Chromatography, 1966* (A. B. Littlewood, ed.), pp. 283–295, Elsevier, New York (1967).
58. D. W. Barber, C. S. G. Phillips, G. F. Tusa, and A. Verdin, *J. Chem. Soc.* **1959**, 18.
59. G. P. Cartoni, A. Liberti, and R. Palombari, *J. Chromatogr.* **20**, 278 (1965).

3

Thermal Conductivity and Diffusivity Measurements

Hiromichi Ohta and Yoshio Waseda

1. Introduction

Heat transfer properties are of importance in designing applications of molten salts as heat transfer fluids for fusion reactors, breeder reactors, and thermal energy storage systems. Salt mixtures of various combinations are especially attractive, because they have lower melting temperatures, compared with the pure salts. For this reason, reliable thermal conductivity data are strongly required to select an optimum composition of salt mixture for the desired application.

Thermal conductivity or diffusivity of a liquid is an important thermophysical property, not only in technological applications, but also in the physicochemical study of liquids. At room temperature, an accuracy of better than 1% is required to examine the reliability and the limits of theoretical predictions based on the physicochemical theory available. However, at elevated temperatures, even measurements with an accuracy of a few percent, which are needed for industrial heat transfer problems, were scarce until quite recently.

In earlier volumes of this series, the techniques used to measure the optical properties and electrochemical properties of molten salts have been described in detail. However, the area of thermal diffusivity measurements of molten salts has not been covered yet. With this in mind, in this chapter some techniques for the measurement of thermal properties of molten salts developed in our laboratory will be presented. It may be noted that the earlier volumes have provided very helpful information for selecting cell materials and for preparing samples.

Hiromichi Ohta • Department of Metallurgy, Faculty of Engineering, Ibaraki University, Hitachi 316, Japan. Yoshio Waseda • Research Institute of Mineral Dressing and Metallurgy (SENKEN), Tohoku University, Sendai 980, Japan.

There have been various attempts to measure thermal conductivities of liquid samples. However, accurate thermal transport measurements on fluids at high temperature are difficult to obtain owing primarily to the following three major difficulties:

1. Onset of convective heat flow is difficult to prevent.
2. At temperatures above 1000 K, heat transfer via radiation becomes significant.
3. Heat leaks to the container, which are difficult to account for in a reliable fashion, are significant.

The transient hot-wire method has long been accepted as one of the most reliable and accurate techniques for determining thermal conductivity of liquid samples and has been widely used because of the simple structure of the measurement cell. A thin metallic wire is dipped into the sample liquid, and the wire is heated by a constant electric current. The temperature rise of the heating wire is determined from the voltage change between two terminals fixed to the terminal wire. Thermal diffusivity is estimated from the rate of this temperature rise. However, short circuit and leakage of electricity from the wire prevents the accurate measurement of the thermal diffusivity of molten salts by the transient hot-wire method, because molten salts are usually electrical conductors. Coating and sheathing of the wire proved unsuccessful at high temperatures. The reason for this failure is the difference in the thermal expansion coefficients of the wire and coating materials. In such a case, thermal resistance between a sheath and a wire is significant.

Nagashima and co-workers[1–4] replaced the wire heat source in the transient hot wire by a mercury thread contained in a quartz glass capillary. Some successful results were obtained. However, they found that current leakage from the heat source to the liquid causes a significant error since the electrical system combines with the metallic cell through the liquid and small voltage signals are distorted and enlarged. Recently, Karasawa *et al.*[5] tested various kinds of coating materials for the hot-wire source, but suitable coating materials at elevated temperatures have not been found yet. In this method, the heating element should be electrically insulated, and perfect contact between the heating element and the solid surface of the insulating sheath should also be ensured. Meeting these requirements appears not to be an easy task at high temperature.

The use of a laser flash method first developed by Parker *et al.*[6] appears to reduce these difficulties. In this method, the energy that causes the temperature gradient is provided by a laser beam (not by an electric current), and this ensures that the electrical properties of the sample liquid do not affect the results. Thus, this method can be applied to ionic melts without any electrical and electrochemical effects. The laser flash method has many advantages and is recognized to be an even better method for determining thermal conductivity of high-temperature materials than the conventional techniques, as shown in our review article concerning several methods for determining thermal conductivity of high-temperature melts.[7] However, the full potential of this laser flash method has been appreciated only recently. One of the major reasons is the difficulty arising from the fact that the radiative heat flow becomes serious at high temperatures, and then the heat transfer should be described as a function not only of thermal diffusivity but also of the radiative properties of the measuring system. The main purpose of this chapter is to present our own developments for overcoming such difficulties in the measurement of thermal conductivity for molten salts at high temperatures by the laser flash method, together with some selected examples. The measurement of thermal diffusivity and heat capacity also is considered.

2. A Brief Background of the Thermal Property Measurements

For conductive heat flow in an isotropic medium, the Fourier equation holds, namely,

$$J = -\lambda \nabla T \tag{1}$$

where J is the quantity of heat flowing in unit time through unit area under influence of a temperature gradient ∇T by conduction, and λ is the thermal conductivity. For the transient condition, Eq. (1) becomes

$$\nabla (\lambda \nabla T) = C_p \rho (dT/dt) \tag{2}$$

where C_p is the specific heat at constant pressure, t is time, and ρ is the density.

For cases where the thermal properties λ, C_p, and ρ are treated as constants, independent of both position and temperature, Eq. (2) becomes

$$\nabla^2 T = (C_p \rho / \lambda)(dT/dt) \tag{3}$$

or

$$\nabla^2 T = (1/\alpha) \, (dT/dt) \tag{4}$$

where $\alpha = \lambda / C_p \rho$ is the thermal diffusivity. Usually, measurements of thermal conductivity or diffusivity are carried out under conditions in which the assumption that the thermal properties of the specimen are independent of temperature and position in the sample liquid is well justified. Of course, particular attention should be given to measurements at temperatures close to any phase transformation, where apparent heat capacity shows strong temperature dependence. Thermal diffusivity can be determined by any experimental methods based on Eqs. (1)–(4). The experimental uncertainties due to heat leak, such as radiation and convection, can be reduced when the experimental time is decreased, and much shorter times are convenient for transient experiments. Steady-state experiments also have a theoretical disadvantage in measurements on a fluid mixture, because the existence of the temperature gradient in the specimen for a long period may cause the separation of the components due to thermal diffusion. Improved electronic devices, which have allowed temperature changes to be accurately determined over short time intervals, have led to an increased interest in transient methods for determining thermal properties, particularly at high temperatures. Owing to these various factors, the laser flash method has presently gained a dominant position.

At low temperatures, heat transfer in liquids is by conduction alone. At elevated temperatures, the transfer of energy through a medium can occur by conduction (phonons) and optical energy waves (photons). In such a case, the heat transfer equation is described as follows:

$$\rho C_p \frac{\partial T}{\partial t} = \lambda \nabla^2 T - \nabla F \tag{5}$$

where F is the radiative heat flux. Equation (5) gives the temperature distribution in the sample liquid, providing that the initial and boundary conditions are specified. As the radiative heat transfer is dependent not only on the local temperature but also on the temperature of the entire surroundings, the energy equation becomes a nonlinear integro-differential equation. There are no general solutions available for this integro-differential equation.

In this report, the following two limiting cases are considered:

1. *Transparent or optically very thin liquid.* In this case, the radiation emitted and absorbed by the medium is negligibly small. In the medium, F in Eq. (5) is almost unchanged, and ∇F is zero. Thus, we can neglect the effect of radiative heat transfer on the temperature distribution in the liquid sample.

2. *Opaque or optically thick liquid.* In this case, the mean penetration distance of an infrared beam is quite small. The radiation emitted by volume elements of the medium is rapidly attenuated. The energy contribution to an arbitrary unit volume then comes from the immediately neighboring media. For these conditions, it is possible to transform Eq. (5) into a diffusion equation, in a similar way to that proposed for the case of heat conduction by Rosseland.[8] The radiative heat flux assumes the same form as Fourier's law of heat conduction. Then, the energy flux q by combined radiation and conduction at any point in the medium can be expressed as

$$q = J + F = -\lambda \nabla T - \lambda_r \nabla T = -(\lambda + \lambda_r)T \tag{6}$$

where λ_r is the radiative (photon) conductivity, represented by the following form:

$$\lambda r = 16n^2 \sigma T^3 / 3k_R \tag{7}$$

where σ is the Stefan–Boltzmann constant, n the refractive index of the medium, and k_R the absorption coefficient of the sample. The effective thermal conductivity λ_{eff}, defined by the following equation, is usually obtained from the experiments.

$$\lambda_{\text{eff}} = \lambda + \lambda_r \qquad (8)$$

It is, however, noteworthy that λ_{eff} becomes λ when the temperature is low enough or the sample liquid is very opaque.

3. Laser Flash Technique for the Determination of Thermal Diffusivity of Transparent Liquid Samples

A two-layered laser flash method was developed by Tada *et al.*[9] in 1981 for application to liquid samples having low thermal conductivity, such as molten salts. Figure 1 shows a schematic diagram of this method. The upper end of a semi-infinite column of liquid sample in contact with a thin metal plate is heated by a pulsed laser beam. The thermal conductivity of the sample liquid is determined from the temperature decay at the top surface of the plate. This technique has the following advantages:

1. The heat leak to the sample cell is eliminated since the sample liquid is sandwiched between a sample holder and a thin metal plate.
2. This cell structure ensures that heat absorbed on the metal plate flows downward one-dimensionally through the sample liquid. The induced temperature distribution in the sample liquid and short measuring time minimize thermal disturbance due to convention in the sample.
3. Measurement of the thickness of the liquid layer is not required; the thermal conductivity of the liquid can be obtained only from the temperature response curve of the metal plate.

This two-layered laser flash method has been applied to some molten salt systems at temperatures above 1000 K.[7, 10–13]

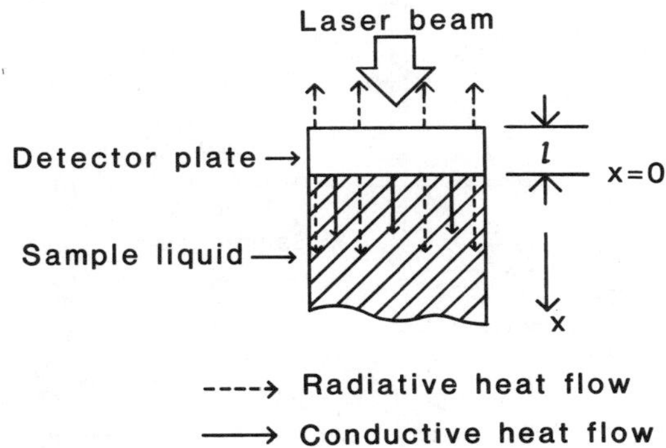

----→ Radiative heat flow

——→ Conductive heat flow

FIGURE 1. Schematic diagram of a cell for the two-layered laser flash method.[11]

3.1. Experimental Setup

The entire cell assembly is enclosed by a bell jar under vacuum of the order of 10^{-4} Pa ($\sim 10^{-6}$ torr).

The two-layered cell system shown in Fig. 2 consists of two parts: a detector metal plate for measuring the temperature response curve, and a column of sample liquid adhered to the metal plate. In experiments with molten salts, a platinum disk (0.2 mm in thickness and 6 mm in diameter) is usually employed as a detector plate. This platinum plate is suspended by three fine platinum wires and contacts the surface of the sample melt. The temperature of the plate is measured by a Pt/Pt–13% Rh thermocouple (0.1 mm in diameter) spot-welded to the center of the platinum detector plate. A platinum sample vessel is placed on the tantalum pedestal with an elevator mechanism to precisely adjust its position. In order to protect the signal from external noise, all circuits connected to the thermocouple are electrically insulated from ground. The sample vessel is also insulated from the tantalum pedestal by a sapphire disk. The cell system is placed in a tungsten mesh heating element. To reduce an inductive effect caused by the heater, direct current was used with a bias of -50 V, at which the noise was found to be minimized.

To estimate the initial temperature rise of the platinum disk and the thermal radiation heat loss from the disk, the temperature response curves for the disk not touching sample liquid were measured at various temperatures.

After removal of gas bubbles, the sample vessel is elevated to just touch the sample surface to the plate, and a liquid column of sample is made by lowering the holder by about 0.65 mm. The length of the liquid column of a sample is dependent upon the geometry of the cell, the surface tension, and the viscosity. However, changes in the vertical distance of the sample holder in the range from

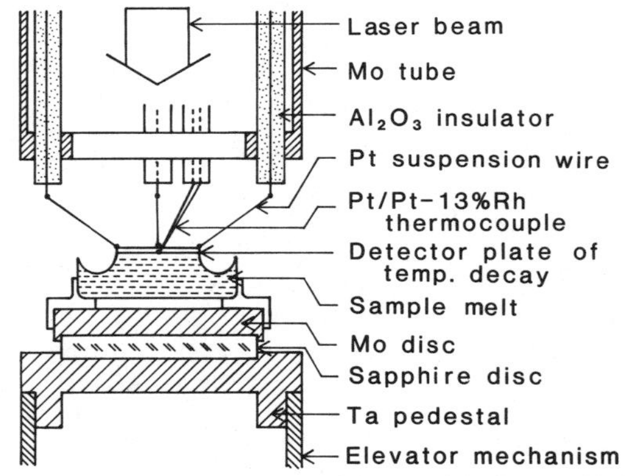

FIGURE 2. Schematic diagram of the sample-cell assembly.[11]

0.5 to 1.3 mm appeared not to affect the temperature response curves under our experimental conditions.

At the desired temperature, a pulsed laser beam is flashed on the top surface of the platinum plate and, to estimate the thermal conductivity value the temperature response curve is stored in a digital transient memory connected to a microcomputer system. After the run, the sample vessel is lowered and detached from the detector plate for direct observation of the liquid sample. The data were discarded when a bubble was present.

3.2. Preparation of Samples

Using this apparatus, the thermal conductivities of silicate melts were determined. Samples were prepared from powder of high-grade silicon dioxide and alkaline carbonates of 99% purity. The weighed reagents of the required composition were mixed well and premelted in a platinum crucible in air to decompose the carbonate and then cooled down to room temperature. After grinding the product into a powder form, the product was again melted at a temperature about 200 K above the liquidus in air to ensure homogeneity. About 1 g of premelted sample is put into the sample vessel, melted, and held for about 2 h under vacuum to remove bubbles. However, it should be noted that the details of the sample preparation depend on the composition.

3.3. Apparatus for Liquids of High Vapor Pressure

A thermocouple has been successfully used to detect the temperature response at relatively low temperature or in high vacuum (less than 10^{-4} Pa) at high temperature. However, certain species volatilize at higher temperatures, especially in measurements on molten salts, and significant noise due to the ionization of the volatilized species by the flashed laser beam, prevents accurate measurement. Such effects appear to be accelerated by the oscillating electric field induced from the heating element. In this case, the noise frequently exceeds the temperature response signal, even when rectified alternating current is supplied to the heater, owing to the small residual ripple current. Since the vapor pressure of most molten salts is relatively high, a detection system for the temperature response based on a thermocouple is strictly limited for molten salts, particularly for measurements at high temperature.

An improved temperature response acquisition system was developed using a germanium infrared detector. Figure 3 shows the block diagram of this apparatus. In order to accurately monitor the supplied energy of the laser beam by a silicon photodetector, a 45°C gold-plated half-mirror also was inserted in the laser beam path.

The schematic diagram for this optical temperature detection system is

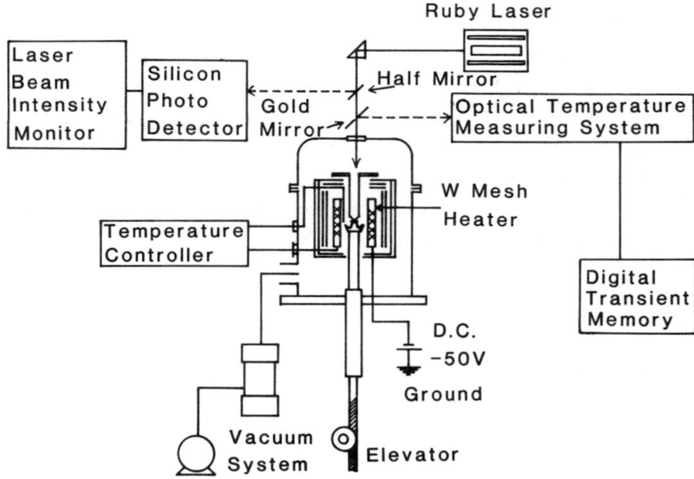

FIGURE 3. Block diagram of experimental apparatus for thermal conductivity measurements by the laser flash method with the optical temperature response acquisition system.[11]

illustrated in Fig. 4. The germanium infrared detector was employed in this system to measure the temperature response curve, and external noise was reduced to a reasonable level. The infrared ray emitted from the plate passed through the upper quartz window and was reflected by a 45° gold mirror with a hole in its center for the laser beam path. The infrared ray was focused on the screen by a quartz lens and the image from the platinum plate transmitted through a quartz fiber-optic guide to a germanium infrared detector. To prevent multiple reflections from the focusing equipment, its inner surface was covered with black felt.

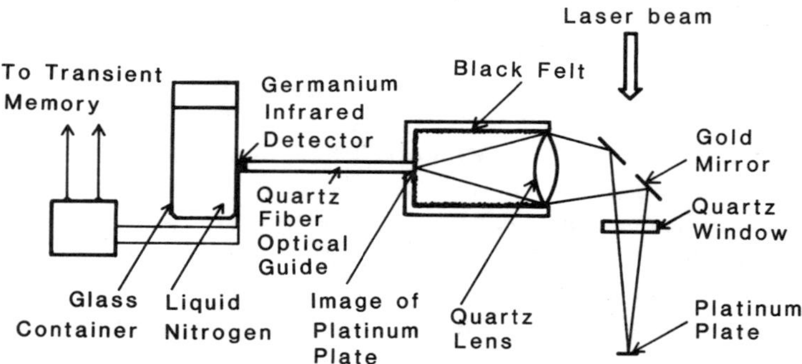

FIGURE 4. Optical temperature response acquisition system.

3.4. Data Analysis

A system composed of a thin metal plate and a transparent liquid, as shown in Fig. 1, is considered in order to construct the theoretical heat transport equations. Before flashing a pulsed laser beam, the detector plate, at temperature θ_0, is in thermal equilibrium with the liquid layer. At $t = 0$, a laser pulse is flashed on the plate and the absorbed heat of the plate discharges into both the sample liquid and the surroundings by both conduction and radiation. This is schematically shown in Fig. 1. It should be noted that Tada *et al.*[9] considered only conduction heat flow, because they measured the thermal conductivity of molten potassium nitrate at temperatures below 750 K. The following six conditions are postulated for higher temperature measurements:

1. The conductive heat flow through the liquid is one-dimensional.
2. The contact resistance of heat flow between the metal plate and the liquid sample is negligibly small.
3. The liquid layer is semi-infinite.
4. The detector plate is thin and has a high thermal conductivity value compared with a sample liquid. Thus, the temperature in the metal plate becomes uniform instantly.
5. The physical properties of the sample are unchanged during the temperature decay.
6. Since the temperature rise of the detector plate due to the pulsed laser beam is of the order of only 10 K under the present experimental conditions, the relation $T_d \ll \theta_0$ holds, where T_d is the increase in temperature of the detector plate from the steady state temperature θ_0, in thermal equilibrium with the liquid sample before triggering of the laser beam. Therefore, the radiative heat flow can be expressed as a linear process with respect to the value of T_d.

For the case involving both radiative and conductive heat flows, the heat balance equation for a metal plate in unit time through unit area is given by Eq. (9). The heat transport equation for a liquid layer, for this case, can be expressed by Eq. (10).

For the metal plate:

$$-\rho_d C_{p_d} l_d \frac{\partial T_d(t)}{\partial t} = -\lambda_s \frac{\partial T_s(t)}{\partial x}\bigg|_{x=0} + q_1(t) + q_2(t) \tag{9}$$

For the liquid layer ($x > 0$):

$$\frac{\partial T_s(x,t)}{\partial t} = \alpha_s \frac{\partial^2 T_s(x,t)}{\partial x^2} \tag{10}$$

where the suffixes s and d indicate the sample and the detector plate, respectively, t is time, Z is the thickness of the detector plate, T is the temperature rise at time t from the equilibrium temperature, x is the distance from the plate, and C_p, ρ, λ, and α_s denote specific heat capacity, density, thermal conductivity, and thermal diffusivity, respectively. The value of the left-hand side of Eq. (9) is the change in heat of the detector plate. The first term on the right-hand side of Eq. (9) is the conductive heat flux from the detector plate to the liquid sample. The terms q_1 and q_2 on the right-hand side of Eq. (9) are the downward and the upward radiative heat fluxes from the detector plate, respectively. They are represented as follows:

$$q_1 = 4n^2\epsilon\sigma\theta_0{}^3 T_d \quad \text{and} \quad q_2 = 4\epsilon\sigma\theta_0{}^3 T_d \tag{11}$$

where σ is the Stefan–Boltzmann constant, ϵ is the total hemispherical emittance, and n is the refractive index of the sample liquid.

On the other hand, the following relation can be readily obtained when assumption 2 is well recognized:

$$T_d(t) = T_s(0,t) \tag{12}$$

Thus, Eqs. (9) and (10) can be rewritten in the following normalized forms with respect to time (t_n) and distance (x_n):

$$-\frac{\partial T_s(0,t_n)}{\partial t_n} = -\frac{\partial T_s(0,t_n)}{\partial x_n} + R_n T_s(0,t_n) \qquad (x_n = 0) \tag{13}$$

$$\frac{\partial T_s(x_n,t_n)}{t_n} = \frac{\partial^2 T_s(0,t_n)}{\partial x_n^2} \qquad (x_n > 0) \tag{14}$$

where

$$R_n = 4(1 + n^2)\,\epsilon\sigma\theta_0^3\,(\rho_d\,C_{p_d}\,l_d/\lambda_s\rho_s\,C_{p_s}) \tag{15}$$

$$h = \left(\frac{\lambda_s\,C_{p_s}\,\rho_s}{\rho_d^2\,C_{p_d}^2\,l_d^2}\right)^{1/2} \tag{16}$$

$$t_n = h^2\,t \tag{17}$$

$$x_n = \frac{C_{p_s}\rho_s}{l_d\,C_{p_d}\,\rho_d}x \tag{18}$$

The initial and boundary conditions are as follows:

Initial boundary conditions:

$$T_s(X,0) = \begin{cases} T_0 & (x = 0) \\ 0 & (x > 0) \end{cases} \tag{19}$$

where T_0 is the initial temperature rise of the detector plate following a flashed laser pulse.

Boundary condition:

$$T_s(\infty, t) = 0 \qquad \text{(for any time)} \tag{20}$$

On the other hand, the temperature response of the plate T_d is obtained from Eq. (12).

The solutions for cases with various radiative effects corresponding to different R_n have been numerically obtained by the finite difference method. Examples of numerical solutions are illustrated in Fig. 5 together with the result of Tada *et al.*[9] for the case where the radiative heat transfer is negligible ($R_n = 0$). The temperature response curves can be given for a wide range of times and for various experimental conditions in terms of three variables—the thermal conductivity (λ_s) of a sample, the radiative heat transfer coefficient (R_n), and the initial temperature rise (T_0). These numerical solutions are transformed into a continuous function using the following orthogonal polynomial equation for convenience of data processing:

$$\frac{T_d}{T_0} = \sum a_i \, t_n^i \, b_j \, R_n^j \tag{21}$$

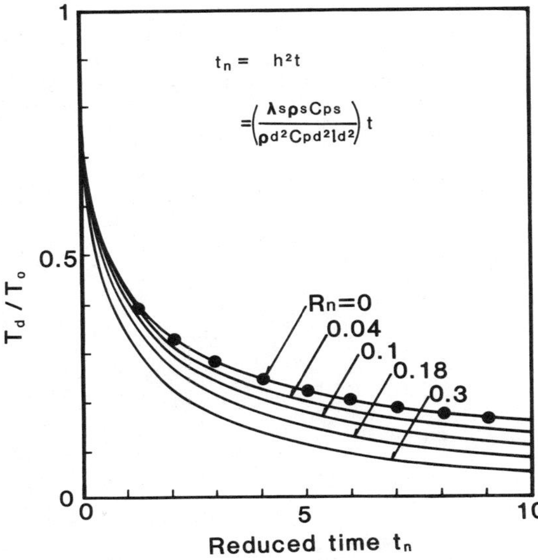

FIGURE 5. Calculated temperature response curves for various values of radiative effect R_n.[10] Experimental points are from the results of Tada *et al.*[9] ($R_n = 0$).

The thermal conductivity (λ) of the sample liquid as well as the radiative component parameter (R_n) and the initial temperature rise (T_0) can be determined by comparing the measured temperature response curves with those of theoretical solutions using the least-squares fitting technique.

3.5. *Validity Check for the Data Analysis Considering the Radiative Component*

Figure 6 provides an example of the curve-fitting scheme in the case of molten calcium aluminosilicate at 1723 K. The present data analysis including the radiative component ($R_n \neq 0$) appears to reproduce well the experimental temperature response curve over the wide time range measured. For the cases where no radiative component is considered ($R_n = 0$), the experimental temperature response curve can be reproduced only in a very limited range of time, as shown in Fig. 6.

The thermal conductivity and the radiative component parameter R_n are

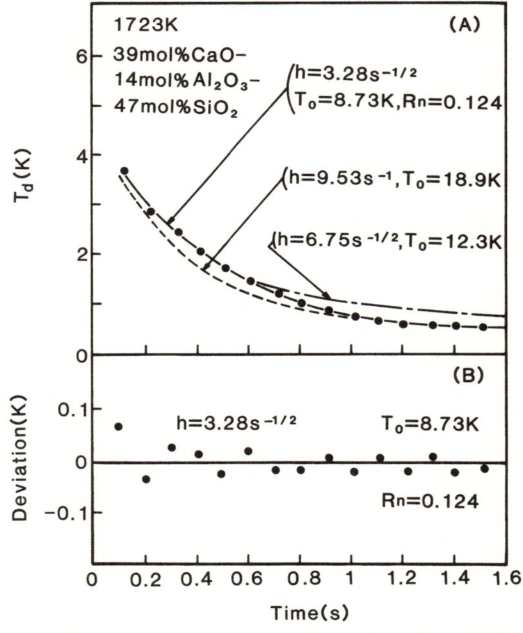

FIGURE 6. Temperature response curve of molten 39 mol % CaO–14 mol % Al$_2$O$_3$–47 mol % SiO$_2$ at 1723 K. (A) Comparison between experimental values (•), values calculated by the present method with thermal radiation effect ($R_n = 0.124$), and values calculated by the method given by Tada *et al.*[9] ($R_n = 0$). (B) Difference between experimental values and values calculated by the present method.

determined simultaneously by a curve-fitting procedure. Hence, the reliability of this data processing may be examined by comparing the theoretical values of R_n with the experimental values. The heat transfer coefficient R_n can be expressed by Eq. (15). The theoretical values of R_n obtained from Eq. (15) for molten sodium silicate as an example are given in Fig. 7, together with the experimental values estimated from the measured temperature response curve. Good agreement was found in the temperature range between 1000 and 1400 K. Thus, the consistency of the present data analysis, which includes the radiative component, could be quantitatively confirmed.

3.6. Single-Parameter Analysis

Multi-parameter fitting schemes are frequently found to be noise sensitive, and a large amount of CPU time is required for computation to determine the parameters. These difficulties have been reduced by a recent development involving the use of an energy monitor system for the incident laser pulse.

When the values of the initial temperature rise, T_0, and the radiative heat transfer component, R_n, are available, the data processing is simplified into a single-parameter fitting analysis. The values of T_0 and R_n themselves can be determined by measuring the temperature responses of the metal plate without a liquid sample. In such a case, the temperature response is expressed as follows:

$$- \frac{d \ln T}{dt} = H \tag{22}$$

where

$$H = \frac{8\epsilon\sigma\theta_0^3}{\rho_d C_{p_d} l_d} \tag{23}$$

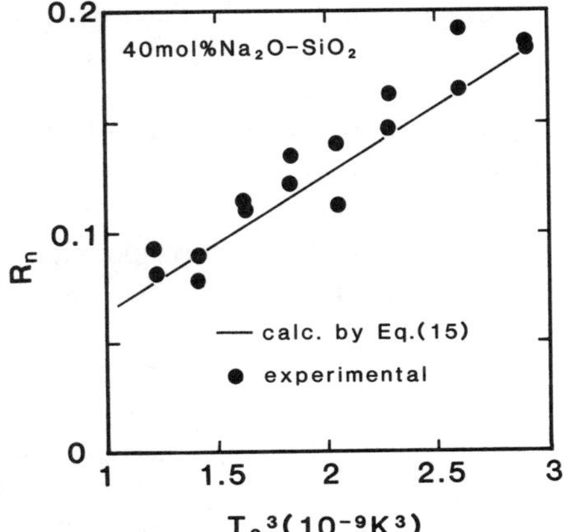

FIGURE 7. Comparison between radiative heat transfer coefficients evaluated from fundamental physical constants of metal plate and sample liquid and the coefficients determined from the measured temperature responses.[11] the solid line was calculated from Eq. (15).

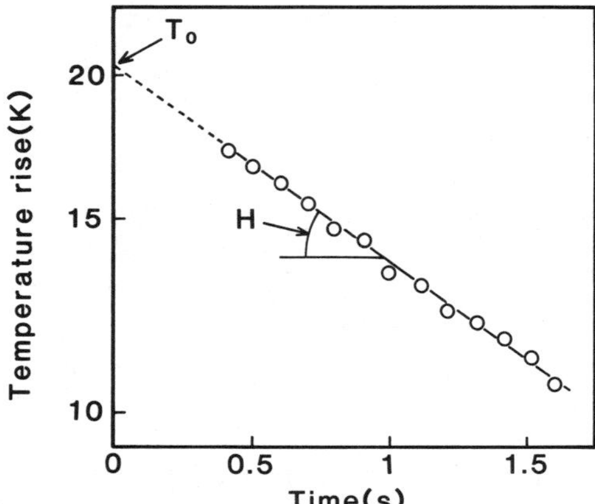

FIGURE 8. Observed logarithmic temperature response without liquid sample.[13]

Figure 8 shows the measured logarithmic temperature response with time. The gradient corresponds to the value of H. Since the relatively large electric current supplied to a laser head occurs in a short time and produces very sharp noise near $t = 0$, the T_0 value is best obtained by an extrapolation to $t = 0$.

The laser energy of each shot appears to be slightly offset. Such an effect can be corrected by the following procedure. The initial temperature rise T_0 is directly proportional to the laser beam intensity, as expressed by

$$T_0 = Q/(\rho_d C_{p_d} l_d) \tag{24}$$

where Q is the heat absorbed by the detector plate from a laser beam. This equation holds for both the plate alone and the plate with a liquid sample. Therefore, the initial temperature rise for each thermal conductivity measurement of sample liquid may be determined from the following equation:

$$T_0 = (I/I') \, T_0' \tag{25}$$

where I is the intensity of the pulsed laser beam measured by the monitor, and I' is the corresponding intensity value for the run without liquid sample; T_0 is the initial temperature value estimated by extrapolation for each run with liquid phase, and T_0' is the corresponding value for the run without liquid sample.

On the other hand, from Eqs. (15), (16), and (23), R_n is expressed as follows:

$$R_n = (\frac{1 + n^2}{2})H/h^2 \tag{26}$$

where h corresponds to the thermal conductivity term of the sample liquid, as defined by Eq. (16). It is also worth mentioning that the relation between the half-decay time, $t_{1/2}$, and the radiative components, R_n, can be readily obtained by numerical calculation using Eqs. (15)–(17) and (24). Here, the half-decay time represents the time to reach 50% decrease of the total temperature change in the temperature response. Therefore, the value of h in Eq. (26) is estimated from the measured half-decay time and the value of H by the simple curve-fitting procedure, when the refractive index, n, of the sample liquid is available. This data analysis has been applied to the measurements for molten carbonate and sodium silicate.[13]

4. Heat Capacity Measurements by the Laser Flash Method

Heat transfer measurements by the laser flash method do not always directly give the thermal conductivity value. For example, in the two-layered method employed in this work, the product of thermal conductivity, λ_s, and specific heat capacity, C_{p_s}, is actually obtained. Thus, the values of the specific heat capacity must be determined when such data are not available.

The laser pulse technique is known to be useful also for the measurement of the specific heat of inorganic materials at high temperature. Several attempts[14–16] have been made to determine the specific heat (C_p) of some materials in comparison with an appropriate reference material.

Specific heat measurements of liquid samples using the laser flash method were carried out by the following procedure. First, the sample is premelted in a measuring cell made of platinum foil (total weight is about 40 mg), as shown in Fig. 9. Then, the cell is suspended upside down to absorb laser energy on the bottom. The temperature response curve of this cell is measured by a Pt/Pt–13% Rh thermocouple spot-welded to the center of the back side of the cell. Fused silica melt is employed as reference material.

The value of the heat (Q) absorbed from a flashed laser pulse was estimated

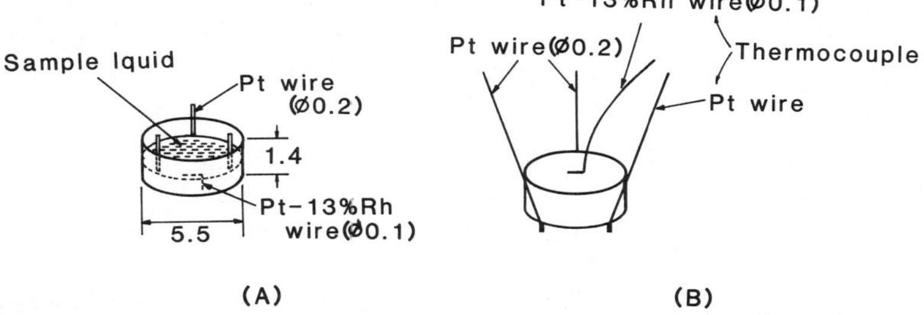

FIGURE 9. Improved cell for the measurement of specific heat capacity of molten sample (all dimensions in millimeters). Sample is melted in the cell (A), and then all is suspended upside down for measurement of heat capacity (B).

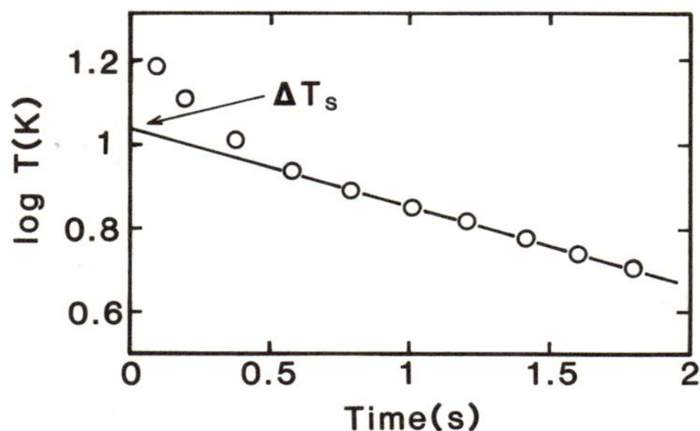

FIGURE 10. Temperature response in specific heat capacity measurement.[11]

from the specific heat data[17] for the reference materials, fused silica and platinum in the present case, using the relation $Q = T_{ref}(C_{silica} + C_{cell})$, where T_{ref} is the temperature rise in the reference measurement with fused silica in the cell, C_{cell} is the heat capacity of the measuring cell, and C_{silica} is the heat capacity of silica glass (reference). When the temperature rise for the cell with a sample, T_s,

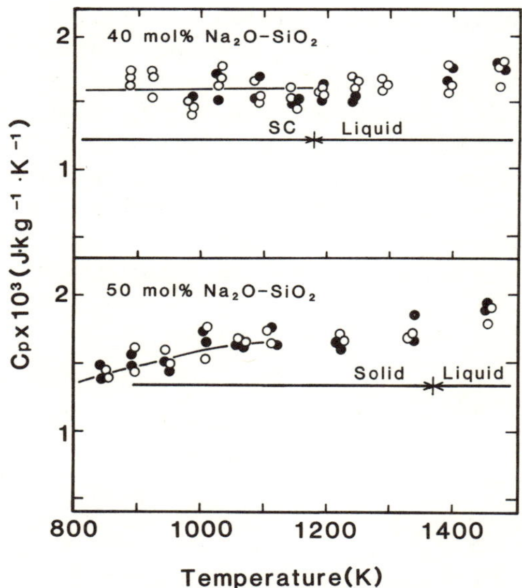

FIGURE 11. Specific heat capacities of Na_2O–SiO_2 system determined by the laser flash calorimeter. The solid lines are the results from conventional drop calorimetry.[18] (a) 40 mol % Na_2O–60 mol % SiO_2[7]: cell 0.0402 g, sample 0.0363 g; cell 0.0440 g, sample 0.0355 g; (b) 50 mol % Na_2O–50 mol % SiO_2[7]: cell 0.0418 g, sample 0.0352 g; cell 0.0421 g, sample 0.0364 g. sc: Supercooled liquid.

is obtained in an adiabatic condition, the specific heat of the sample liquid can be determined from the following equation:

$$C_{p_s} = (Q - C_{\text{cell}}\Delta T_s)/M\Delta T_s \tag{27}$$

where M is the mass of the sample liquid. The extrapolation method, in which $\ln T$ is plotted against time, is shown in Fig. 10. It is used to obtain the value of T_s and to eliminate the effects due to heat loss from the sample by radiation and a heat transfer process which occurs at the initial stage of the run before an isothermal condition is achieved through the sample.

Examples of specific heat capacity measurements are given in Fig. 11, using the results for sodium silicates. The values obtained by the laser flash method[11] agree well with the results determined by conventional drop calorimetry[18] within an error of less than $\pm 3\%$.

5. Selected Examples of Measured Thermal Conductivity of Transparent Molten Salt Systems

The laser flash method was applied to determine the thermal conductivity of several molten silicate melts at temperatures between 800 and 1700 K. Three-parameter fitting of the temperature response measured with a thermocouple was employed. Figures 12 and 13 show some representative results together with the available data from other techniques.[19–22] All data for sodium silicates fall within the same order of magnitude, and this implies that the agreement is rather good except for the difference in the temperature dependence. The reason for this discrepancy could not be identified with the presently available information.

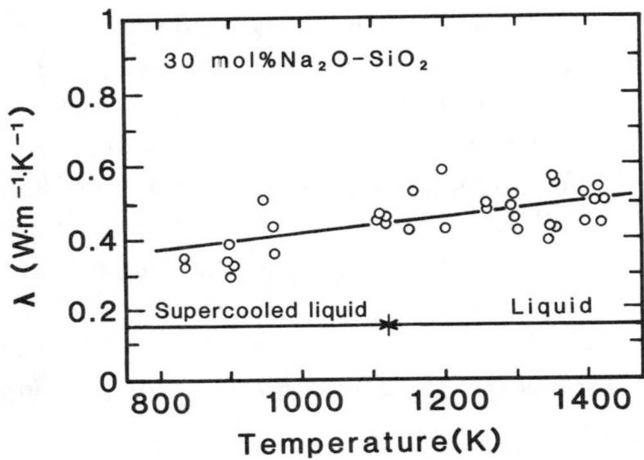

FIGURE 12. Thermal conductivity of 30 mol % Na_2O–70 mol % SiO_2 system.[11] (The points are the measured values; the solid line represents the values determined from the measured ones by the method of least squares.)

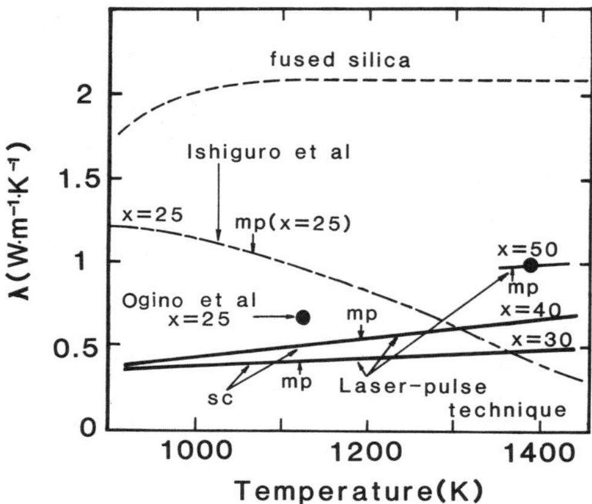

FIGURE 13. Thermal conductivity of Na_2O–SiO_2 system determined by several techniques: ——, two-layered laser flash method[11]; ——·——, transient hot-wire method[19–21]; •, concentric cylinder method[22]; sc, supercooled liquid; *x*, mol % Na_2O.

The experimental uncertainty of these results is difficult to estimate. However, the following comments may be made:

1. Random errors are mainly caused by the electrical noise in the measurement of the temperature response curve.
2. The uncertainty due to the measurement of the thickness of a detector metal plate is 0.1% at most. This gives an error in thermal conductivity of 2% at most.
3. Theoretical temperature response curves estimated by numerical calculation of the heat transport equations are found to include an uncertainty of less than 1%.

A newly devised system with an infrared detector and a laser intensity monitor is described in Section 3.3, and the data processing for this method is given in Section 3.6. The thermal conductivities of molten sodium silicate and carbonate have been determined using this new laser flash method. The values are 0.428 W/(m·K) for sodium silicates at 1133 K and 0.478 W/(m·K) for sodium carbonate at 1273 K, respectively. The former value agrees well with the value obtained from an experiment with a thermocouple.

6. Laser Flash Method for the Determination of Thermal Diffusivity of Opaque Liquids

For optically thick or opaque liquids, one can use a simpler cell system than the multilayered type of cell assembly, such as the two-layered sample cell system[23, 24] shown schematically in Fig. 14. The front surface of a small disk-

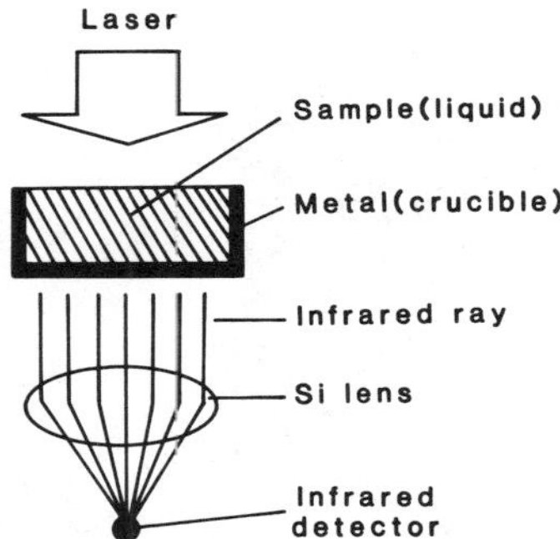

FIGURE 14. Schematic diagram of the cell assembly for opaque liquids.[24]

shaped cell is subjected to a single pulse from the laser beam source. The resulting temperature rise of the back surface of the sample cell is detected, and the thermal diffusivity value is determined from the temperature rise versus time data. This method can be used for molten ferrites with high iron oxide content. The method appears to be useful for the measurements of opaque liquids, such as metal-salt solutions at elevated temperatures.

6.1. Experimental Setup

The basic arrangements of the experimental setup for determining thermal diffusivity of opaque liquids are almost the same as those used for transparent liquids. A schematic diagram of the apparatus employed is shown in Fig. 15. A platinum sample cell containing the sample liquid is placed on four small alumina pins to minimize the conductive heat leak. The cell system is heated by alternating current using a platinum wire heating element around an alumina tube. The temperature of the sample before flashing a pulsed laser beam is measured by a Pt/Pt–13% Rh thermocouple located just above the front surface of the sample. The temperature response of the back surface of the cell is measured by a remote optical sensing device. The infrared ray from the back surface of the sample cell passes through the pedestal of the alumina tube to a gold mirror, which directs the infrared ray to a silicon lens, which focuses it onto the InSb infrared detector. Since the alumina tube is penetrated by radiation at high temperatures, the radiative heat flux through the alumina tube is found to fluctuate due to the alternating current from the main power source. Multiple reflections of the radiative heat flux also take place between the inner wall of the alumina tube and the back surface of the sample cell. These factors sometimes affect the detection of the infrared ray emitted from the back surface of the cell. Then, significant sinusoidal noise, in phase with the power source, is frequently induced in the measured temperature response curve. However, such fluctuations of the measured temperature response curve could be reduced to the level of

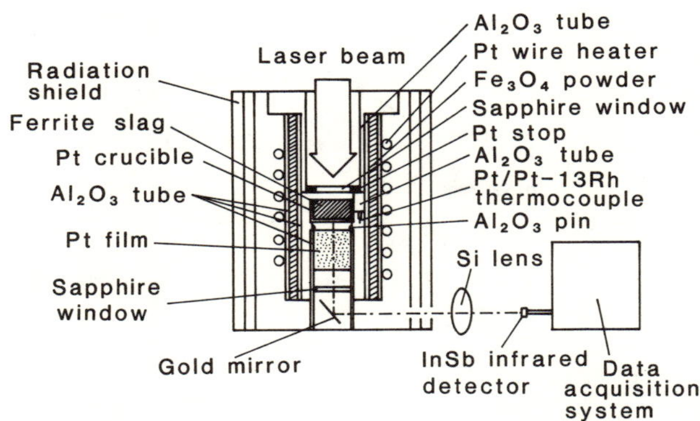

FIGURE 15. Schematic diagram of the experimental apparatus for the determination of thermal diffusivity of opaque liquids.[24]

insignificance by putting magnetite powder as a radiation absorber between the alumina tube that is wound by heating elements and the inner alumina tube.

The present cell system is designed for measurements in air atmosphere. Therefore, the top and bottom of the sample chamber were covered with sapphire windows to decrease the heat loss by convection of air.

6.2. Data Analysis

Schriempf,[25] Bates,[26] and Taylor and Lee[27] have made several attempts to measure the thermal diffusivity of liquid samples by the laser flash method. However, the quantitative accuracy of their results is not well established because of the leakage of heat from the specimen to the container, as reported by Ang *et al.*[28, 29] On the other hand, the $t_{1/2}$ method, first proposed by Parker *et al.*,[6] has been widely employed as a simple method for determining thermal diffusivity of solid and liquid materials. Thermal diffusivity values can be calculated from the time to reach 50% of the maximum value of the back surface temperature. Thus, the $t_{1/2}$ method requires the time to reach its maximum value in the temperature response curve. When the times necessary to obtain the thermal diffusivity values become long, heat losses by radiation and conduction increase. Therefore, it is desirable to analyze the initial short-time region of the temperature response curve.

For this purpose, the logarithmic analysis proposed by James[30] in 1980 is considered to be a useful one. According to James, the temperature response of the back surface can be represented as follows:

$$T/T_{max} = 2(l^2/\pi\alpha t)^{1/2}\Sigma \exp[-(2n + 1)l^2/4\alpha t] \tag{28}$$

where T is the temperature rise of the back surface of the sample, α and l are the thermal diffusivity and the thickness of the sample, respectively, and t is the time elapsed after flashing the laser beam. The series on the right-hand side of Eq. (28) can be approximated by the first term in the range from $t = 0$ to the maximum temperature rise. Taking the logarithms of both sides of Eq. (28), the following equation is readily obtained.

$$\ln(t^{1/2} \cdot T) = \ln[2(l^2/\pi\alpha)^{1/2} \cdot T_{\max}] - l^2/4\alpha t \tag{29}$$

A plot of $\ln(t^{1/2} \cdot T)$ as a function of $1/t$ gives a straight line with slope $-l^2/4\alpha$, and the thermal diffusivity value can be determined from the slope.

In the case of the two-layered sample cell system shown in Fig. 14, the slope of the straight line is $-(n_1 + n_2)^2/4$, where n_i is the square root of the diffusion time, or $n_i = l_i/\alpha_i^{1/2}$, and the subscript i indicates the ith layer. The assumption is made that the interfacial thermal contact resistance between the two layers is negligibly small.

Figure 16 shows the theoretical temperature responses calculated from Eq. (29); t_n is the nondimensional time, defined as $\alpha t/l^2$. The experimental data employed for analysis should be determined so as to compensate for the effect of heat leakage. For example, logarithmic analyses can be employed in the time region corresponding to 10–30% of the maximum temperature rise. Since the temperature rise is small in the region corresponding to less than 10% of the maximum temperature rise, the signal-to-noise ratio is usually not enough to estimate the thermal diffusivity with high accuracy. The slope of the straight-line plot of $\ln(t^{1/2}T)$ versus $1/t$ is determined by the least-squares method.

Yamamoto *et al.*[31] estimated the effect of the pulse width on thermal diffusivity measurements of solid metals. The thermal diffusivities of molten

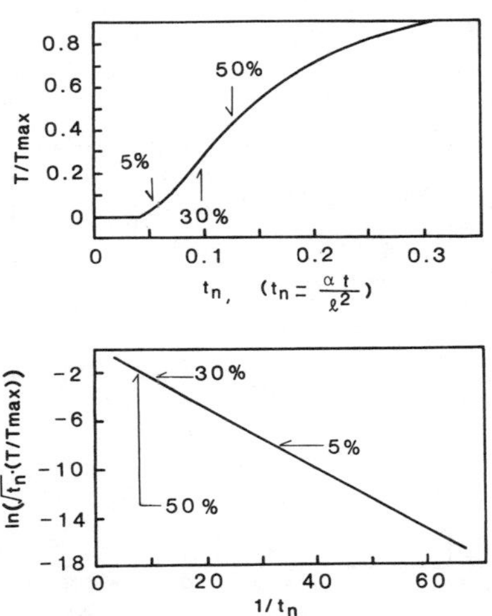

FIGURE 16. Theoretical temperature–time response curve for choosing the best time region analyzed by the logarithmic method.[24]

salts are relatively low. Thus, the rate of temperature rise is slow compared with the duration of the laser beam pulse. Therefore, the pulse width effect is found to be negligibly small under such experimental conditions.

6.3. Effects of Radial and Radiative Heat Flows on Thermal Diffusivity

To evaluate the effect of the heat leak due to radial heat flow on thermal diffusivity, two series of experiments were done. Figure 17 shows the results obtained with three solid samples of calcium ferrite with diameters of 17, 13, and 10 mm. (The diameter of laser beam used was 10 mm.) As is easily seen from Fig. 17, the diameter of the sample size does not affect the result and the thermal diffusivity values under these experimental conditions.

Figure 18 indicates the effect of the laser beam diameter on thermal diffusivities both in air and in vacuum. Almost constant values were obtained when the laser beam diameter was larger than 7 mm. These results lead one to suggest that logarithmic analysis is one way to minimize the effect of radial heat flow due to conduction to the crucible wall. Figure 18 also shows that the heat leak to the atmosphere by convection is almost negligible. Nevertheless, the effect of radiative heat loss on thermal diffusivity in the logarithmic analysis was estimated for the case of a sample containing 50 mol % CaO at temperatures of 1500 and 1673 K.

According to Heckman,[32] the back surface temperature responses, when heat losses from both front and back surfaces are considered, are as follows:

$$T(l,t) = T_\infty \sum_{n=1}^{\infty} Y_n(0)Y_n(l) \exp(-\beta_n^2 \alpha t/l^2) \tag{30}$$

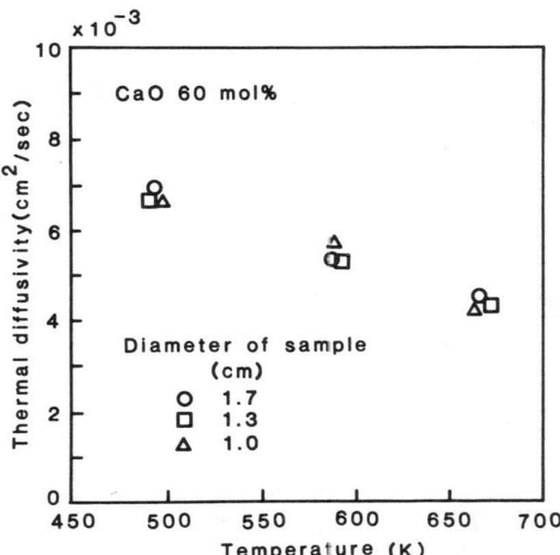

FIGURE 17. Effect of sample diameter on thermal diffusivity.[24]

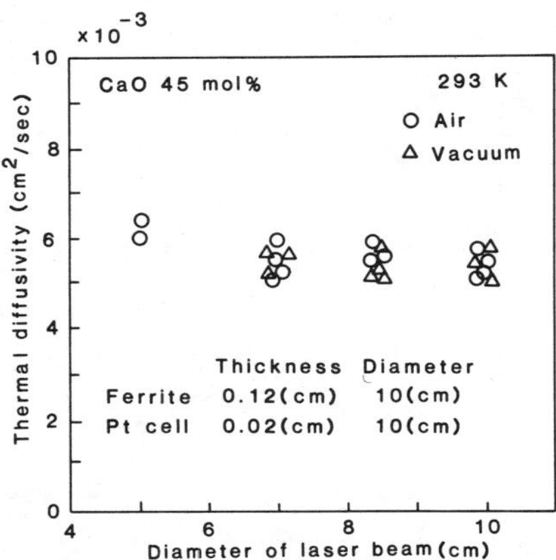

FIGURE 18. Effect of laser beam diameter on thermal diffusivity.[24]

where

$$Y_n(x) = \frac{2^{1/2}(\beta_n^2 + L_2^2)^{1/2}[\beta_n\cos(\beta_n x/l) + L_1\sin(\beta_n x/l)]}{[(\beta_n^2 + L_1^2)(\beta_n^2 + L_2^2 + L_2) + L_1(\beta_n^2 + L_2^2)]^{1/2}} \qquad (31)$$

and the βn ($n = 1,2,3,\ldots$) are the positive roots of

$$(\beta^2 - L_1 L_2)\tan\beta = \beta(L_1 + L_2) \qquad (32)$$

In Eq. (32), L_1 and L_2 are the radiative heat loss factors from the faces at $x = 0$ (front surface) and $x = l$ (back surface), respectively, and are given by

$$L_i = 4l\sigma\epsilon_i T_A^3/\lambda \qquad (i = 1,2) \qquad (33)$$

where σ is the Stefan–Boltzmann constant, l is the sample cell thickness (in the present case, for the calcium ferrite sample, $l = 1.3$ cm; for Pt, $l = 0.02$ cm), ϵ_i is the emissivity ($\epsilon_1 = 0.8$, $\epsilon_2 = 0.2$), and T_A is the temperature of the environment (1500 and 1673 K). The thermal conductivity value, λ, can be obtained from the experimental data for the heat capacity, density, and thermal diffusivity. As previously mentioned, the back-surface temperature responses of the two-layered sample cell system, with thermal diffusivities α_1 and α_2 and thicknesses l_1 and l_2, can be obtained by substituting the diffusion time for l^2/α in Eq. (30). Under the present experimental conditions, the square roots of the diffusion times are $l_1/\sqrt{\alpha_1} = 2.20$ and $l_2/\sqrt{\alpha_2} = 0.04$. Since the platinum plate is very thin and its thermal diffusivity is very high, compared with that of molten calcium ferrite, the platinum layer scarcely affects the shape of the temperature responses, i.e., by less than 2%. Thus, a single layer of calcium ferrite with different emissivities at the two surfaces was assumed in order to investigate the radiative heat loss. Figure 19 provides information about the influence of radiative heat loss on the thermal diffusivity measurement. The solid line is the experimental temperature

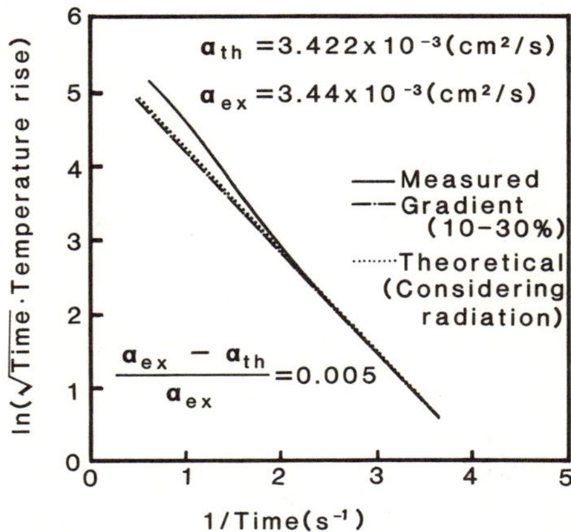

FIGURE 19. Effect of radiative heat loss on thermal diffusivity.[24]

response at 1500 K. The dotted line is the calculated temperature response considering radiative heat loss using Eq. (30) and physical constants for the cell at 1500 and 1673 K. The results imply that for the thermal diffusivity values determined by logarithmic analysis, the uncertainties due to radiative heat loss are less than 0.5% at 1500 K and 0.6% at 1673 K. Thus, the logarithmic method enables us to determine the thermal diffusivity even at high temperatures without uncertainties due to radiative heat loss.

6.4. Permeation of the Laser Beam

The pulsed laser beam is intended to be absorbed at the surface of the sample in this experiment. The effect of permeation of the laser beam can be easily examined by experiments with samples of different thicknesses. It may be noted that the optical properties of calcium ferrite samples are not significantly changed in the temperature range that has been investigated. As shown in Fig. 20, the results are not significantly affected by the thickness of the sample in both single-layered and two-layered systems. This proves that the presently used calcium ferrite samples are sufficiently opaque.

6.5. Selected Examples of Measured Thermal Diffusivity of Opaque Molten Salt Systems

Using the method described herein, the thermal diffusivities of calcium ferrites containing 33–55 mol% CaO were determined over the temperature range 1003 to 1673 K.[23, 24]

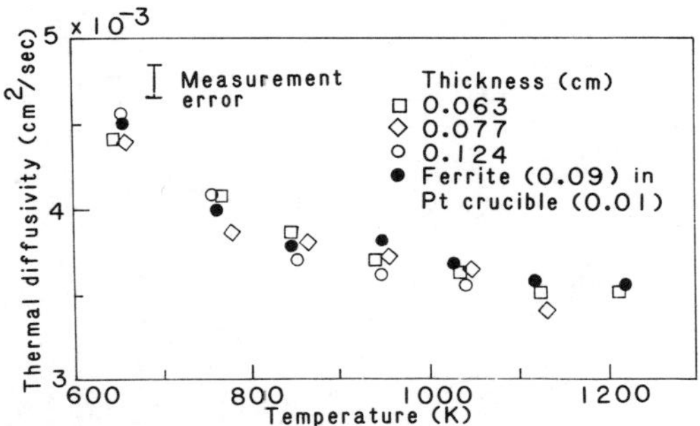

FIGURE 20. Effect of sample thickness on thermal diffusivity.[23]

Figure 21 shows the reproducibility of the measurements, using the results for a ferrite sample containing 55 mol % CaO in both solid and liquid phases. The same degree of reproducibility has been obtained in measurements for samples of other compositions.

Figure 22 illustrates the temperature dependence of thermal diffusivity for the calcium ferrite system. The thermal diffusivity of this system decreases as the temperature increases in both solid and liquid phases.

The thermal diffusivities of the molten calcium ferrite system at 1633 K are plotted in Fig. 23 as a function of the CaO content. The minimum value is found at a composition of 40 mol % CaO. In the CaO-rich region, an increase in the thermal diffusivity is evident, but its composition dependence is not apparent in this system.

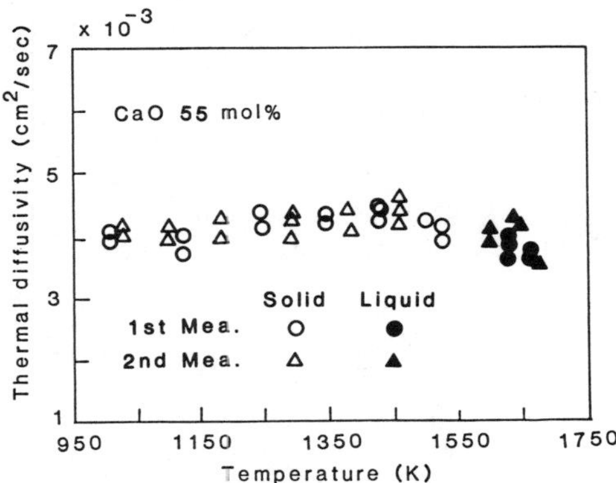

FIGURE 21. Thermal diffusivity of ferrite containing 55 mol % CaO.[24]

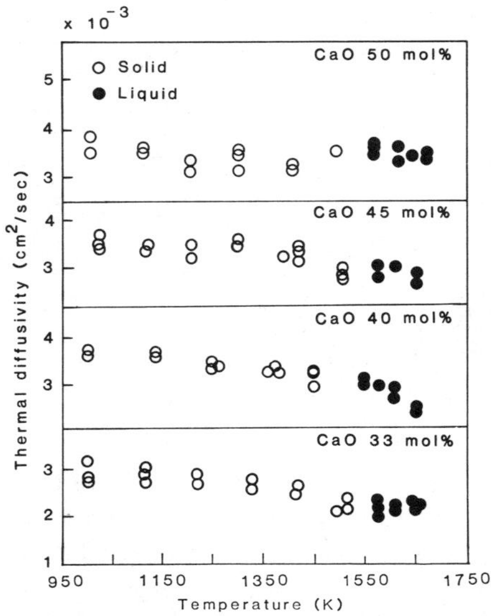

FIGURE 22. Thermal diffusivity of ferrite containing 33, 40, 45, and 50 mol % CaO.[24]

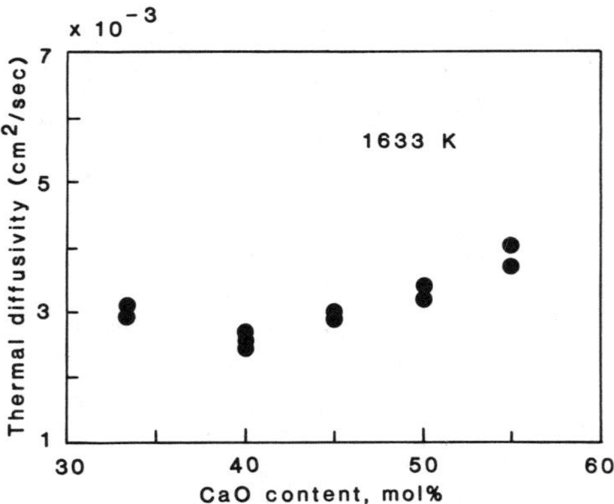

FIGURE 23. Compositional dependence of thermal diffusivity of molten calcium ferrite at 1633 K.[24]

7. Concluding Remarks

This chapter demonstrates the validity and usefulness of the laser flash method for determining thermal conductivities of molten salts at high temperature, with some selected examples. It includes also a description of recent improvements in remote sensing of the infrared rays and in data processing. The laser flash method requires small, easy to fabricate samples, and the results can be obtained within a few seconds, which reduces some of the difficulties encountered in more conventional methods. The laser flash method also is a useful and simple way to estimate the specific heat capacity of molten salts at high temperatures. No other technique has the versatility of the laser flash method. As shown in this chapter, the laser flash method basically works very well, and its potential power, in our view, cannot be overemphasized. However, some further developments are required for measurements of heat transport properties of molten salts by this method, for example:

1. Data processing for the determination of the thermal conductivity of semitransparent liquids at high temperature
2. Criteria to distinguish the sample liquid into optically transparent, semitransparent, and opaque media.

On the basis of the preliminary results,[33, 34] it can be predicted that if further effort and priority is assigned to addressing the above points, the laser flash method will be recognized as one of the most powerful tools for measuring thermal properties of molten salts.

Acknowledgments

We would like to acknowledge the important technical assistance of In-KooK Suh (SENKEN, Tohoku University). This work was funded in part by a Grant-in-Aid from the Ministry of Education, Science and Culture (#62850122). The support and encouragement of Professors A. Yazawa and Y. Shiraishi (SENKEN, Tohoku University) and Y. Tomota (Department of Metallurgy, Faculty of Engineering, Ibaraki University) are also greatly appreciated.

References

1. M. Hoshi, T. Omotani, and A. Nagashima, *Rev. Sci. Instrum.* **52,** 755 (1981).
2. T. Omotani, Y. Nagasaka, and A. Nagashima, *Int. J. Thermophys.* **3,** 17 (1982).
3. T. Omotani and A. Nagashima, *Trans. Jpn. Soc. Mechanical Eng.* **48B,** 2034 (1982).

4. Y. Nagasaka and A. Nagashima, *J. Phys. E: Sci. Instrum.* **14,** 1435 (1981).

5. T. Karasawa, Y. Nagasaka, and A. Nagashima, *Trans. Jpn. Soc. Mechanical Eng.* **52B,** 940 (1986).

6. W. J. Parker, R. J. Jenkins, C. P. Butler, and G. L. Abbott, *J. Appl. Phys.* **32,** 1979 (1961).

7. Y. Waseda and H. Ohta, *Solid State Ionics* **22,** 263 (1987).

8. S. Rosseland, *Astrophysik auf Atom—Theoretische Grundlagen,* pp. 41–44, Springer–Verlag, Berlin (1931).

9. Y. Tada, M. Harada, M. Tanigaki, and W. Eguchi, *Ind. Eng. Chem. Fundam.* **20,** 333 (1981).

10. T. Sakuraya, T. Emi, H. Ohta, and Y. Waseda, *Jpn. Inst. Metals* **46,** 1131 (1982).

11. H. Ohta, Y. Waseda, and Y. Shiraishi, in: *Proceedings of the Second International Conference on Metallurgical Slags and Fluxes,* Lake Tahoe, Nevada, p. 863, The Metallurgical Society of AIME, New York (1984).

12. H. Ohta, Y. Waseda, and Y. Shiraishi, Proceedings of the First International Symposium on Molten Salt Chemistry and Technology, Kyoto, Japan, April 1983, p. 261.

13. H. Ohta and Y. Waseda, Proceedings of the Joint International Symposium on Molten Salts, Honolulu, Hawaii, October 1983, p. 353.

14. J. B. Moser and O. L. Kruger, *J. Appl. Phys.* **58,** 3215 (1967).

15. Y. Takahashi, *J. Nucl. Mater.* **51,** 17 (1974).

16. Y. Takahashi, H. Yokokawa, H. Kadokura, Y. Sekine, and T. Mukaibo, *J. Chem. Thermodyn.* **11,** 379 (1979).

17. Y. S. Touloukian (ed.), *Thermophysical Properties of High Temperature Solid Materials,* Vol. 4, Macmillan, New York (1967), p. 1655.

18. Y. Takahashi and T. Yoshino, *Yogyo Kyokai Shi* **81,** 524 (1973).

19. K. Nagata M. Susa, and K. S. Goto, *Testu-to-Hagane* **69,** 1417 (1983).

20. S. Ishiguro, K. Nagata, and K. S. Goto, *Testu–to–Hagane* **66,** S669 (1980).

21. K. Nagata and K. S. Goto, in: *Proceedings of the Second International Conference on Metallurgical Slags and Fluxes,* Lake Tahoe, Nevada, p. 875, The Metallurgical Society of AIME, New York (1984).

22. K. Ogino, K. Nishiwaki, and T. Yamomoto, *Testu-to-Hagane* **65,** S683 (1979).

23. H. Ohta, In-Kook Suh, and Y. Waseda, *Proceedings of the Third International Symposium on Metallurgical Slags and Fluxes,* University of Strathclyde, p. 223, The Institute of Metals, London (1989).

24. In-Kook Suh, H. Ohta, and Y. Waseda, *High Temp. Mater. Processes* **8,** 231 (1989).

25. J. T. Schriempf, *Rev. Sci. Instrum.* **43,** 781 (1972); *High Temp.-High Pressures* **4,** 411 (1972).

26. J. L. Bates, Thermal Diffusion and Electrical Conductivity of Molten and Solid Slags, Battelle Pacific Northwest Lab. Report (1976).

27. R. E. Taylor and H. J. Lee, Report on the 14th Thermal Conductivity Conference, p. 135 (1974).

28. C. S. Ang, H.S. Tan, and S. L. Chen, *J. Appl. Phys.* **44,** 687 (1973).

29. C. S. Ang, S. L. Chen, and H. S. Tan, *J. App. Phys.* **45,** 179 (1974).

30. H. M. James, *J. Appl. Phys.* **51,** 4666 (1980).

31. O. Yamamoto, M. Yamawaki, T. Ido, and Y. Takahashi, The Eighth Japan Symposium on Thermophysical Properties, p. 229 (1987).

32. R. C. Heckman, *J. Appl. Phys.* **44,** 1455 (1973).

33. H. Ohta, Y. Waseda, and Y. Shiraishi, *J. High Temp. Soc. Jpn.* **11,** 81 (1985).

34. In-Kook Suh, H. Shibata, and H. Ohta, *High Temp. Mater. Processes* **8,** 135 (1989).

4

Magnetic Measurement Techniques

W. W. Warren, Jr.

1. Introduction

The magnetic properties of condensed matter are determined by the nature of valence and ion-core electronic states. Investigation of magnetic properties can therefore provide insight into such important issues as the character of chemical bonding in a particular material, the formation of molecular species, the state of nonbonding valence electrons, and the presence of phase transitions. Magnetic measurements on solids and liquids have a long history, and the general experimental techniques are well developed. It is the purpose of this chapter to discuss the execution of magnetic measurements under the special experimental conditions required for the study of molten salts. The emphasis is on techniques developed for measurements well above room temperature on materials which may be highly reactive. Thus, we are not concerned here with studies of "low-temperature" molten salts close to room temperature. Those materials are accessible using more or less standard techniques as they are applied to other classes of chemical systems. The general theory of magnetism and the interpretation of magnetic data are discussed at only a very basic level, mainly to suggest the kind of information available from magnetic studies and to define the quantities one might wish to measure.

1.1. Paramagnetic and Diamagnetic Susceptibilities

A fundamental property of any substance is the magnetic susceptibility, the quantity expressing the response of the material to application of a magnetic field that is constant in time and uniform in space. In a field $\mathbf{H}$, a volume v of material acquires a magnetization (dipole moment per unit volume) $\mathbf{m}$ given by

$$\mathbf{m} = \chi_v v \mathbf{H} \tag{1}$$

W. W. Warren, Jr. • AT&T Bell Laboratories, Murray Hill, New Jersey 07974 where χ_v is the volume susceptibility. The magnetization also can be expressed in terms of the moment per unit mass, in which case the corresponding susceptibility is the mass (gram) susceptibility $\chi_g = \chi_v/\rho$, where ρ is the density of the substance. If the gram molecular weight M is known, the susceptibility can be expressed equivalently as a molar susceptibility $\chi_M = \chi_g M$.

The susceptibility may be either positive or negative. A positive or *paramagnetic* susceptibility indicates that the magnetization develops parallel to the applied magnetic field while a negative or *diamagnetic* susceptibility corresponds to magnetization opposing the field.

Although it is relatively simple to define, the magnetic susceptibility is, in fact, a highly complex physical quantity.[1] The observed susceptibility of a substance is often the net sum of several contributions, diamagnetic and paramagnetic, whose relative magnitudes vary from material to material. The contribution common to all materials is the ionic diamagnetism of closed electron shells, $\chi_v^{d,i}$. In any ion, circulation of electron charge in the magnetic field produces a moment which tends to oppose the applied field. For simple closed-shell compounds such as NaCl, this is the dominant susceptibility contribution. The ionic diamagnetism is proportional to the number of ions per unit volume, n, and the mean square of the distance $\bar{r}^2$ of the electrons from the nucleus:

$$\chi_v^{d,i} = n(e^2/6m)\sum \bar{r}^2 \tag{2}$$

Electron orbits are larger in heavier ions, so the magnitude of the ion-core diamagnetism increases significantly with atomic number.

The susceptibility becomes more complex (and perhaps more informative) when the electronic states include unfilled shells. This occurs in transition metal salts or, in some cases, when elemental metals are dissolved in a molten salt. Unpaired electrons in incomplete shells produce spin paramagnetism. If the electrons are *localized* and obey classical Boltzmann statistics, the spin susceptibility $\chi^{p,s}$ at a given temperature is given by the Curie law:

$$\chi^{p,s} = n\mu^2/3kT \tag{3}$$

where μ is the paramagnetic moment of the ion. If, on the other hand, the electrons are completely *delocalized,* so that they approximate a free-electron gas and obey Fermi–Dirac statistics, the susceptibility is given by the Pauli value:

$$\chi^{p,s} = 2\mu_B^2 N(E_F) \tag{4}$$

where μ_B is the Bohr magneton and $N(E_F)$ is the density of electronic states per unit energy at the Fermi level E_F. For the volume susceptibility, the density of states is defined per unit volume. In general, if $T \ll E_F/k$, the Pauli susceptibility is much smaller than the Curie value, Eq. (3), because of the high degree of electron spin pairing imposed by the Fermi–Dirac statistics. Highly localized and completely delocalized states represent ideal limiting cases, and there are many interesting and important intermediate cases for which the susceptibility is described by neither Eq. (3) nor (4). Moreover, interactions among the electrons can modify $\chi^{p,s}$ with respect to the free-electron (Pauli) value.

Adding to the effects of spin paramagnetism, the *orbital* motion of electrons in unfilled shells also contributes to the susceptibility. Orbital (Van Vleck) paramagnetism $\chi^{p,o}$ is particularly important in transition metal systems for which the d-electron states are only partially occupied. If electrons are delocalized, there is also a diamagnetic contribution $\chi^{d,e}$ from the orbital circulation of the electron gas. This is the Landau diamagnetism and it makes a significant contribution to the magnetic susceptibility of metals.

In a susceptibility measurement, the experiment yields the total susceptibility, which is the sum of the various contributions:

$$\chi^{\text{total}} = \chi^{d,i} + \chi^{p,s} + \chi^{p,o} + \chi^{d,e} \tag{5}$$

It is often difficult to interpret χ^{total} without independent information about some of the terms in Eq. (5). In many interesting cases, however, progress can be made by means of resonance experiments that either provide information about specific susceptibility contributions or, at least, combine them with a different weighting than given by Eq. (5). We now turn to a brief introduction to the resonance methods.

1.2. Resonance Techniques

Nuclear magnetic resonance (NMR) and electron spin resonance (ESR) are spectroscopies based on the resonant absorption of energy by the nuclear and electronic magnetic moments, respectively. Let us consider first NMR. The paramagnetism of atomic nuclei *per se* is rarely very informative about the characteristics of a particular chemical system. The static nuclear paramagnetic susceptibility is determined by the concentration of the various nuclear species in the material, that is, the chemical and isotopic composition, together with the magnitude of the moments. In material of natural isotopic abundance and known chemical composition, these quantities are known *a priori*. However, the behavior of nuclear moments under the influence of combined static and *time-dependent* magnetic fields is usually very sensitive to the microscopic environments of the nuclei.[2] Although there are many contributions to the local magnetic environment, nuclei are particularly responsive to the effects of magnetic hyperfine fields produced by electron spin and orbital moments. These hyperfine interactions establish NMR as an indirect, but very powerful, probe of electronic magnetism.

The basic resonance effect is a consequence of the *Zeeman interaction* of an assembly of nuclear moments μ_n in a magnetic field H_0. The nuclear spins develop an energy E_{Zeeman} given by

$$E_{\text{Zeeman}} = -\sum \mu_n^i \cdot \mathbf{H}_0 = -\sum \gamma_n^i \hbar \mathbf{I}_i \cdot \mathbf{H}_0 \tag{6}$$

where γ_n^i is the gyromagnetic ratio, and $\hbar \mathbf{I}_i$ is the nuclear angular momentum of the ith nuclear species. The energy levels corresponding to the quantum-mechanically allowed orientations of the nuclei in the magnetic field are separated by an energy $\hbar\omega_0$, where ω_0 is the Larmor precession frequency

$$\omega_0 = \gamma_n H_0 \tag{7}$$

For typical laboratory fields of a few tesla (1 T = 10 kG), the frequencies are in the radio frequency range, 1–100 MHz. NMR is based on the resonant absorption of radio frequency energy when, in addition to a static field H_0, the system is also subjected to a time-dependent magnetic field $H_1 \cos \omega t$, with $\omega \sim \omega_0$.

It is rare that the frequency for exact resonance is precisely that given by Eq. (7). This is because the electrons in the material generate additional local magnetic fields at the nuclei which are superimposed on the externally applied field H_0. The local-field contributions can be either positive or negative depending on the sign of the hyperfine interaction between electrons and nuclei and on whether the local fields are associated with diamagnetic or paramagnetic electronic magnetization. Shifts of the NMR frequency due to local fields can be very informative regarding the electronic structure of the system under study. Not only is the information derived from NMR highly local, that is, determined by electronic structure on the atomic scale, but the method is specific to particular chemical elements. In CsI, for example, the local-field shifts can be measured at both the Cs and I sites since ^{133}Cs and ^{127}I have resonances that can be studied separately.

The local magnetic fields are time dependent, in general, and so carry information about electron and ion dynamics. These fluctuating fields reveal themselves in the NMR experiment by affecting the linewidth of the resonance and causing nuclear spin relaxation. The latter is the process by which the nuclei come to equilibrium after having been disturbed by an abrupt change of magnetic field or, more commonly, by manipulation of the nuclear spins with radio frequency pulses.

Electron spin resonance is directly analogous to NMR but is based on the Zeeman interaction of unpaired electron spins. It is the most direct probe of electron spin paramagnetism. In an applied magnetic field $\mathbf{H}_0$, the energy difference between states with spin parallel and antiparallel to the field is

$$E_{\text{Zeeman}} = -\gamma_e \hbar \mathbf{S} \cdot \mathbf{H}_0 \tag{8}$$

in complete analogy to Eq. (6). The electronic gyromagnetic ratio γ_e is related to the "g-factor" by $\gamma_e \hbar = g\mu_B$. Because the magnetic moment of the electron is ~ 1000 times that of typical nuclear moments, the resonance frequencies for ESR fall in the gigahertz rather than the megahertz range for typical laboratory fields of a few kilogauss.

The integrated intensity of an ESR line is proportional to the paramagnetic susceptibility of the spins, $\chi^{p,i}$. If an appropriate theoretical expression for $\chi^{p,i}$ is available, measurement of the absolute ESR intensity provides a "spin count," that is, a measure of the number of unpaired spins in the sample. This is possible, for example, in situations where the Curie law, Eq. (3), is applicable. Important information is also provided by the position of the resonance, the so-called "*g*-shift," which is determined by interactions of the electrons with their environment, mainly through the spin–orbit interaction. Relaxation and linewidth effects are related to the dynamics of this interaction as well as dipolar spin–spin couplings to other electrons and hyperfine interactions with nuclei.

1.3. Relevance to Studies of Molten Salts

Even limited to the domain of molten salt science, magnetism is a rich and detailed subject. Static and dynamic susceptibilities are governed by a diversity of physical mechanisms and can provide detailed microscopic information about the chemistry and physics of these systems. The diamagnetic susceptibility and NMR shift, for example, are quite sensitive to variations in the character of the chemical bond in a given class of salts. This was exploited in an early NMR study of thallium halides by Hafner and Nachtrieb,[3] who used systematics of the NMR shift to evaluate the ionicity of the Tl–halogen bond.

Magnetic measurements can be especially informative when there is the possibility of spin paramagnetism from unpaired electron states. The state of excess electrons introduced by solution of metal in molten salts is the classic example of this type of situation. Magnetic susceptibility and magnetic resonance studies provide information about the state of unpaired electrons as well as the formation of diamagnetic species containing paired electrons. The dynamic susceptibility is strongly affected by the metal–nonmetal transition at higher concentrations of dissolved metal.

These examples and a few others introduced later in this chapter are not inclusive, but serve merely to illustrate typical motivations for carrying out magnetic measurements in molten salts and molten salt mixtures. More complete discussions of experimental data and their interpretation may be found in various review articles[4–8] and, of course, in the original literature. We turn now to the central focus of this chapter, the techniques required for measurement of the magnetic properties of molten salts well above room temperature.

2. General Experimental Considerations

The actual practice of magnetic measurements is simplified by the fact that the experiments do not require physical contact to the sample materials. This is particularly helpful for studies of highly reactive liquid samples at high temperatures. It is possible to determine the susceptibility of virtually any molten salt or molten salt solution that can be enclosed in a suitable cell, that is, a cell which is chemically resistant to the sample material at the measurement temperature and which can be sealed against the vapor pressure of the sample. Similar requirements apply to resonance measurements, with the added condition that the cell must be transparent to radio frequency radiation in the megahertz range for NMR and in the gigahertz range for ESR. Thus, the general experimental requirements for magnetic measurements reduce to the problems of (i) providing a high-temperature environment in an appropriate magnetic field configuration and (ii) confining the liquid sample in a container that is chemically compatible with the sample material and, if necessary, transparent at radio or microwave frequencies.

2.1. High-Temperature Environments and Thermometry

Magnetic measurements on molten salts require sample environments at temperatures that may reach well above 1000°C, depending on the system, and magnetic fields ranging up to more than 100 kG. A generic arrangement providing these conditions is shown schematically in Fig. 1. Heating is provided by a cylindrical resistive element whose diameter is a compromise—small enough to fit into a laboratory magnet (with space for insulation), yet large enough to provide an adequate internal volume at a homogeneous high temperature. Thermocouples in the heated space measure the temperature and can be used in a feedback circuit to regulate the temperature. The external container is formed from a cylinder of nonmagnetic metal, such as brass or nonmagnetic stainless steel.

In designing and operating the heating element, it is necessary to consider the magnetic effects of the heating current. It is usually advisable that the heater be noninductive. This can be ensured to a relatively high degree with the use of either a bifilar winding (Fig. 2a) or a "zigzag" configuration (Fig. 2b). The use of alternating current (AC) to power the heater eliminates any residual static magnetic fields that might result from the lack of perfect symmetry in the winding. However, the element may vibrate at the AC frequency due to the interaction of the current with the applied field and, in the case of a resonance measurement, there may be electrical "pickup" in the sensitive receiver circuitry. Direct current (DC) power is preferable for such measurements. Good practice requires that the residual field produced by the heater be checked at the sample position at high temperature.

An important characteristic of any high-temperature system is the speed with which the temperature can be changed. This is governed mainly by the "thermal mass" of the system, that is, the quantity of material that must be brought to the experimental temperature and the heat capacity of that material. Systems of the basic type shown in Fig. 1 tend to have relatively high thermal masses because of the size of the element, its supporting structure, and the thermal insulation around the heater. Thus, even if substantial excess power is available to speed the heating process, cooling may be so slow that the time required to bring the system to equilibrium is inconveniently long. For typical systems of the type described in this chapter, this is not a serious problem above

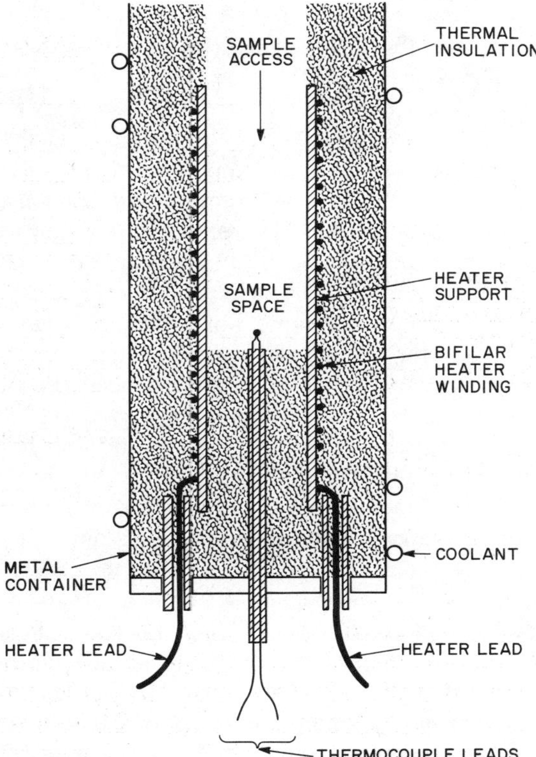

FIGURE 1. Essential elements of a system for magnetic studies on molten salts.

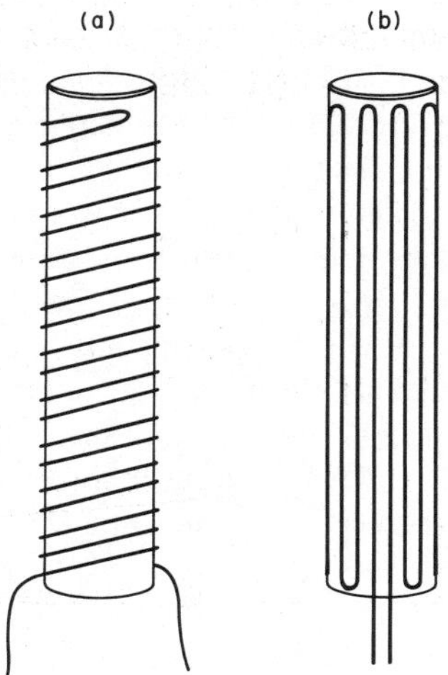

FIGURE 2. Noninductive windings for heater elements: (a) bifilar style; (b) "zigzag" style.

500–600°C, where radiative heat transfer begins to play a role. At these higher temperatures, it is not difficult to achieve equilibration times of one minute or less for a temperature change of 100°C. However, at lower temperatures, thermal conduction dominates heat transfer, and a typical system may require 30 minutes or more to equilibrate at, say, 100–200°C. For these more moderate conditions, a gas flow system may be much more effective. In such a system, a stream of hot gas is directed on the sample. It is only necessary to heat the gas itself and the relatively small mass of the sample. The response is thus much faster than with a "static" system of the type shown in Fig. 1. For reasons discussed later, gas flow systems are not suitable for magnetic susceptibility measurements, but they can be very useful for resonance experiments at moderately high temperatures. An actual gas flow system for NMR is described in Section 4.

The materials used in apparatuses for high-temperature magnetic measurements are relatively standard, except that strongly magnetic materials must be avoided. Ni metal, for example, is often used as an insert in high-temperature heating elements to help homogenize the temperature distribution. However, it is not suitable for magnetic measurements since it is ferromagnetic up to 358°C and strongly paramagnetic above that temperature. Similarly, Fe–constantan thermocouples may cause problems if they are close to the region of measurement.

For heating elements to be operated in air, nichrome and related alloys such as Kanthal may be used up to about 1000°C. Platinum or platinum–rhodium alloy are good choices for higher temperatures, and the latter may be used up to more than 1500°C. These materials are ductile and are easily formed into the desired configurations, such as those shown in Fig. 2. It is sometimes necessary that the high-temperature apparatus be enclosed and operated in a protecting atmosphere of inert gas, or in vacuum. While this obviously complicates the overall design of the experimental system, it has the advantage of permitting the use of additional high-temperature materials which would oxidize rapidly in hot air. In nonoxidizing environments, molybdenum is a particularly good choice for heater construction as it can be used to very high temperatures ($\sim$1650°C) and is easily formed into coils or zigzags. Tungsten may be used to even higher temperatures, but its brittleness makes it very difficult to bend and shape. Tungsten is less likely to break if it is heated slightly during working, and it is possible (with practice!) to make zigzag tungsten elements by heating the wire to $\sim$100°C at the point of bending.

Temperatures can be measured conveniently with the standard high-temperature thermocouples such as chromel–alumel or, in an oxidizing environment, Pt/Pt–Rh. For very high temperatures in a nonoxidizing environment, W/W–Re is a good choice. These thermocouples are available with coaxial metallic shields, which are especially advantageous for resonance measurements. A thermocouple located close to a radio frequency coil can act as an "antenna," delivering unwanted electrical interference to the measurement circuitry. Such problems are greatly reduced if the thermocouple is enclosed in a grounded shield.

A final consideration in the choice of heater and thermocouple materials is the nature of the sample material itself. As we discuss in the next section, it is preferable that the sample be sealed tightly in a suitable cell, and, in this case, the sample obviously has no influence on the components of the high-temperature system. However, depending on the temperature range and the sample involved, it may be impossible to avoid leakage of sample material vapors into the region of the heater and thermocouples. If this is a possibility, it is necessary to take into account the corrosive effects of the sample material on the external components and to choose materials for which these effects are minimal.

2.2. Sample Containment Requirements

Because magnetic measurements *can* be made on samples in fully sealed cells, it is advisable to use sealed containers if at all possible. Not only does a sealed cell eliminate loss of sample material by evaporation (accompanied by a change of composition in a binary or multicomponent system), but also the sample is protected from the corrosive effects of the ambient atmosphere in the high-temperature apparatus. Even in nominally controlled environments such as argon, it is difficult to keep the atmosphere so pure that alkali metals, for example, will not react with residual oxygen or water. Furthermore, as discussed in the previous section, vapors of the sample material may corrode heating elements, thermocouples, and other components of the high-temperature apparatus, shortening their useful lives.

The simplest solution of the containment problem is the use of sealed glass cells, namely, Pyrex or fused-silica cells. If there is no chemical reaction between the sample and silica, and if the internal pressure is less than 2 or 3 atm, fused silica may be used at temperatures up to about 1100°C without difficulty. Pyrex and other soft glasses are limited to temperatures below 500°C, but they may be preferable to silica in this range for reasons of chemical compatibility. Glass cells enjoy the advantage of being easily formed into special configurations. For example, inclusion of a reentrant well for a thermocouple in the middle of the sample volume can be readily achieved.

The situation becomes more complex if glass cells are not acceptable, and a special cell design may be necessary for each experiment. Magnetic susceptibilities may be measured in (nonmagnetic) metallic cells, and a material such as tantalum can be used in many cases. Tantalum cells can be formed from tubing and sealed under vacuum with electron-beam welding. Other refractory metals and oxides such as sapphire can be closed by means of a mechanical seal in a "header" assembly attached to the body of the cell. Two cells based on this principle, developed at Marburg University for susceptibility measurements [9,10] on alkali metals, are illustrated in Fig. 3. In these cells, a niobium cone is forced against a sharp edge on the opening of a molybdenum or molybdenum–rhenium cap, which is electron-beam welded to the molybdenum cell body. A similar arrangement with a sapphire cell body is shown in Fig. 4.[11] In this case, the

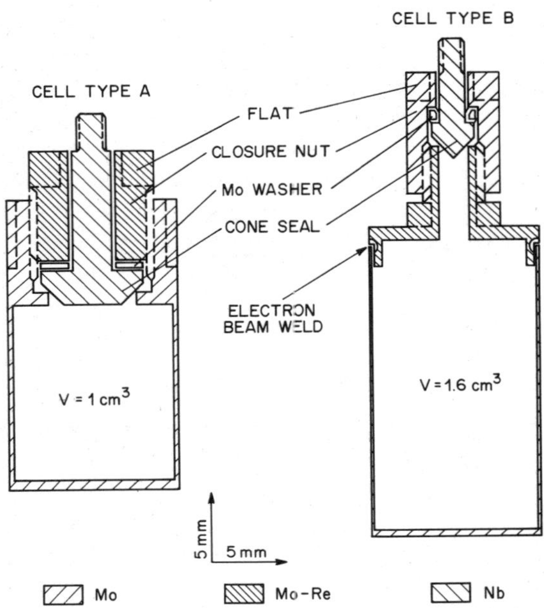

FIGURE 3. Metallic cells for magnetic susceptibility measurements.[9,10] "Type A" has one-piece molybdenum body; "type B" has thin-walled molybdenum–rhenium tube welded to top portion. Vacuum-tight seal is achieved by mechanically forcing niobium cone against sharp metal edge.

metal closure assembly is sealed to the sapphire by means of a metal (Ni–Ti) brazing alloy.[†] Oxide glazes also may be used if they are compatible with the sample material. The niobium–sapphire seal is less reliable than an electron-beam metal–metal seal, but the latter is unacceptable for resonance studies since the cell body must be electrically insulating.

2.3. Sample Purity and Handling Considerations

The procedures required for preparation of molten salt samples for magnetic measurements vary widely depending on the properties of the specific material. Only a few generalizations can be made. Good practice obviously dictates that the sample material be loaded and sealed in the measurement cell in the purest state attainable with a reasonable level of effort. It is fortunate, in this regard, that magnetic measurements often impose less severe demands on sample quality than techniques such as optical spectroscopy or some electrical transport measurements. This is not to encourage careless sample handling or the use of material of low quality when better samples are readily at hand. However, it is

---

† The possibility of sealing sapphire to niobium using a Ni–Ti brazing alloy was first brought to my attention by N. H. Nachtrieb.

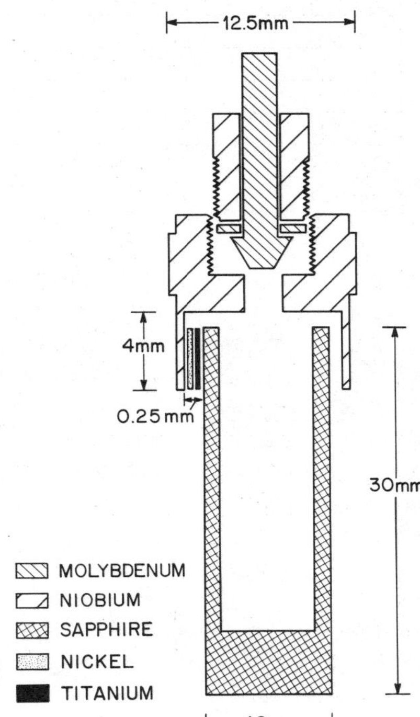

FIGURE 4. Nonmetallic (sapphire) cell for resonance measurements.[11] Sapphire cell body is sealed to niobium portion using nickel–titanium braze.

useful to know that it may be unproductive to expend great effort to reduce impurities to the ppm level in a particular case if the measurements are insensitive to an impurity concentration of 1%.

For measurements of the susceptibility above room temperature, for example, it is usually sufficient that the concentration of magnetic impurities be kept below about 0.1%. This is because the strong $1/T$ dependence of the Curie law, Eq. (3), suppresses the impurity susceptibility at high temperatures. Nonmagnetic impurities, if they do not react with the host molten salt, can be present at levels of 1% before they begin to introduce serious errors.

NMR measurements are even more forgiving if the impurities contain no nuclei of the species of interest in the molten salt. Such impurities are simply invisible to NMR unless they react chemically with the host and thus change its properties. However, if the impurities carry magnetic moments, they can affect the nuclear spin relaxation rates even if they do not contribute directly to the NMR signal. Whether or not this presents a problem depends on the strength of the host nucleus–impurity coupling compared with other, intrinsic interactions. There is no general rule for this.

ESR measurements also are quite immune to the effects of nonreacting, nonmagnetic impurities, but they are seriously affected by magnetic impurities. These are more troublesome than in NMR because *all* electron spin resonances, including those of impurities, fall in the same narrow frequency range. In contrast, NMR frequencies for different isotopes vary widely. It can be quite difficult in some cases to distinguish the ESR signal of impurities from the intrinsic resonance associated with the host molten salt. In any ESR experiment, it is important to ensure that the observed ESR signals are actually characteristic of the pure sample material.

After purification of the sample material to the desired level, it is important to load and seal the containers without loss of sample quality. Often, this requires protection of the sample from the ambient atmosphere, especially oxygen and water. The cells described in the previous section can usually be loaded in the controlled atmosphere of a dry box or glove bag. Those with mechanical closure can be sealed with the internal gas being that of the dry box or, if necessary, under vacuum. Glass cells can be closed temporarily, then transferred to a pumping line for evacuation and back-filling with gas, before being sealed with a torch.

3. Magnetic Susceptibility

3.1. Measurement Principles

There are many ways to measure the magnetization of a sample and hence, according to Eq. (1), the susceptibility.[12] Modern methods include the vibrating sample magnetometer and techniques employing superconducting quantum interference devices (SQUIDs). However, the classical techniques of Faraday and Gouy are most readily adapted to studies of molten salts and other liquid samples at elevated temperatures.

Both the Faraday method and the Gouy method are based on measurement of the force on a sample in an inhomogeneous magnetic field. The force on an element of material dv in, say, the z-direction is given by

$$dF_z = (1/2)dv[\chi_v(\text{sample}) - \chi_v(\text{medium})]\partial H^2/\partial z \tag{9}$$

where $\chi_v(\text{sample})$ and $\chi_v(\text{medium})$ are the susceptibilities of the sample material and the surrounding medium, respectively. Since the medium is most frequently a gas such as air or argon, or even vacuum, its susceptibility is usually negligible compared with the susceptibility of the sample. The two methods differ in that for the Faraday method, the sample volume is assumed to be small enough that the field–gradient product

$$\partial H^2/\partial z = 2\mathbf{H}\cdot\partial \mathbf{H}/\partial z \tag{10}$$

is uniform over the sample. For the Gouy method, one uses samples large enough that the field varies significantly over the volume v. For both methods, the sample is suspended from a sensitive microbalance and the apparent change in weight is measured as the magnetic field is applied. These two methods are described in more detail in the following sections.

3.1.1. Faraday Method

The total force on a small sample in an inhomogeneous magnetic field is easily seen from Eqs. (9) and (10) to be

$$F_z = \chi_v(\text{sample})\mathbf{H}\cdot\partial\mathbf{H} / \partial z \tag{11}$$

where we neglect the susceptibility of the medium. Special magnet poles (Fig. 5) are available which give a uniform value of $\mathbf{H}\cdot\partial\mathbf{H}/\partial z$ over the 10- to 20-mm distance needed to accommodate typical samples.

For liquid samples, it is necessary that the material be contained in a cell of susceptibility $\chi_v(\text{cell})$, so that the total weight change of sample and cell in a Faraday apparatus is

$$\Delta w = (1/g)\,[\chi_v(\text{sample})v_s + \chi_v(\text{cell})v_{\text{cell}}]\mathbf{H}\cdot\partial\mathbf{H}/\partial z \tag{12}$$

where g is the acceleration of gravity. In order to obtain the sample susceptibility, it is necessary to determine the contribution of the empty cell. In principle, this could be done by a simple calculation if the susceptibility of the cell material and its volume are known. However, it is better in practice to measure the empty cell contribution directly. This not only eliminates the problem of determining the "active" volume of the cell, which may be quite difficult if the cell extends outside the region of uniform $\mathbf{H}\cdot\partial\mathbf{H}/\partial z$, but it corrects for possible variations in the susceptibility of the cell material and other systematic errors.

After subtracting the contribution of the empty cell, one obtains from a Faraday susceptibility measurement the quantity $\chi_v(\text{sample})v_s\mathbf{H}\cdot\partial\mathbf{H}/\partial z$ (assuming that we are on Earth and g is known!). One might determine $\mathbf{H}\cdot\partial\mathbf{H}/\partial z$ by a careful mapping of the magnetic field, but it is much more common to calibrate the apparatus by measuring the susceptibility of standard substances of known

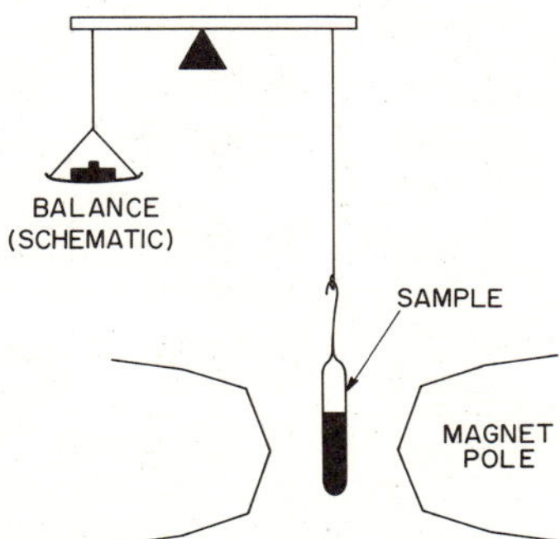

FIGURE 5. Schematic diagram of Faraday susceptibility measurement. Magnet poles are shaped to provide uniform value of $\mathbf{H}\cdot\partial\mathbf{H}/\partial z$ over the sample volume. The Faraday method is best suited to relatively small samples.

susceptibility. Calibration at room temperature may be made with water or with any number of well-characterized organic substances or elemental metals. To allow for possible effects of heater currents at high temperatures, calibrations can be made over a wide temperature range using various metals (Pd, Au, Mo, Nb, etc.).

It is usually easier to determine the mass of a sample than its volume, so a Faraday measurement of a known weight of material normally yields the specific (gram) susceptibility, $\chi_g = \chi_v(\text{sample})v_s/m$, where m is the mass in grams. The specific susceptibility can be converted easily to volume or molar units if the density or molecular weight, respectively, is known.

3.1.2. Gouy Method

The Gouy method is appropriate when relatively large samples are available. In a typical experiment, a cylindrical sample of cross section A is suspended from a balance so that one end is in the magnetic field of strength H_1 and the other remains outside the field or, at least, is in a region of significantly different field value, H_0. The experimental arrangement is shown schematically in Fig. 6. The total force on the sample is obtained by integrating Eq. (9) over the length of the sample to obtain

$$F_z = (1/2)A\chi_v(\text{sample})(H_1^2 - H_0^2) \tag{13}$$

where we have again neglected the effect of the medium. One advantage of the Gouy method is that the force depends only on the field values, H_1 and H_0, and

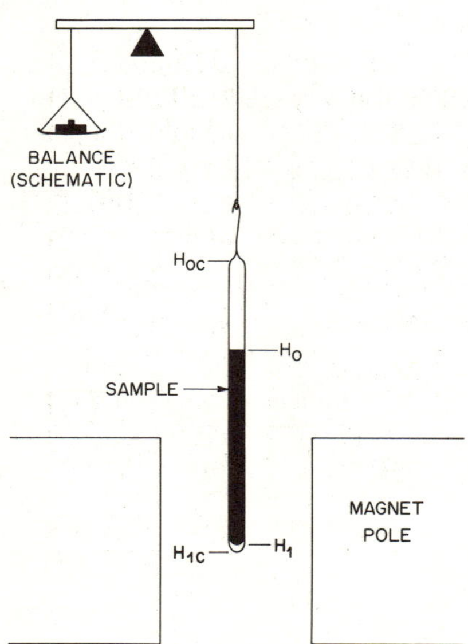

FIGURE 6. Schematic diagram of Gouy susceptibility measurement. Best for relatively large samples, the measurement is based on the difference between the field value at the two ends of the sample, H_1 and H_0, respectively. H_{1c} and H_{0c} are the field values at the corresponding ends of the cell.

not on the detailed configuration of the field. Thus, unlike the Faraday method, the Gouy method does not require special magnet pole design, and the measurements may be done with a variety of permanent magnets or electromagnets.

For a sample contained in a cell made of material with susceptibility $\chi_v(\text{cell})$, the apparent weight change on applying the magnetic field is

$$\Delta w = (A/2g)[\chi_v(\text{sample})(H_1^2 - H_0^2) + \chi_v(\text{cell})(H_{1c}^2 - H_{0c}^2)] \qquad (14)$$

where the subscripts $1c$ and $0c$ indicate that the fields at the two ends of the cell may not be the same as at the two ends of the sample material (Fig. 6). The cell contribution [second term in Eq. (14)] can be determined by a measurement on the empty cell placed in the same position as in the measurement on the loaded cell. Rather than calibrate the fields H_1 and H_0 directly, it is more common to employ a standard sample, as discussed in the previous section.

Since the cross section A is determined by the dimensions of the cell, the Gouy susceptibility measurement yields directly the volume susceptibility. Determination of the specific or molar susceptibilities thus requires independent knowledge of the density. This may be a disadvantage for certain experiments, such as studies of molten salt solutions, for which the mass susceptibilities may be more easily interpreted than the volume susceptibility and density data may not be available.

3.2. Apparatus Design

A general approach to provision of a high-temperature environment and containment of samples for magnetic studies was described in Section 2. The requirements of the Faraday and Gouy methods of susceptibility measurement impose additional conditions. The measurement is essentially a delicate weighing operation (microgram to milligram range), and it is important to avoid mechanical disturbance of the suspended cell. It is usually necessary to provide at least some mechanical isolation of the apparatus from building vibrations. In any case, it is common practice to evacuate the space containing the balance and the suspended cell to avoid disturbance by convection currents. Thus, heating a susceptibility sample with hot flowing gas is clearly unacceptable despite the convenience of rapid thermal response that such a system offers.

Molten salt samples can be contained in any cell of nonmagnetic material that is chemically compatible with the sample and stable at the measurement temperature. In addition to fused silica, it is possible to use various refractory metals, oxide ceramics such as alumina (Al_2O_3), or single-crystal sapphire. The main problem with nonglassy cell materials is that of obtaining a tight closure seal at the experimental temperature. Some techniques for sealing were discussed in Section 2. The presence of paramagnetic impurities in the cell material is not a serious problem at elevated temperatures because of the $\chi \propto 1/T$ temperature dependence of the Curie susceptibility of the impurities. Nevertheless, for the most accurate measurements it may be necessary to determine the cell suscepti-bility as a function of temperature over the experimental range.

A complete high-temperature susceptibility system,[9,10] suitable for mea-surements on molten alkali halides with added alkali metal, is shown in Fig. 7. An apparatus of this design has been used at temperatures up to > 900°C. The vacuum space is defined by a cylinder of nonmagnetic stainless steel which is connected via a flexible bellows to a pump-out manifold and enclosure for the balance (not shown). Heating is provided by a radiative heater consisting of a bifilar Mo winding inside a Mo tube. (Heater windings on metallic support structures can be insulated by means of ceramic beads strung along the wire in "necklace" fashion.) Thermal insulation is provided by a cylindrical Mo radia-tion shield on the sides and by plates of zirconia which close the ends of the high-temperature region. The entire high-temperature portion is supported on a water-cooled base plate by vertical Mo rods.

Sensitive microbalances that may be operated in an enclosed evacuated

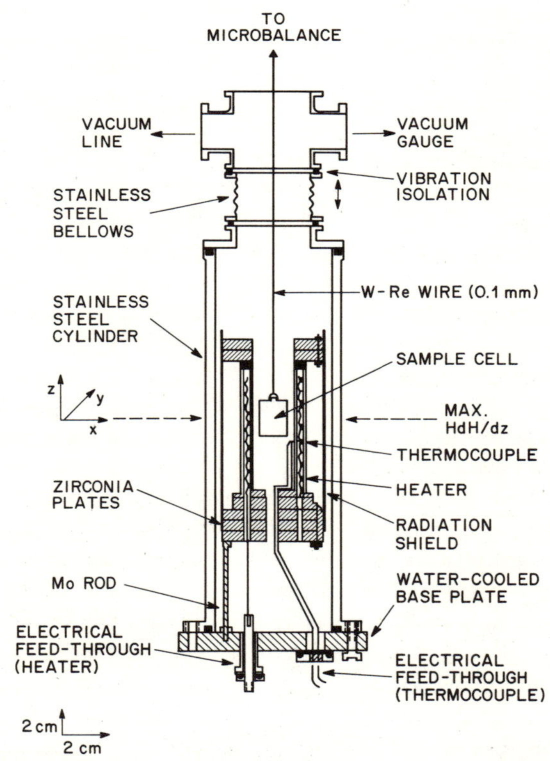

FIGURE 7. High-temperature Faraday susceptibility system.[9,10]

space are available commercially. The balance should have a resolution of a few micrograms for a total load consisting of some tens of grams. The following explicit example should help define the requirements of the balance and the weighing operation. A typical cell, especially one with a relatively heavy metal closure arrangement such as those illustrated in Figs. 3 and 4, may weigh 10 to 20 g. For a typical molten salt, CsI, the weight change Δw of the loaded cell in a field of a few kilogauss would be about 30 mg for a 1-ml sample volume. The weight change of the empty cell for this case might be about 40 mg. The precision of the susceptibility value depends on the precision of the *difference* in weight change (-10 mg). (Note that for this example the apparent weight change of the sample is negative, indicating that the susceptibility is diamagnetic.) Thus, a susceptibility measurement with a precision of 1% requires that a given weight be determined to much better than 100 μg in order that the the final difference in weight changes be determined to within 100 μg.

3.3. Illustrative Experimental Results

In the space of the present chapter, it is not possible (or even desirable) to review the experimental literature on susceptibility measurements in molten salts. The purpose of this section is rather to demonstrate, using specific examples, the quality of data obtainable and some of the physical information that can be extracted from magnetic susceptibility measurements. The same approach will be taken subsequently in Sections 4.3 and 5.3, in which examples of NMR and ESR data, respectively, are presented.

The first example, shown in Fig. 8, presents the molar susceptibility of K_x–KCl_{1-x} solutions as a function of temperature for compositions covering the full range from pure salt ($x = 0$) to pure metal ($x = 1$).[10] These data illustrate several basic points:

1. The magnetic susceptibility of pure molten KCl is diamagnetic and essentially independent of temperature. The diamagnetism reflects the filled-valence-shell electronic configuration K^+—Cl^- of the compound. There are no unpaired spins or partially filled shells that could provide any paramagnetic contribution. One would not expect this closed-shell ionic configuration to be affected by changes in temperature so it is not too surprising that the temperature dependence of the susceptibility is rather uninteresting.

2. As potassium metal is added to the pure salt, the susceptibility changes rapidly in the paramagnetic direction, showing that the species associated with the dissolved metal carry unpaired electron spins. It is not straightforward to identify the species from susceptibility data alone, but, in fact, extensive studies have shown that the valence electron of the added metal localizes to form F-center analogues in the dilute limit ($0 <$

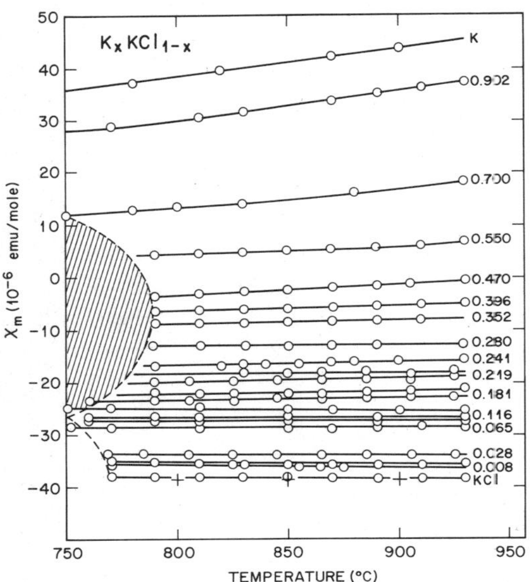

FIGURE 8. Molar magnetic susceptibility χ_m versus temperature for K_x–KCl_{1-x} metal–molten salt solutions.[10] Curves are labeled by mole fraction (x) of potassium. Shaded area denotes region of liquid–liquid immiscibility.

$x \ll 1$). For sufficiently high metal concentrations ($x > 0.5$), the total susceptibility becomes paramagnetic. It is important to recognize that the composition where $\chi = 0$ has no special physical significance, because the susceptibility is the sum of *independent* diamagnetic and paramagnetic contributions.

3. The susceptibility of pure liquid potassium is dominated by the Pauli paramagnetism of the conduction electrons. There is a weak temperature dependence, reflecting, mainly, the effects of temperature and thermal volume expansion on the density of states $N(E_F)$. The diamagnetic susceptibility of K^+ ions is still present in the pure metal, of course, and in order to extract the electronic contribution to the susceptibility, it is necessary to subtract the (negative) ionic susceptibility. This is usually determined independently from studies of ionic compounds and solutions.

Nicoloso and Freyland[10] analyzed the concentration dependence of the paramagnetic susceptibility in K_x–KCl_{1-x} and found a rapid decrease in the susceptibility *per excess K* in the range of a few percent added metal. The paramagnetic contribution only corresponds to the Curie value given by Eq. (3) for the most dilute solutions. At higher concentrations of excess metal, the spins become partially paired, reducing the susceptibility. This can be due to formation of spin-paired species like K_2 or K^- in nonmetallic solutions. In the metallic

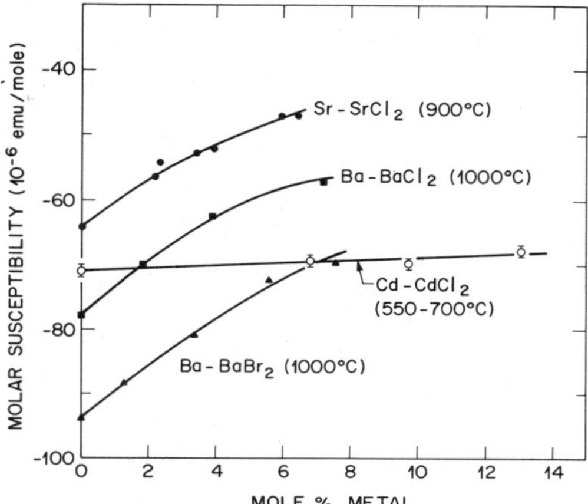

FIGURE 9. Molar magnetic suscepti-bility versus mol % metal for divalent metal–metal halide solutions.[13–15] Rapid rise of χ_m for the alkaline earth metal solutions reveals formation of paramagnetic species. Species formed in Cd–CdCl$_2$ are mainly diamagnetic.

range, we expect pairing because of the Fermi–Dirac statistics leading to the Pauli susceptibility of Eq. (4). The data of Fig. 8 led to the conclusion[10] that the metallic limit is reached for $x > 0.2$, where the susceptibility per added K is essentially the same as in the pure metal.

The second example is that of divalent metals dissolved in their molten halides. Figure 9 shows the concentration dependence of the magnetic suscepti-bility of solutions of Cd, Sr, and Ba in their respective halides (mainly chlo-rides).[13–15] These measurements are restricted to relatively low metal contents because of limited metal solubility in these salts at the temperatures investigated. As is the case for the alkali halides, we see that the total susceptibilities of the pure divalent metal salts are diamagnetic. However, the effect of excess metal is quite different for different metals. For the alkaline earth metals, Sr and Ba, there is a strong paramagnetic contribution associated with the dissolved metal. Since the electronic configuration of the neutral atoms is s^2 and is therefore spin-paired, the data clearly demonstrate that these metals do not enter the solution in the atomic state (M^0). The most likely possibility is that the paramagnetic ion M$^+$ is formed.

It is evident from Fig. 9 that the chemistry of Cd–CdCl$_2$ solutions is quite different from that of Sr and Ba solutions. There is almost no change in suscepti-bility with addition of Cd. Grantham[16] proposed formation of the diamagnetic species Cd$_2^{2+}$. However, it would be necessary to invoke additional arguments to exclude other diamagnetic species, including the neutral atom Cd^0. These data and the issues raised by their interpretation illustrate both the high sensitivity of the magnetic susceptibility to formation of paramagnetic species and the diffi-culty of identifying a unique species actually responsible for the observed effects.

4. Nuclear Magnetic Resonance

Nuclear magnetic resonance (NMR) is a powerful experimental technique whose reach extends far beyond the subject of this chapter. The great versatility of NMR is due, in part, to the more than 100 nuclear isotopes representing 80 chemical elements which can be utilized. Because these include hydrogen (^{1}H and ^{2}H), carbon (^{13}C), and oxygen (^{17}O), NMR has found widespread applicability in organic chemistry, biology, and medicine. However, suitable isotopes may be found in all parts of the periodic table, and NMR has long been established as one of the key experimental techniques for inorganic matter with many applications in such areas as condensed matter physics, inorganic chemistry, materials science, and geology.

Our concern here is the application of the NMR technique to investigations of molten salts. This section begins with a brief review of the general concepts underlying the technique, but this can serve only to summarize the quantities one might wish to measure and to suggest what can be learned about molten salts using NMR. It is necessary to assume that the reader contemplating studies on molten salts has some prior knowledge of the technique and some experience with NMR in, say, organic chemistry or solid-state physics. The application of NMR to the study of molten salts extends the technique to unusual limits of temperature, sensitivity, and sample reactivity. The NMR novice therefore is advised to turn first to one of the many excellent books on the theoretical basis of NMR[2,17–20] and to acquire some experience with one of the more common applications of NMR.

4.1. General Principles

4.1.1. Nuclear Zeeman Interactions and Resonant Absorption

A collection of nuclear magnetic moments μ_n in an applied magnetic field acquires a Zeeman energy (Eq. 6), which depends on the orientation of the moments with respect to the field. For nuclei with spin I, the Zeeman interaction produces a quantized set of $2I + 1$ energy levels separated by the energy $\Delta E = \hbar\omega_0 = \hbar\gamma_n H_0$ (Fig. 10a). Transitions between adjacent levels can be induced by application of a time-dependent magnetic field with a component H_1 rotating with frequency ω in the plane perpendicular to the applied field. This is achieved experimentally by placing the sample in a coil of inductance L whose axis is perpendicular to the field (Fig. 10b). The coil is tuned to *electrical* resonance at the operating frequency with a parallel or series capacitance obeying the usual relation $\omega^2 = 1/LC$. When the *magnetic* resonance condition $\omega = \omega_0$ is met, NMR is observed either by absorption of energy from the external circuit or by detection of the weak current induced in the coil by the magnetization precessing at frequency ω_0 (Fig. 10c).

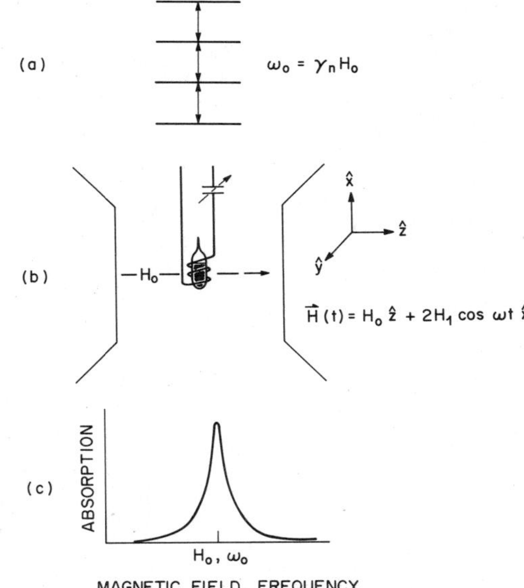

FIGURE 10. Nuclear magnetic resonance (NMR). (a) Zeeman energy levels of nuclei in absence of additional perturbing interactions. Separation between levels $\Delta E_{\text{Zeeman}} = \hbar\omega_0$, where ω_0 is the Larmor frequency. (b) Schematic representation of NMR sample and tuned rf coil in magnetic field H_0; H_1 is the rotating component of the rf magnetic field. (c) Resonant absorption when $\omega = \gamma_n H_0$.

In the simplest class of experiment, low levels of radio frequency power are applied, and the frequency dependence of the absorbed power obeys the relation

$$P(\omega) = \frac{\chi_n H_1^2 \omega\omega_0 T_2}{1 + (\omega - \omega_0)^2 T_2^2} \tag{15}$$

where T_2, the spin–spin relaxation time, determines the linewidth of the resonance at $\omega = \omega_0$, and χ_n is the nuclear magnetic susceptibility:

$$\chi_n = N\gamma_n^2\hbar^2 I(I+1)/3kT \tag{16}$$

The magnitude of the nuclear magnetic moment is given by $\mu_n = \gamma_n\hbar[I(I+1)]^{1/2}$. The resonance can be observed by sweeping either the magnetic field or the applied frequency through the resonance condition. Like any resonant physical system, nuclear moments in a magnetic field can be probed by their response to either steady-state or transient driving perturbations (applied rf power). The responses are related by Fourier transformation. We discuss transient and steady-state methods further in Section 4.1.3.

4.1.2. NMR Parameters: Resonance Shifts and Relaxation Rates

A collection of nuclear spins isolated except for their interactions with each other and with the applied magnetic fields would yield a simple resonance at the frequency ω_0 with a linewidth $1/T_2$ determined by the spin–spin couplings. Such a resonance would be of limited interest, however, except to measure the nuclear gyromagnetic ratio, a property which is fixed for each nuclear isotope and generally known to high precision. The real power of the NMR technique is due to modifications of the resonance parameters by internally generated electric and magnetic fields in condensed matter. For example, the diamagnetic circulation of electronic charge in the applied field produces a local magnetic field opposing the applied field, and this shifts the resonance to lower frequency compared with that expected for a "bare" nucleus. The strength of the local field and the frequency shift depend on details of the electronic states, and these differ for the same element in different chemical environments. This is the so-called "chemical shift," and it forms the basis for many applications of NMR, including the standard spectroscopy familiar to many chemists. The structure of a complicated organic molecule, for example, can be revealed by its rich ^1H spectrum, containing many lines, each representing a different nonequivalent proton.

In paramagnetic substances, there are also resonance shifts associated with local magnetic fields produced by the electron spins. These *spin hyperfine fields* produce dramatic effects when large concentrations of unpaired electrons are present as, for example, in solutions of metals in molten salts.

The local fields in condensed matter are not static, but fluctuate in time according to the dynamic properties of the electrons and ions. The shifts merely reflect the time average of these fluctuating fields. The time dependence of the local fields are responsible for another important class of NMR effects, nuclear spin relaxation. Nuclear *spin–lattice relaxation,* for example, is the process whereby the nuclei achieve thermal equilibrium with the "lattice," that is, the reservoir of thermal motions of the ions. Thus, if the spin system is prepared in a nonequilibrium state by a rapid change of the applied field or, more commonly, by application of pulses of radio frequency power (Section 4.1.3), the fluctuating local fields induce transitions between nuclear Zeeman states to restore the equilibrium magnetization. Relaxation rates can sometimes be inferred from the NMR linewidths while in other cases transient techniques are required.

4.1.3. Steady-State and Transient Methods

In steady-state or "cw" NMR, rf power is applied with constant amplitude, usually at a relatively low level. One detects the response of the tuned sample circuit as the frequency or magnetic field is scanned through resonance. The nature of the signal depends on specific characteristics of the actual circuitry, but it is, in general, a mixture of absorption and dispersion. A particularly simple circuit is represented by the so-called "marginal oscillator,"[21] an rf oscillator in which the sample coil provides the inductive component of the resonant tank circuit. With feedback in the oscillator adjusted to a "marginal" level, barely enough to maintain oscillation, the circuit is very sensitive to power absorption in the tank circuit at resonance. After detection of the rf amplitude, such circuits yield the derivative of the absorption lineshape as the frequency or field is scanned while the external field or the frequency is modulated.[21] More sophisticated steady-state systems often employ a bridge arrangement with two orthogonal rf coils, a "transmitter" and a "receiver." The resonant nuclei are driven by the rotating rf fields generated by power applied to the transmitter coil, and the precessing nuclear magnetization generates a signal in the receiver.

While they are relatively easy to use under normal circumstances, steady-state NMR techniques can present serious experimental difficulties under the less-than-ideal circumstances presented by molten salt research. Chief among these is the instability of detected rf level and bridge balance conditions associated with slight variations of temperature in a high-temperature probe. Another problem plaguing steady-state measurements is unwanted modulation of the rf amplitude by microphonic vibrations of the apparatus. The result is a noise level on the detected output that is far in excess of the fundamental Johnson noise background associated with the temperature of the sample coil and adjacent circuit elements. These problems develop mainly because, in steady-state NMR, one is detecting a very small NMR signal in the presence of a much larger cw rf level.

Many difficulties encountered in steady-state measurements at high temperatures are avoided through the use of transient or "pulsed" NMR techniques.[20,21] Short, intense, rf pulses are applied to the sample coil, or a separate transmitter coil in some arrangements, and the nuclear precession signal is detected *in the absence of any applied rf power*. The only currents flowing in the sample coil are those generated by precessing nuclei and thermal noise. Typical pulse conditions are 1–3 kW of rf power applied for durations of a few microseconds.

The pulse conditions are determined by the desired rotations of the nuclear magnetization. For example, if the applied rf produces a linearly polarized magnetic field $2H_1$ oscillating at the resonance frequency ω_0 in a plane perpendicular to the applied static field, a pulse of duration δt will rotate the net nuclear magnetization away from its equilibrium polarization along the (z) direction of the static field (Fig. 10b) by an angle $\theta = \gamma_n H_1 \delta t$. When $\theta = \pi/2$, one has a so-called 90° pulse, and, after the pulse is "turned off," the magnetization is left precessing in the x–y plane with an initial value equal to the equilibrium magnetization. The signal induced in the sample coil by the precessing magnetization decays because the individual nuclear moments lose their phase coherence, and eventually their orientations are equally distributed in all directions in the perpendicular plane. This decay is known as the "free induction decay" (FID), and it is characterized by a time T_2^*, the "spin-phase memory time." One of the commonest sources of spin-phase loss is simply the inhomogeneity of the applied magnetic field, which causes nuclear moments in different parts of the sample to precess at slightly different rates. However, phase loss can result from many different effects, including spin–spin interactions and spin–lattice relaxation. The latter causes the net magnetization to leave the transverse plane and return to its equilibrium orientation along the direction of the applied field.

Observation of the FID after a 90° pulse is only one of many transient NMR methods, of which some involve elaborate sequences of rf pulses with various rotation angles and phase relationships. A second pulse applied after a time τ with rotation angle 180°, for example, reverses the dephasing of nuclei in an inhomogeneous applied field, leading to refocusing of the transverse magnetization to form a "spin echo" at time 2τ. More complicated multipulse sequences are of great importance for the study of many solids and molecular liquids with complex structures and interactions, but, for reasons discussed in the next section, they are frequently of little value for the study of molten salt systems.

A particularly important application of transient methods is the direct measurement of the spin–lattice relaxation time, T_1. In the most straightforward type of measurement, one applies an initial 180° pulse, which generates no transverse magnetization but inverts the nuclear magnetization with respect to the field direction. The amplitude of the FID following a 90° pulse applied at a subsequent time, t, reflects the value of the longitudinal magnetization component, M_z, at time t. In many important cases, M_z relaxes to its equilibrium value with a single time constant T_1:

$$M_z(t) - M_z(\infty) = [M_z(0) - M_z(\infty)]\exp(-t/T_1) \tag{17}$$

The problems facing the experimentalist interested in transient NMR studies of molten salts can by summarized by the following essential requirements. One must provide a spectrometer of suitable power capability and time resolution, a tuned sample coil operating at the necessary elevated temperature, and sample cells that are both electrically insulating and chemically compatible with the molten salt system of interest. With such a system, one must display the resonance in a form that enables the position (shift) of the line to be measured with respect to an appropriate reference. In addition, because the dynamic information contained in relaxation rates can be very informative, it is desirable to be able to measure the spin–lattice relaxation time T_1 and, in some cases, the linewidth or FID lifetime. We turn now to a discussion of some specific strategies for meeting these requirements.

4.2. Apparatus Design

4.2.1. Spectrometer

The spectrometer is the central (and most expensive) element of any system for NMR experiments on molten salts. The realities of science funding are such that the experimenter considering molten salt studies may have to adapt an available spectrometer to high-temperature work rather than purchase or build a system that is ideally matched to the task. This is relatively easy for salts with low melting points, since the spectrometer requirements are quite close to those imposed by many other chemical systems studied at or near room temperature.[8,22] Suitable spectrometers are rather widely available and need only be adapted to the moderately elevated temperatures required.

In this chapter, we are more concerned with the needs of investigations of inorganic salts melting at higher temperatures. Because of the large number of chemical constituents that may be encountered among different salts, the spectrometer should be capable of operation at a variety of frequencies and magnetic fields. Ideally, the spectrometer should permit relatively easy changes of frequency and field so that, if necessary, more than one nuclear species can be investigated in a single experiment.

Depending on the nuclei and details of the electronic and ionic dynamics of the particular salt, the resonance lines may become quite broad and the spin relaxation times very fast. The NMR conditions in these cases are more akin to those encountered in applications of NMR to problems in solid-state physics than to those of typical liquid-state NMR. A pulsed NMR spectrometer may have to produce rf pulses of relatively high power (1–3 kW) in order that the nuclear magnetization be rotated in a time short compared with the dephasing time T_2^*. Moreover, the receiver must recover from the overload of the rf pulse in no more than a few microseconds. Spin echo techniques cannot be used to recover the nuclear magnetization after the FID if the decay is due to irreversible processes such as spin–lattice relaxation. (These can be as short as 1 or 2 μs in some liquids at high temperatures; if they are shorter, the resonance is generally unobservable with the best spectrometers.)

Pulsed NMR spectrometers may be purchased commercially or constructed by the experimenter from a combination of "homemade" circuits and commercial components. Modern commercial spectrometer systems are usually designed for use with high-field superconducting magnets. The high fields, typically 50–100 kG, offer high sensitivity due to the large nuclear polarization achieved, and they yield high resolution of spectral features that are proportional to the magnetic field, such as chemical shifts. Many commercial systems also provide sophisticated sequencing of rf pulses to increase resolution and discriminate among various nuclear sites for the same chemical species. However, because of the rapid relaxation times often encountered in molten salts, these features may not be applicable.

The experimenter interested in developing a pulsed spectrometer from individual components should consult the classic paper of Clark [23] for an introduction to the basic concepts of these systems. The fundamentals have survived the transition from vacuum tube to solid-state technology. Numerous articles have appeared in recent years describing the adaptation of solid-state microelectronics to spectrometer design.[†] Solid-state technology has greatly simplified the construction of logic and low-level analog circuitry. The high-power pulse generation in the transmitter may require the use of vacuum tubes in the final stages,

---

† See, for example, Ref. 20 and the references therein.

depending on the power level desired. In our laboratory, we use the combination of a commercial wide-band power amplifier and a homemade final stage with ceramic power tubes to obtain pulse power in the range 2–3 kW.

A superconducting magnet for molten salt NMR must have a bore large enough to accommodate the high-temperature furnace, insulating material, and cooling. A magnet system with a room temperature access space of about 50-mm diameter is minimal, while 75 mm would be adequate for studies well above 1000°C. It is highly desirable that the magnet have a sweep capability of several hundred gauss in order to measure broad lines and large shifts. Although one can, in principle, sweep the frequency of the spectrometer while the field remains constant, a continuous-frequency sweep is difficult in practice because of the various tuned elements in the spectrometer.

The alternative to a superconducting solenoid is an iron-core electromagnet. Standard laboratory magnets with a 75-mm gap are limited to about 20 kG, so there is a substantial loss of sensitivity compared with a high-field superconducting magnet. (Signal-to-noise ratio is roughly proportional to the square of the field). However, the iron-core magnet has compensating advantages, including easy access to the air gap and the fact that no liquid helium is needed for its operation. Moreover, the orientation of the field with respect to the direction of external access gives an important advantage to the iron-core magnet. As we discuss in more detail shortly, the field in an iron-core electromagnet is perpendicular to the gap opening so that sample containers of arbitrary length and complexity along the axis of the rf coil are permitted. In a superconducting solenoid, the field is oriented along the cylindrical axis of the bore, and since the rf coil must be aligned perpendicular to the field, a sample must be placed transverse to the direction of easy access. Thus, the length of a sample container and any associated assembly such as a mechanical closure (see Figs. 3 and 4) must fit within the *diameter* of the high-temperature experimental space. This constraint might be decisive in experiments employing complex sample cells.

4.2.2. Probe Design

The purpose of the NMR probe is, first, to generate the rf magnetic field H_1 in the region of the sample and, second, to receive the transient nuclear induction signals after the pulse is turned off. The probe must be designed to concentrate as much rf energy as possible on the sample while minimizing the pulse energy that reaches the input circuitry of the sensitive receivers. At the same time, the transmitter circuitry must be isolated from the receiver circuit during reception of the NMR signal in order that the signal-to-noise ratio not be degraded by this excess load. These goals can be achieved with varying degrees of perfection using either the *crossed-coil* or *single-coil* configurations. In the crossed-coil scheme, a receiver coil is closely wound around the sample while a second, transmitter coil is aligned with its axis perpendicular to that of the receiver coil.

The axes of both coils are perpendicular to the static magnetic field. Isolation of the transmitter and receiver circuitry is achieved by the orthogonality of the two coils. Perfect isolation is never possible in practice, of course, and is particularly difficult to achieve within the constraints on spatial dimensions and construction materials of a high-temperature apparatus. Furthermore, the transmitter coil inevitably has a larger volume than the receiver coil so that the fraction of rf power applied to the sample volume is less than is the case if the transmitter and receiver coils are the same.

A single-coil scheme in which one coil performs both functions is much more suitable for high-temperature experimentation. The probe requires only a small solenoid that fits closely around the sample cell. There is need for just one rf lead into the high-temperature apparatus; one side of the coil can be grounded to the metal enclosure. Transmitter–receiver isolation is more complex than with the cross-coil scheme and is achieved by external circuitry.

A single-coil scheme that we have used very successfully is illustrated in Fig. 11.[24] The circuit is basically a "T" in which the central leg consists of a series resonant circuit formed by the sample coil and a variable capacitor. The latter is a vacuum variable capacitor with a voltage rating of 2–5 kV. In the transmitter leg, the transmitter output is connected to the tuned circuit through a matching circuit in series with a bank of "back-to-back" diodes. The receiver leg contains a matching circuit and a parallel configuration of diodes. The diodes present a high impedance to low-level signals and a low impedance to high levels. Thus, those in the transmitter leg pass the high-voltage pulse but isolate that portion of the circuit during reception of the low-level NMR signal. The parallel diodes in the receiver leg help protect the input of the preamplifier by presenting a lower impedance to ground during the pulse.

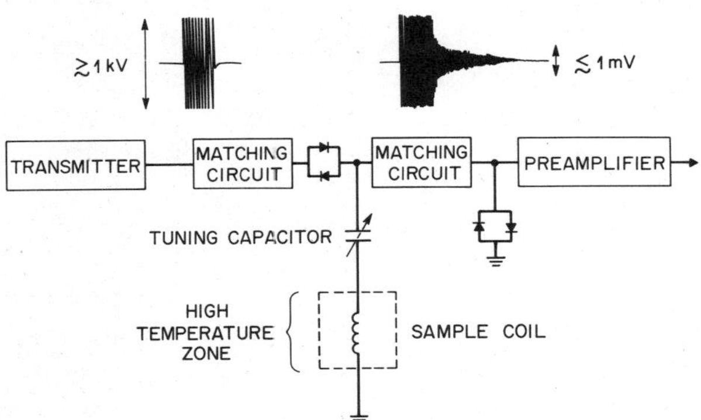

FIGURE 11. Single-coil probe and matching arrangement for pulsed NMR at elevated temperatures. Waveforms represent (*left*) rf pulse at transmitter output and (*right*) residual pulse followed by weak NMR free induction decay at preamplifier input.

The simplest form of matching circuit consists of quarter-wavelength coaxial cables. These transform the relatively low impedance of the series tuned resonant circuit Z_s to a higher value Z_i according to

$$Z_i = Z_0^2/Z_s \qquad (18)$$

where Z_0 is the characteristic impedance of the cable. In a typical case, Z_s is a few ohms and $Z_0 = 50\ \Omega$ so that the impedance "seen" by the transmitter and preamplifier is a few hundred ohms. This matching scheme is simple to implement, but it is rather inflexible in that a different set of cables is needed for optimum matching at each frequency. Moreover, at low frequencies ($<\sim 5$ MHz) the necessary cable lengths become inconveniently long. On the other hand, for operation at a few fixed frequencies in the range 10–50 MHz, the quarter-wavelength scheme is quite adequate. For a more flexible arrangement, the cables can be replaced by appropriate tuned resonant circuits, as discussed by Clark and McNeil.[24]

In the scheme shown in Fig. 11, it is important that the tuning capacitor be as close as possible to the sample coil. Otherwise, additional *parallel* capacitance is introduced by the long leads, and the performance of the circuit deteriorates. This requirement is relatively easy to meet at frequencies below 30 MHz when the experiment is done in an iron-core electromagnet. When a superconducting magnet is employed, however, this scheme may become impractical because of the combined effects of the higher frequencies and the distance from the coil position in the solenoid to the space external to the magnet dewar where the tuning capacitor can be located. One approach with a superconducting magnet is to locate some of the tuning capacitance as close as possible to the probe. Alternatively, as shown by McNeil and Clark,[25] it is possible to use transmission line techniques and incorporate the leads to the sample in the tuning of the coil.

We now consider the actual construction of rf coils for use in high-temperature studies of molten salts. As mentioned previously, the coil configurations are different for iron-core and superconducting magnets. These are illustrated schematically in Figs. 12a and 12b, respectively. For the iron-core magnet, the size of the gap places a limit on the *diameter* of the coil, but there is no constraint on access along the direction of the coil axis. Samples may be exchanged easily in this geometry. The coil in the superconducting magnet is transverse to the cylindrical magnet bore, and this limits the *length* of the coil. All samples sealed in high-temperature materials, even those sealed in glass, require extra length for the "sealing" portion of the cell, and the coil must be short enough to allow the sample cell to fit within the diameter of the furnace. Although one can compensate to some extent by making the sample "fatter," it is clear that the geometry of the superconducting solenoid may require sacrifices in sample size that partially offset the advantages of working at higher field values.

High-temperature rf coils for NMR should be wound with wire of the lowest

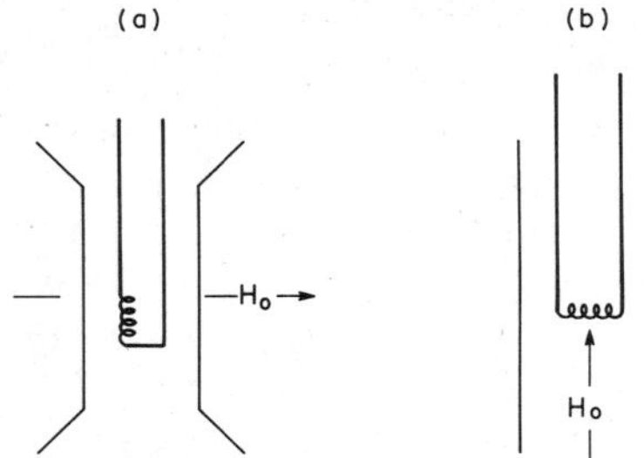

FIGURE 12. Orientations of sample coils for NMR in (a) iron-core magnet, and (b) superconducting solenoid.

resistance material compatible with the environment. Thus, gold is the material of choice for operation in air at temperatures below 1000°C. At higher temperatures, platinum or platinum–rhodium alloy is best. If the coil is to be located in a nonoxidizing atmosphere, molybdenum may be used to at least 1500°C. Figure 13 illustrates two simple rf coil designs for high-temperature NMR in an iron-core magnet. The coil in Fig. 13a is imbedded in a form made from castable alumina-based ceramic. It is constructed by winding the wire on a glass tube or rod which is then inserted into a mold of the desired outer diameter. After the ceramic has set, the rod can be withdrawn or broken if necessary. The coil in Fig. 13b is wound internally on a form machined from alumina ceramic and coated with alumina cement to hold it in place. In choosing cements and castable ceramics, it is important to avoid the presence of zirconia if the coil is to be used above 900°C, since the ionic conductivity of ZrO_2 will effectively short-circuit the coil at higher temperatures.

4.2.3. NMR Furnace

The basic requirement of a furnace for high-temperature NMR on molten salts is provision of a stable and homogeneous zone, at the required temperature,

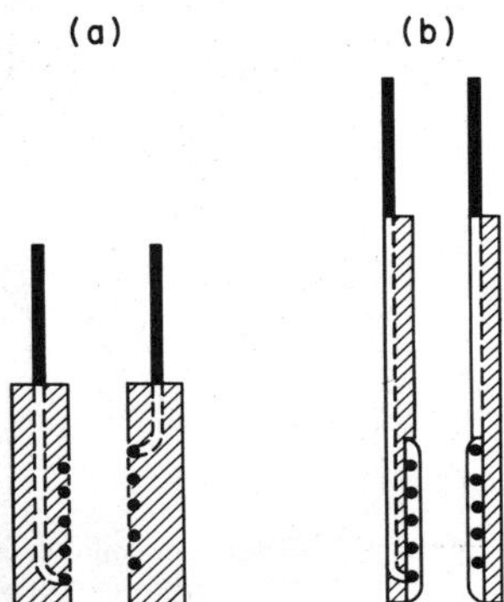

FIGURE 13. Sample coils for high-temperature NMR: (a) cast ceramic form, and (b) machined ceramic form.

large enough to accommodate the sample and rf coil. The entire assembly, heating element, thermal insulation, enclosure, and cooling must fit within the space allowed by the magnet gap or bore.

In Fig. 14, we illustrate a basic design used in our laboratory for many years in an iron-core magnet with a 76-mm gap.[26] This design has been used at frequencies up to 50 MHz, but most experiments have been done at frequencies below 20 MHz, where the majority of nuclei resonate in the 20-kG magnet. The system shown in Fig. 14 is designed for operation in air and employs a bifilar heater winding of platinum wire with a 0.5-mm diameter wound on an alumina tube and coated with alumina cement. The overall length of the element is 150 mm. For thermal insulation, materials such as "Fiberfrax" (Carborundum Co.), low-density zirconia fiberboard, or alumina "bubbles" have been used. The thermal efficiency varies somewhat depending on the type of insulation and the degree to which unused internal portions of the heater tube are filled with insulation. A typical power requirement is 150 W for a stable temperature of 1000°C. The enclosing brass cylinder and end pieces are cooled by water flowing in

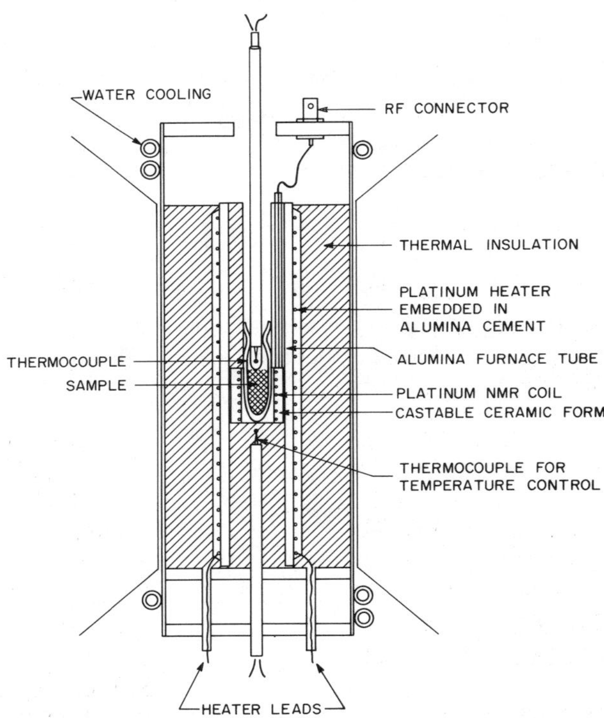

FIGURE 14. Furnace for molten salt NMR studies in an iron-core magnet.[26] Sample is shown in fused-silica cell sealed to closed-end tube containing thermocouple.

copper tubing soldered to the enclosure at the ends and along one side facing the gap opening. A flow of 15°C water at about 1 liter/min suffices to keep the enclosure cool to the touch.

The basic design of Fig. 14 is readily adapted to many special requirements. It can be used for high-field studies, for example, by replacing the longitudinal rf coil by a transverse coil assembly and reducing the diameter to fit in the bore of a superconducting magnet. In the constrained geometry of the superconducting magnet bore, cooling can be provided by a thin, concentric water jacket extending over the full length of the enclosure. A different requirement, that of providing a protected atmosphere for the sample cell, can be met by replacing the end pieces with vacuum flanges containing 0-ring seals. In such a case, a hermetic-type coaxial BNC connector is used for the rf lead and standard vacuum feedthroughs are used elsewhere. A nonoxidizing atmosphere is essential when the cells have closure assemblies made of refractory metals such as molybdenum or niobium. Such a furnace can be operated with the enclosure containing an inert gas such as argon, or, if desired, the enclosure can be evacuated.

A frequent problem with a furnace of the type shown in Fig. 14 is the introduction of rf interference into the receiver circuitry via the heater leads and thermocouple. We have found it necessary in most cases to provide filtering. A simple π network (i.e., series inductance and parallel capacitance) can be mounted on the bottom of the furnace enclosure and is quite effective for filtering the heater circuit. To obtain an accurate measurement of the sample temperature, it is desirable to position the thermocouple as close as possible to the sample position. The rf noise that would otherwise be radiated by this antenna and picked up by the NMR coil can be greatly reduced if the thermocouple is enclosed in a coaxial metal shield. The shield can be grounded directly to the furnace enclosure at the point of entry.

The problem of slow thermal response at low temperatures was discussed in Section 2.1. A hot gas flow system provides much shorter thermal equilibration times at temperatures below 500°C. A gas flow system using heated air is illustrated in Fig. 15.† This system employs nichrome heaters located well away from the sample coil. This has the added advantage of reducing the rf interference problems discussed above. A gas flow apparatus has a rapid thermal response, but it is also inefficient since much of the heat is carried away in the gas stream. The system shown in Fig. 15 requires roughly 500 W to obtain a steady temperature of 500°C.

4.3. Illustrative Experimental Results

Our first example of NMR data is drawn from a study of the system In–InI$_3$.[27] This system contains three stoichiometric salts—InI, InI$_2$, and InI$_3$—

---

† This system was designed and constructed by R. W. Schmutzler.

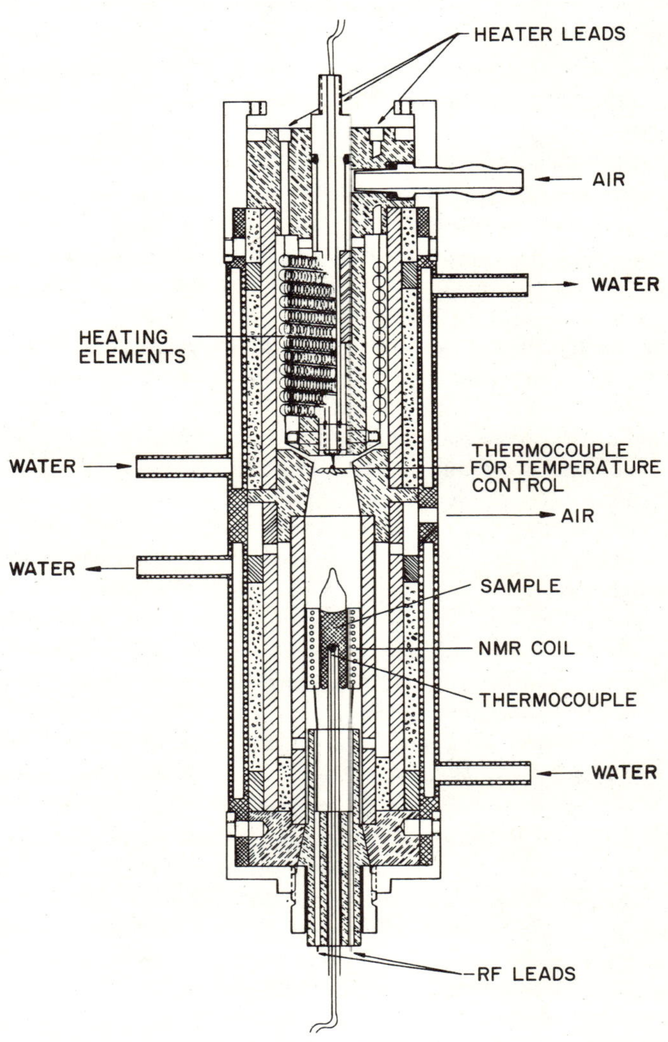

FIGURE 15. Hot air flow furnace for molten salt NMR studies below 500°C. Heating element shown in upper portion is duplicated in lower portion of furnace (not shown). Sample in fused-silica cell is inserted from below.

and a wide range of liquid–liquid phase separation between the pure liquid metal, In, and InI. Figure 16 shows a series of ^{113,115}In spectra for the three molten salts and one phase-separated solution, $In_{0.66}I_{0.34}$. The data show how the resonance shift (chemical shift) changes in the paramagnetic direction; that is, the resonance shifts to lower magnetic field with decreasing covalency in the sequence InI_3, InI_2, InI. There is, in addition, a complex variation in the linewidth associated with changes in the relative nuclear spin relaxation rates. At the indicated temperatures, InI_2 exhibits the narrowest line, but it was found that this resonance broadens dramatically as the temperature is raised. This behavior was explained[27] in terms of the dynamic formation of the paramagnetic species In^{2+} according to the reaction $In^+ + In^{3+} \rightarrow 2In^{2+}$.

The data shown in Fig. 16 also illustrate the distortion of the absorption line introduced by incomplete Fourier transform[23] when the FID is short. For each In iodide, there are minima on either side of the main peak, and the minima are deepest when the line is broadest. This simply reflects the fact that the broader the absorption line is, the shorter the FID (and hence the more severe the distortion).

The spectrum of the phase-separated sample shows a strong narrow resonance at the position of InI and a small peak shifted far downfield, which comes from the metal-rich component. The apparent intensity of the metal-rich reso-

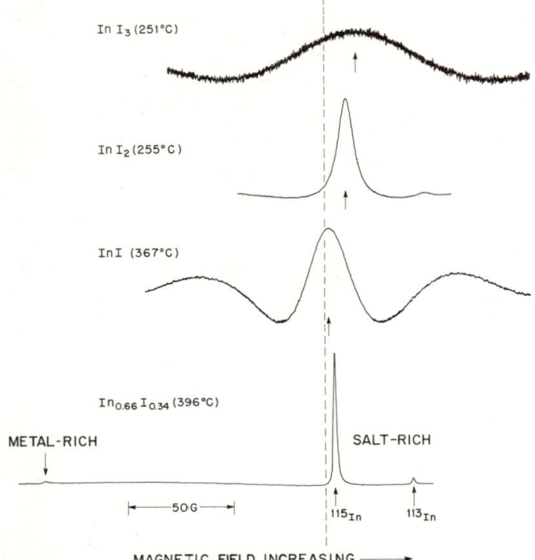

FIGURE 16. ^{113,115}In NMR spectra for In–InI_3 solutions.[27] Distortion (oscillations) in wings of resonance line is due to incomplete Fourier transform. $In_{0.66}I_{0.34}$ sample is phase-separated in liquid–liquid immiscibility range between In and InI.

nance is reduced by the exclusion of rf fields from the highly conducting material ("skin effect"). The large shift of the metal resonance is the Knight shift associated with Pauli paramagnetism.[28] The value of the metallic shift in this phase-separated solution is nearly equal to that of pure liquid In, showing that the compositions of the respective separated components are quite close to In and InI. Also visible in the spectra for InI_2 and $In_{0.66}I_{0.34}$ is a weak line due to the minority isotope [113]In, whose gyromagnetic ratio is fortuitously close to that of [115]In. The resonances of InI_3 and InI are too broad to resolve the unusually small splitting of the In isotopes.

In Fig. 17, we present an example of NMR data for a *homogeneous* metal–molten salt solution. This figure shows the nuclear spin–lattice relaxation rate $1/T_1$ for [133]Cs in the system Cs–I. The relaxation rate of pure CsI is extremely low because of the absence of unpaired electron spins and the weakness of other potential processes such as nuclear quadrupole relaxation. However, addition of *either* excess Cs or excess I increases the relaxation rate dramatically. The effect is largest for solution of metal whereby the hyperfine interactions with unpaired spins raise the rate by $\sim 10^5$ with only 1% excess Cs. This rapid rate of increase does not continue indefinitely at higher concentrations of metal, however, and at about 5% excess Cs, the rate passes through a maximum. This effect is a consequence of the delocalization of the excess electrons at the metal–nonmetal transition. Delocalized (metallic) electrons are far less effective in relaxing nuclei than are localized electrons because, with their high mobility, they have less opportunity to undergo a mutual spin flip with a particular nucleus. The sensitivity of

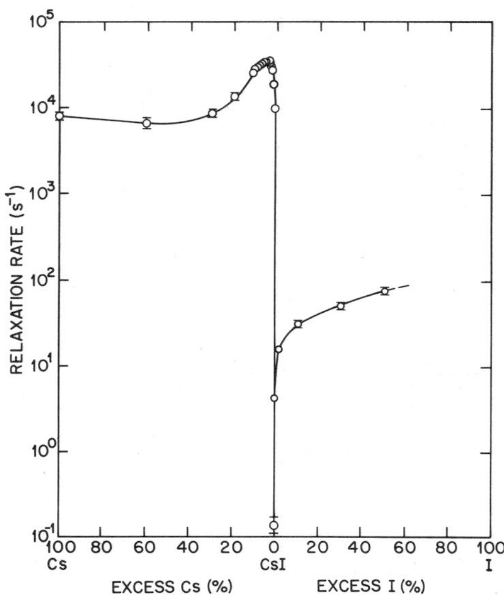

FIGURE 17. Cesium-133 nuclear spin–lattice relaxation rate (logarithmic scale) versus composition for Cs–I solutions.[29] Rate observed for pure molten CsI is sharply enhanced by addition of small amounts of excess Cs or I. Peak in rate on metal-rich side is due to delocalization of electrons at metal–nonmetal transition.

the nucleus–electron interaction to the dynamics of the electron state is one of the powerful features of the NMR technique.

The increase in relaxation on the I-rich side is also very strong, roughly a factor of 100. This effect has been attributed to formation of the paramagnetic species I_2^-,[29] and this is supported by magnetic susceptibility measurements.[30] The concentration of paramagnetic species increases quite strongly with temperature, apparently due to a shift to the right of the equilibrium

$$I_3^- + I^- \rightarrow 2I_2^- \tag{19}$$

The NMR experiments permit measurement of the enthalpy of reaction for this equilibrium and its evolution as the I content is increased. Combined with other data, including data from susceptibility and optical studies, NMR data have revealed much of the rich chemistry of these interesting solutions.

5. Electron Spin Resonance

Like NMR, electron spin resonance (ESR) is a versatile experimental technique with many applications in physics, chemistry, materials science, and biology.[31] However, ESR is limited in principle to systems exhibiting electron spin paramagnetism, and so its use is excluded for the large class of diamagnetic substances (including many molten salts) in which all the valence electron spins are paired. Even in paramagnetic materials, including many metals, the electron spins may interact so strongly with their local environments that the ESR is unobservably broad. Moreover, as we discuss in this section, ESR experiments are relatively difficult at elevated temperatures. Thus, compared with the systems accessible to study by NMR, the number of materials for which ESR is observable is quite restricted. Nevertheless, because ESR is a direct probe of the electron spin system, the experiments are highly informative when they are feasible. The most ambitious ESR studies of molten salts are probably those of Nicoloso and Freyland[32,33] on solutions of alkali metals in alkali halides. The discussion of ESR techniques in this section draws heavily from their experience.

5.1. General Principles

The basic principles of ESR are identical to those of NMR except that we now consider the Zeeman interactions and resonant absorption of electron rather than nuclear spins. Thus, we replace the nuclear moment μ_n by the electronic moment $\mu_e = g\mu_\beta[S(S+1)]^{1/2}$, where g is the "g-factor", μ_β is the Bohr magneton, and S is the electron spin. For single electrons, $S = \frac{1}{2}$, but higher values are possible for species, such as transition metal ions, in which the electronic moment is formed from more than one electron. In an applied magnetic field H_0, the ESR frequency is given by $\hbar\omega_0 = g\mu_\beta H_0$, in complete analogy with Eq. (7). Because of the large electronic moment, the resonant frequencies fall in the microwave (gigahertz) range rather than the radio frequency (megahertz) range, but the resonant absorption of power from an external oscillator is still given by Eq. (15).

Most ESR experiments are conducted in the steady-state mode in which the resonance is displayed as the frequency, or magnetic field, is swept. The position of the resonance is then described by a "g-shift," that is, a shift of the resonance from the nominal $g = 2.0023$ position of free electrons. The value of the g-shift, Δg, is a measure of the interactions of the electrons with their local environments. The other important experimental parameters in an ESR experiment are the linewidth, related to the transverse, or spin–spin, relaxation time, and the total integrated intensity of the resonance.

5.2. Apparatus Design

Although the principles of ESR and NMR are essentially identical, the experimental techniques are quite different because of the much higher resonant frequencies normally employed in ESR. Thus, the tuned wire sample coil of NMR is replaced by a resonant cavity in ESR, and wire rf leads and coaxial cables are replaced by a waveguide. The principal technical problem posed by ESR study of molten salts is provision of a cavity which can contain the high-temperature sample. A commercial ESR spectrometer constitutes the rest of the system including microwave generation, modulation, and signal detection and processing. The magnetic field requirements for ESR are modest: 3200 G for the commonly used X-band frequency (9 GHz).

Cavity designs for high-temperature work up to 1200°C have been reviewed elsewhere.[31,34] Three methods of heating a sample in a cavity have been developed. One approach is to heat the sample in a stream of hot gas in an otherwise cool cavity. This relatively simple method may be suitable for some experiments, but it suffers from the difficulty of providing a uniform temperature over the sample volume. This can lead to unacceptable concentration gradients in multicomponent systems. A second method is to heat the cavity, or some region within, with an electrical heating element. This technique provides much more uniform temperature distributions, but great care must be exercised in the design to ensure that resonant properties of the cavity are disturbed as little as possible by the heater. The use of electrical resistive heating is most closely related to the techniques described previously for high-temperature susceptibility and NMR experiments. The third method is to heat the sample locally with a high-powered laser.[34]

A high-temperature X-band cavity based on the second method, resistive heating, is shown in Fig. 18.[32,33] In this system, a heater of the "zigzag" type is inserted into the cavity together with the sample cell and thermocouple for temperature measurement (detail, Fig. 18). The insert is completely isolated

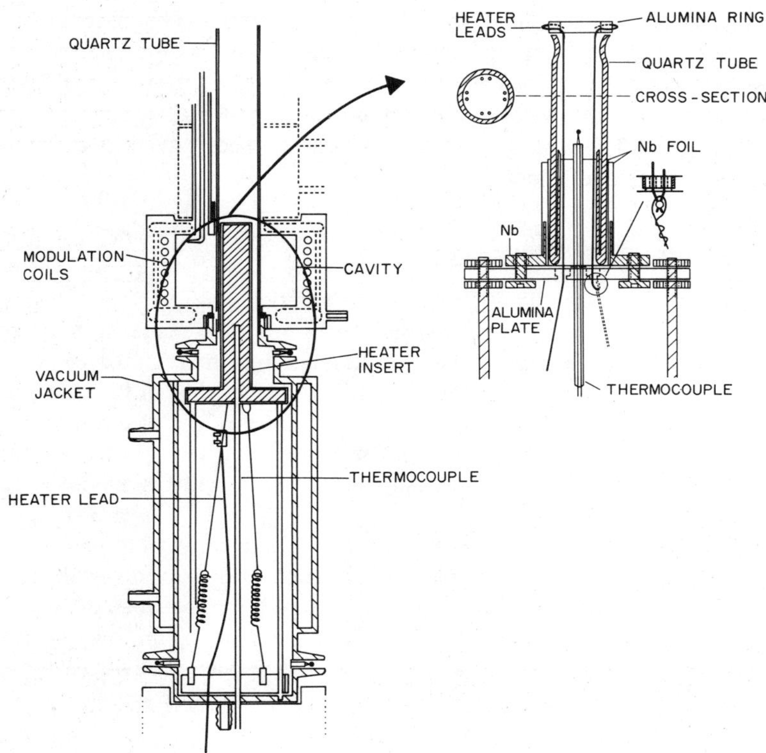

FIGURE 18. Microwave cavity with high-temperature insert (detail) for ESR studies of molten salts.[32,33]

within a larger concentric quartz tube in order that material vaporized from the sample or outgassed from the hot components will not be deposited on the highly reflecting internal surfaces of the resonant cavity. Another necessary precaution is provision of solid mechanical support for the apparatus in order to minimize signals generated by microphonics. (The steady-state ESR experiments suffer from some of the same problems as steady-state NMR, discussed in Section 4.1.3.) The heater windings are connected mechanically to chromel wire below the insert, and tension on the leads is maintained with tungsten springs. In order to protect the metallic components from oxidation, the entire high-temperature apparatus is held in an inert gas flow which is carefully regulated to ensure a constant and reproducible temperature distribution within the cavity.

Materials selected for ESR sample containers must meet the same basic requirements as those discussed in the NMR case. The most important of these, of course, is that the cell body must be electrically insulating. An additional restriction imposed by ESR studies is the necessary absence of interfering ESR signals from paramagnetic impurities in the cell material. This requirement extends also to the heater support and to any other construction material located within the cavity. Thus, for example, Nicoloso and Freyland[32,33] found that boron nitride, an otherwise useful high-temperature material, was not acceptable for ESR samples. Those workers found sapphire to be best suited for their studies of alkali metal–halide solutions.

5.3. Illustrative Experimental Results

ESR spectra for the alkali metal-halide solution $Na_{0.04}[(Na,Rb)Cl]_{0.96}$ are shown in Fig. 19 for three temperatures.[32] In this study, eutectic hosts rather than pure alkali halides were chosen for dissolution of metal in order to reduce the experimental temperature. These data reveal a clear shift of the resonance to higher magnetic field relative to the field $H_{f.e.}$ of the free electron spin resonance. This negative g-shift is quite similar to those of excess electrons localized in alkali halide crystals and provides support for the F-center model of electrons in the molten salts.

The integrated intensities of the resonances typically yield spin densities well below those expected from the concentration of dissolved metal. This led Nicoloso and Freyland to propose a chemical equilibrium between the F-centers, with their unpaired spins, and an unspecified diamagnetic, spin-paired species. The strong temperature dependence of the resonance intensity evident in Fig. 19

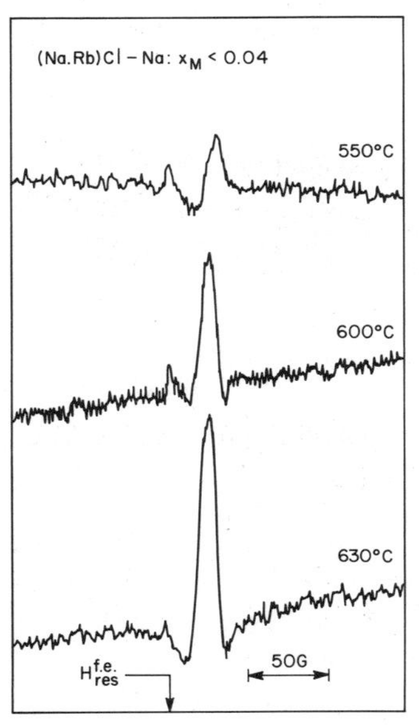

FIGURE 19. ESR spectra for metal–molten salt eutectic solution $Na_{0.04}[(Na,Rb)Cl]_{0.96}$.[32] Arrow denotes position of free electron resonance visible to left of main resonance line. Spectra show negative g-shift and strongly temperature-dependent spin concentration.

supports this interpretation and implies a shift of the equilibrium toward the paramagnetic centers as the temperature increases. The temperature dependence is consistent with a thermal activation energy of about 12 kcal/mol.

Of the three magnetic techniques discussed in this chapter, ESR has been the least exploited as a tool for the study of molten salt systems. Some of the factors and experimental problems which have inhibited researchers from widely employing this technique have been discussed. However, the example presented in Fig. 19 suffices to show that ESR experiments are possible on molten salt systems at temperatures of several hundred degrees Celsius and that very respectable spectra can be obtained. It is to be hoped that workers in the molten salt field will be encouraged to extend ESR studies to other systems and still higher temperatures.

Acknowledgments

I am deeply indebted to G. F. Brennert, whose continuing contributions to the development of our apparatus for high-temperature NMR have been invaluable. This work also has benefited from incorporation of various ideas and suggestions from a number of colleagues and collaborators, including W. G. Clark, R. Dupree, U. El-Hanany, W. Freyland, G. Schäfer, F. Hensel, R. Schmutzler, and S. Sotier. Their contributions are acknowledged with pleasure. I am particularly grateful to W. Freyland and his former students N. Nicoloso and G. Steinleitner for permission to reproduce figures describing their susceptibility and electron spin resonance systems.

References

1. J. H. Van Vleck, *Electric and Magnetic Susceptibilities*, Oxford University Press, London (1932).
2. A. Abragam, *Nuclear Magnetism*, Oxford University Press, London (1961).
3. S. Hafner and N. H. Nachtrieb, *J. Chem. Phys.* **40**, 2891 (1964); **42**, 631 (1965).
4. M. Blander (ed.), *Molten Salt Chemistry*, Interscience, New York (1964).
5. W. W. Warren, Jr., in: *Advances in Molten Salt Chemistry* (G. Mamantov and J. Braunstein, eds.), Vol. 4, p. 1, Plenum Press, New York (1981).
6. W. W. Warren, Jr., in: *The Metallic and Nonmetallic States of Matter* (P. P. Edwards and C. N. R. Rao, eds.), p. 139, Taylor and Francis, London (1983).
7. W. W. Warren, Jr., in: *Molten Salt Chemistry* (G. Mamantov and R. Marassi, eds.), NATO ASI, Ser. Series C, Vol. 202, p. 237, Reidel, Dordrecht (1987).
8. J. Wilkes, in *Molten Salt Chemistry* (G. Mamantov and R. Marassi, eds.), NATO ASI, Series C, Volume 202, p. 405, Reidel, Dordrecht (1987).
9. G. Steinleitner, Thesis, University of Marburg (1978).
10. N. Nicoloso and W. Freyland, *Z. Phys. Chem. N. F.* **135**, 39 (1983).
11. R. Dupree, D. J. Kirby, and W. W. Warren, Jr., *Phys. Rev. B* **31** 5597 (1985).

12. P. W. Selwood, *Magnetochemistry*, Interscience, New York (1956).

13. K. Grjotheim, H. A. Ikeuchi, and J. Krogh-Moe, *Acta Chem. Scand.* **24**, 985 (1970).

14. N. H. Nachtrieb, *J. Phys. Chem.* **66**, 1163 (1962).

15. K. Grjotheim, S. Dhabanandana, and J. Krogh-Moe, *Acta Chem. Scand.* **26**, 3427 (1972).

16. L. F. Grantham, *J. Chem. Phys.* **44**, 1510 (1966).

17. C. P. Slichter, *Principles of Magnetic Resonance*, 2nd ed., Springer, Berlin (1978).

18. R. T. Schumacher, *Magnetic Resonance*, Benjamin, New York (1970).

19. J. Mason (ed.), *Multinuclear NMR*, Plenum Press, New York (1987).

20. E. Fukushima and S. B. W. Roeder, *Experimental Pulse NMR, A Nuts and Bolts Approach*, Addison-Wesley, Reading, Massachusetts (1981).

21. Reference 2, Chapter III.

22. W. Go, M. Sakai, T. Zawodzinski, J. Osteryoung, and R. A. Osteryoung, *J. Phys. Chem.* **92**, 6125 (1988).

23. W. G. Clark, *Rev. Sci. Instrum.* **35**, 316 (1964).

24. W. G. Clark and J. A. McNeil, *Rev. Sci. Instrum.* **44**, 844 (1973).

25. J. A. McNeil and W. G. Clark, in: *Pulsed Nuclear Magnetic Resonance and Spin Dynamics in Solids* (J. M. Hennel, ed.), Proc. 1st Specialized Colloque Ampere, p. 159, Institute of Nuclear Physics, Krakow (1973).

26. U. El-Hanany and W. W. Warren, Jr., *Phys. Rev. B* **12**, 861 (1975).

27. K. Ichikawa and W. W. Warren, Jr., *Phys. Rev. B* **20**, 900 (1979).

28. C. H. Townes, C. Herring, and W. D. Knight, *Phys. Rev.* **77**, 852 (1950).

29. W. W. Warren, Jr., S. Sotier, and G. F. Brennert, *Phys. Rev. B* **30**, 65 (1984).

30. B. Nakowsky, N. Nicoloso, and W. Freyland, *Ber. Bunsenges. Phys. Chem.* **88**, 297 (1984).

31. C. P. Poole, Jr., *Electron Spin Resonance*, Wiley, New York (1967).

32. N. Nicoloso and W. Freyland, *J. Phys. Chem.* **87**, 1997 (1983).

33. N. Nicoloso, Thesis, University of Marburg (1982).

34. T. A. Yager and W. D. Kingery, *Rev. Sci. Instrum.* **51**, 464 (1980).

5

Calorimetric Methods

Marcelle Gaune-Escard

1. Introduction

For a complete thermodynamic description of reactions involving inorganic phases, calorimetric measurements are indispensable. Heats of reaction may also be obtained from the temperature coefficients of electromotive forces, dissociation pressures, and other equilibrium measurements, but the enthalpies thus determined are often not sufficiently accurate and generally not entirely reliable when one or more solid phases take part in the reaction. This shortcoming of equilibrium measurements is not always appreciated.

At the time when most calorimeters were room temperature apparatuses, the temperature of the calorimetric reaction needed not rise much above room temperature except for brief periods.

This was true, for instance, for the most common calorimeter, the combustion bomb,[1] in which the temperature may rise momentarily to more than 2000°C because of the heat produced by the reaction itself. All that is required is a high pressure of oxygen and ignition by a comparatively small local heat input. The technique may be used to measure heats of formation, or as a difference method. As such, it is ideal for the investigation of organic compounds but must be used with care for metal oxides. The equilibrium state of the product must be ascertained and its nature determined (e.g., Al_2O_3).[2] The bomb has also been used for the chlorination of metals.[3] Similar in principle are the calorimeters for reactions of metals with nonmetals, such as sulfur.[4]

There were, however, other calorimeters in which use was made of a relatively high equilibrium temperature to ensure that a reaction of comparatively weak driving force went to completion, and these will be briefly described in this chapter.

Advances in high-temperature calorimetry during the past 20 years will be extensively discussed and the techniques applicable to particular molten salt mixtures stressed.

Marcelle Gaune-Escard • S.E.T.T., Université de Provence, 13397 Marseille Cedex 13, France.

2. General Considerations

Generally speaking, any kind of classification is somewhat arbitrary, and this is particularly true in the case of calorimetry. Classifications may be based on:

- apparatus (e.g., adiabatic, isothermal, heat flow calorimetry)
- properties to be determined (e.g., enthalpy of mixing, enthalpy of reaction, heat capacity)
- temperature range (low temperature, room temperature, high temperature)
- substances under investigation (e.g., fluid, molecular, metallic, ionic).

For convenience, a distinction will be made in the following between reaction calorimetry and nonreaction calorimetry, although there is not always a clear difference between the two types of calorimeters used for the determination of heats of reaction and heats of transformation; one apparatus may even be used for both purposes. The distinction lies rather in posing the problem. Application of calorimetry to the determination of phase diagrams will also be discussed.

The heat of a reaction changes with temperature, and, according to Kirchhoff's law, this change is given by ΔC_p, the molar heats of the products minus those of the reactants, as follows:

$$\Delta H_T = \Delta H_0 + \int_0^T \Delta C_p \, dT \tag{1}$$

provided that phase changes are not involved. For the thermodynamic calculation of a practical system, heat values are required at a certain temperature. It is, however, not necessary to measure the heat value at this particular temperature. It is experimentally more convenient to determine the heat of reaction at a temperature most suitable for attaining equilibrium at a reasonable rate, to measure separately the molar heats of the substances involved, and to apply Eq. (1). This is the basis for the distinction between calorimeters for the determination of molar heats and those for the determination of heats of reaction.

The simplest and most common method for the measurement of molar heats is the drop method using room temperature calorimeters (see, e.g., Ref. 5). This method yields curves for the mean specific heats which must be differentiated in order to obtain true specific heats. It is sufficiently accurate, but cannot be relied upon when the substance undergoes transformations, in particular, slow transformations, in the temperature range studied. In such cases, direct methods for measuring true specific heats, that is, high-temperature methods, become indispensable. These methods consist essentially in the determination of the temperature increase of the substance caused by a measured energy input. In view of the problems which may be encountered by the thermochemist, the methods should be applicable up to as high temperatures as possible. These direct methods may also be used to measure small heats of transformation.

The measurement of enthalpies of reaction requires other types of calorimeters. Bomb calorimeters or corresponding room temperature apparatuses may be used if the driving force of the reaction is strong and an initial small heat input suffices to ensure completeness of reaction. Reactions with weak driving forces and with kinetic steps, for example, formation of compounds, may be studied either by difference methods (involving a solution in acid of the compound and the component salts) or directly in a calorimeter at high temperatures. Difference methods require high accuracies of measurement to obtain moderate accuracies in the resulting values, in particular if these constitute small differences between high measured values. Frequently, it is found preferable to use high-temperature calorimeters, the aim being to increase the reaction rate. This may sometimes be achieved by choosing the minimum temperature at which the reactants and products are liquid or react spontaneously in the solid state or at which diffusion rates are sufficiently high to allow a solid-state reaction to go to completion within a few hours.

The first attempts to apply high-temperature calorimeters consisted in the investigation of moderately exothermic reactions, using simple apparatus and providing limited accuracy. More accurate calorimeters have now been constructed. At the beginning of its development, the term high-temperature calorimetry applied to apparatuses which were used up to 500°C. Considerable advances in high-temperature calorimetry have been made during the past 30 years, and calorimetric investigations are currently being made up to 1500°C. The difficulties encountered at high temperatures include the increase in the rate of heat transfer by radiation, the onset of thermionic emission, and the decrease in the thermal and electrical resistance of ceramics. The major difficulty, however, arises from limitations in the choice of constructional materials imposed by the increase in reactivity of all materials with temperature. Since there are other limitations on the choice of a refractory such as price and availability, the design of a high-temperature calorimeter can involve considerable compromise. The calorimeters described below may be regarded as representing interim developments in design.

3. Reaction Calorimetry at High Temperature (200 < T < 1000°C)

3.1. Direct Methods for Liquids

3.1.1. Approximate Methods

It is often more important to quickly obtain a rough value for a heat of reaction than to measure it more accurately after building a complicated apparatus. The following method has been used (see, e.g., ref. 5) to give approximate values for heats of spontaneous, exothermic reactions. A powder mixture (e.g., Mg plus Bi) was placed in a steel crucible (45 mm long, 15 mm wide), closed by a conical lid, which was hammered in tightly. A steel sheath carrying a thermocouple was fixed to the lid and embedded in the powder. The complete calorimeter was suspended in a furnace. When a certain temperature was reached, the mixture reacted. The temperature change was observed as in normal isothermal calorimetry, the heat of reaction being calculated from the rise in temperature and the masses and specific heats of the compound and the crucible. When this primitive calorimeter was calibrated with a known reaction, an accuracy of $\pm$ 15% was attainable. The water equivalent was 2 to 4 cal/°C. The heat changes measured in the example given were 500–2000 cal, in the temperature range 300 to 600°C.

In measuring the heats of formation of liquids, complete reaction within a short time is usually ensured, particularly when an efficient stirring device is used. One or both components may initially be molten with the calorimeter, and the reaction is started either by dropping in the other component from room temperature or by allowing the two liquids to flow together.

A primitive calorimeter for liquids was used by Sauerwald and Fleischer.[6]

3.1.2. Precision Methods

In each of the foregoing calorimeters for liquids, the temperature measured has been that of the melt, and the reaction vessel has been of low water equivalent. Thus, the accuracy has depended to a large extent on an accurate knowledge of the specific heats of the products. In addition, where ceramic reaction vessels are used, the temperatures of their outer surfaces are uncertain, and it is on these temperatures that heat losses to the enclosure depend. It is significant, therefore, that in the only high-temperature heat-of-mixing calorimeters to attain accuracies as good as ± 1%, in addition to the use of superior thermostats, the reaction vessel is enclosed in a metal block, and the temperatures measured are those of the block. The fact that the temperature rise is smaller than in ceramic reaction vessels is offset by the possibility of using multiple thermocouples and by the fact that the rate of heat loss to the surroundings is lower. In addition, these methods permit direct electrical calibration and calculation of compensation for heat losses.

In a calorimeter for use at temperatures up to 500°C, Kleppa[7] introduced a considerable amount of thermal insulation and heat capacity between the furnace and the uniform temperature enclosure, so that the cyclic temperature changes in the latter were very small. This design was based on work done by Tian[8] on multiple enclosure thermostats, which have served as the basis for microcalorimeters operating at room temperature. The hollow cylindrical constant-temperature enclosure (30 cm high, 18 cm external diameter), made of Al, was enclosed in a heavy steel shield within a furnace. Additional shields of Al sheet were placed between the furnace and the steel shield, and between that shield and the jacket. The furnace was equipped with additional top and bottom heaters and kept at nearly constant temperature by means of a regulator. The heat changes occurred within a removable graphite crucible inside a cylindrical aluminum block within the constant-temperature enclosure. The temperature measurement was made by means of eight differential chromel–alumel thermocouples connected in series, with one set of junctions in the block and the other set in the enclosure. The calorimeter was calibrated electrically. The whole assembly was maintained in an atmosphere of argon. Through a central opening extending from the top of the enclosure to the outside of the furnace insulation, normally closed by removable plugs, crucible and samples could be introduced without seriously disturbing the thermal equilibrium. Both samples (e.g., Pb and Sn) were contained in the graphite crucible, one being held in a special stirring and charging device. From 5 to 22 ml of metal were used in each experiment; an accuracy of $\pm1\%$ was claimed. The heat changes measured were in the range 70 to 300 cal.

Although the amplitude of temperature fluctuations in the enclosure was reduced, this was done at the expense of using a thermostat of high heat capacity, which therefore required a long time to reach equilibrium. In addition, it was not possible to directly control the temperature of the enclosure, which thus contributed to some extent to the heat in the heat change following the reaction.

Several versions of this calorimeter were constructed by Kleppa and co-workers. They were designed to operate at higher temperatures and, therefore, differed mainly in the materials used for construction. The enthalpies of mixing in binary liquid alkali metal iodides,[9] for instance, were measured in a new single-unit microcalorimeter suitable for work at temperatures up to about 1100°C.

The temperature-sensing device of this calorimeter was a 60+60 junction Pt/Pt–13% Rh thermopile. The output of this thermopile, which measured the temperature difference between the external surface of the calorimeter proper and the internal surface of a heavy SiC jacket surrounding the calorimeter, was amplified and recorded.

The furnace assembly, which surrounded the calorimeter on all sides, consisted of a cylindrical main heater and two separately wired end heaters. The furnace was maintained at temperature by means of a stable proportional temperature controller system which was activated by a single shielded Pt/Pt–13% Rh thermocouple. The power supplied to the two end heaters was voltage-stabilized.

In spite of repeated attempts to optimize the performance of the control system by varying proportional band, rate, and reset time, it was found that the controller was not quite adequate to maintain the "baseline" of the calorimeter completely stable at very high calorimetric sensitivities. Small temperature fluctuations in the main heater were reflected in fluctuations in the baseline of the calorimetric emf–time curves. This introduced an additional uncertainty in the determination of the area between the observed emf–time curve and the baseline.

An isothermal solution calorimeter capable of detecting heat changes of the order of 0.1 cal at high temperatures has been developed.[10] This instrument was designed for measuring heats of solution as a function of composition and temperature in liquid systems, where the data can be used without any correction for melting effects. Moreover, very reactive molten salt mixtures could be investigated since the separate components were brought together and mixed within a completely enclosed, vacuum-sealed cell. The apparatus consisted of an inner spherical compartment containing the reaction cell, surrounded by a number of concentric spherical shells made of Inconel alloy.

The main features of the calorimeter are shown in Fig. 1. The sensing junctions of a four-couple Pt/Pt–13% Rh thermopile are positioned on the walls of the alumina crucible. The cylindrical quartz calorimetric cell contains two compartments, X and Y. When the calorimeter is rotated by 180° and then brought back to its original position, the separate components in compartments X and Y are both trapped in compartment Y, where they mix. Calibration is performed electrically with a heater made of platinum wire. It was concluded, however, that the accuracy of the results obtained with this instrument was difficult to assess on an absolute basis. While the sensitivity of the temperature-measuring device allows the detection of temperature differences of the order of 0.002°C, the accuracy of the measured heat changes depended entirely upon the calibration procedure. Enthalpy of mixing experiments were carried out on the molten salt mixtures of AgCl with NaCl, KCl, RbCl, and CsCl and on the melts

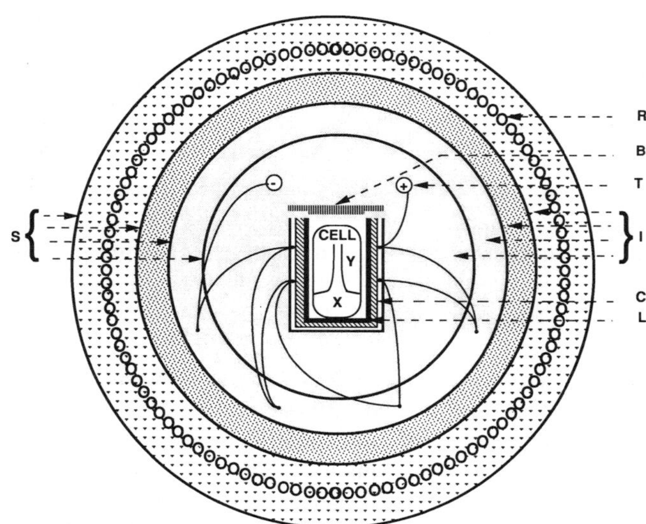

FIGURE 1. Schematic diagram of a high-temperature solution calorimeter. R, Spherical heating element; B, boron nitride cover; T, thermopile; I, fibrefrax insulation; C, alumina crucible; L, platinum liner; S, Inconel spheres.

NaCl–KCl and $CdCl_2$–$PbCl_2$; the results obtained were in fair agreement with literature values.[11–13]

However, as mentioned above, single-unit calorimeters are sensitive to external temperature fluctuations and yield moderately accurate results. Twin microcalorimeters brought about an improvement in these problems.

The principle of the microcalorimeter was proposed in 1923 by Tian[8], but only some 20 years later, thanks to Calvet,[14] did this apparatus become easy to use over a very large temperature range (100–1300 K). This isoperibolic calorimeter has been described many times, so only its essential characteristics are reiterated here.

A microcalorimeter is composed mainly of the following parts (Fig. 2):

- An external steel enclosure surrounding a cylindrical furnace; heating is provided by four resistors, one on the bottom, one on the top, and two on the cylindrical walls.
- A calorimeter block, made of either aluminum or copper for the medium-temperature version (M.T.) and of either Kanthal or alumina for the high-temperature version (H.T.). This block generally contains two cavities carrying the thermopiles (some early versions were designed with four thermopiles, but this was not generally adopted, especially in the H.T. apparatus). The block is surrounded by many enclosures acting as thermal and electric shields. Each thermopile is composed of a cylindrical bottom-closed tube, a so-called "calorimeter cell" of silver (M.T.) or alumina (H.T.) on which the "hot" thermocouple junctions lie, the "cold" junctions being in contact with the calorimetric block. The type of thermocou-

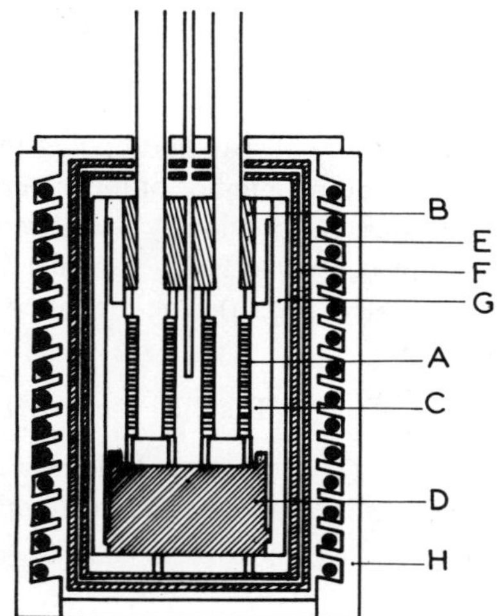

FIGURE 2. Schematic diagram of a Calvet microcalorimeter. B, C, and D, Top, middle, and bottom parts of the alumina calorimetric block; E and F, alumina shields; G, alumina enclosure; H, electrical furnace.

ple depends on the operating temperature of the calorimeter (chromel–alumel for M.T. and Pt/Pt–10% Rh for H.T.).

The size of the cell, generally 17 mm in diameter and 80 mm high, and the shape of the thermocouple supports depend on technical considerations, but one condition has to be met: the maximum thermal flux should be integrated by the thermocouples.

In order to obtain good stability of the apparatus with respect to time and temperature, the two thermopiles are connected in opposition; any exterior thermal perturbation is thus eliminated.

The furnace is maintained at constant temperature with an electronic regulator; the sensing element is either a thermistor (M.T.) or a thermocouple (H.T.), located closest to the heating element.

The experimental temperature is controlled by means of a thermocouple situated in the calorimetric block center.

3.1.3. Experimental Techniques

The experimental techniques devised in calorimetry for the determination of enthalpies of mixing are numerous. For all of them, the essential condition to be met is elimination of all the effects arising generally from material–atmosphere interactions (e.g., with oxidizing or hygroscopic substances), from solvent or solute–crucible interactions, from difficulties of mixing (A and B with very different densities), from stirring necessary to homogenize the final product, from introduction of a solute at a temperature different from that of the solvent, and so on. In order to solve these specific problems, some researchers not only created devices but sometimes built special calorimeters.[15–17]

The main techniques[18] for systems whose components melt above ambient temperature are indicated here.

The first and also simplest method to achieve the mixing of two substances is the so-called "drop method," described at length by Kubaschewski and Evans[19]; it consists in dropping into the liquid substance A, considered as the solvent and maintained at the experimental temperature T_E, a precise amount of substance B, previously stabilized at temperature T_0 (generally room temperature). This technique, which is very simple, has been applied to microcalorimetry (Fig. 3). The measured heat (Q_p) corresponds to the enthalpy of the reaction:

$$n_B B(s,T_0) + n_A A(l,T_E) \rightarrow (n_A + n_B) AB(l \text{ or } s, T_E)$$

The heat Q_p is the sum of the heat of mixing (Q_M), the heat of fusion (Q_F), and the heat necessary to raise the temperature of the sample B (mass m) from T_0 to T_E:

$$Q_p = Q_M + Q_F + \int_{T_0}^{T_E} C_p dT$$

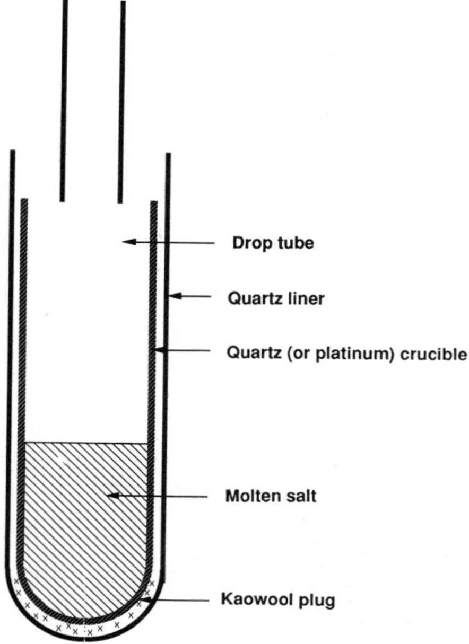

FIGURE 3. Drop method.

Unfortunately, for some systems, the uncertainty in the last two terms on the right-hand side of the above equation is of the same order of magnitude as Q_M and can, therefore, lead to meaningless results of the enthalpy of mixing measurements. However, when the enthalpy of formation of AB referred to the liquid state is negative (exothermic), the positive (endothermic) enthalpy increment of B reduces the overall measured change, and the method then reduces to a null method.

In order to improve the accuracy of the previous method, an indirect drop method was developed (Fig. 4), which involves preheating of the sample B. The mixing experiment is thus carried out in two steps. First, the sample B is dropped into the funnel, where preheating takes place; this drop is guided by the drop tube, the lower end of which acts as a stopper by obstructing the funnel aperture. Then, when B is thermally equilibrated, a simple vertical shift of the drop tube allows the mixing to be performed. In some cases, a stirring device was added in order to ensure homogeneity of the mixture.[20] In order to keep the samples either under vacuum or under inert atmosphere (argon U), the upper part of the system involves a set of taps and rings, and all the manipulation of the drop tube and of the stirrer is electromagnetically operated.

The break-off ampoule method[21] (Fig. 5) has often been used with molten salt mixtures. Inside the Pyrex (or quartz) liner, a cylindrical crucible contains the liquid salt A while salt B is contained in an ampoule. This Pyrex (or quartz)

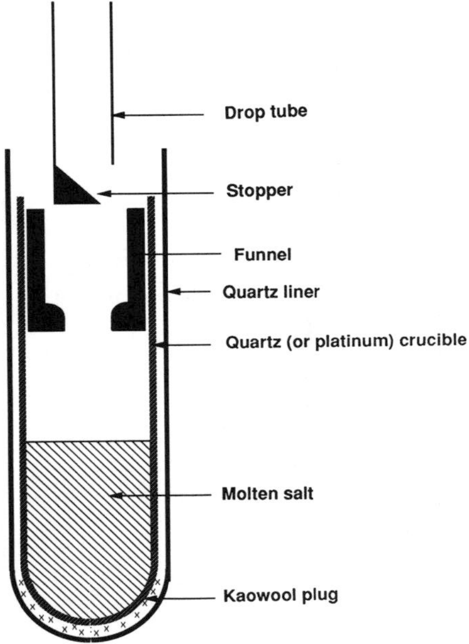

FIGURE 4. Indirect drop method.

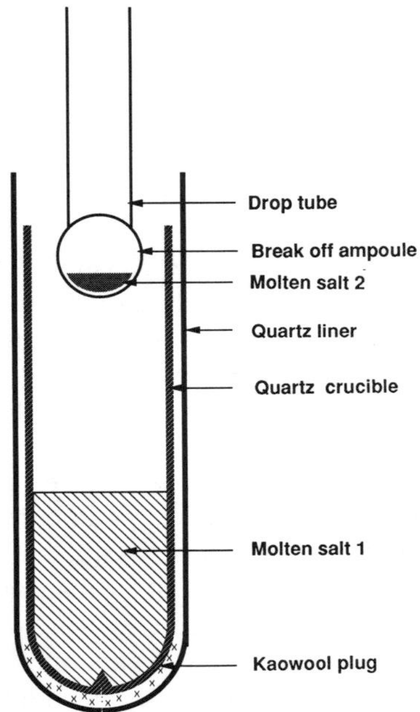

FIGURE 5. Break-off ampoule method.

ampoule has to be thin enough to be broken with a single stroke. The mixing of the two liquids is initiated by crushing the ampoule against the bottom of the crucible. The thermal change arising from the ampoule break-off was found to be very small and reproducible.

This method, of course, cannot be employed when the melt reacts with Pyrex (or quartz). For the investigation of molten alkali metal hydroxide mixtures a suspended cup method[22] (Fig. 6) was developed. Inside the Pyrex (or quartz) liner, a cylindrical silver crucible contained the liquid hydroxide A while the salt B was contained in a small silver cup. This silver cup was held by tongs and could be released by external operation of the tongs. The mixing of the two liquids was initiated by dropping the silver cup into the crucible. The thermal excursion arising from this drop (Fig. 6) was very small and reproducible.

Enthalpies of mixing may also be determined using the method of continuous injection of the liquid component A, contained in a special syringe, into the liquid component B (Fig. 7). This technique has mostly been used for aqueous solutions[23]; its main advantage lies in the fact that the enthalpy of mixing is obtained over the whole concentration range from only a few experiments. The application of this method to high temperatures is possible, but it has been little employed so far since it is rather difficult to find cell materials suitable both to contain molten salts and to construct a syringe.

The methods of mixing described above are general, but the choice of a

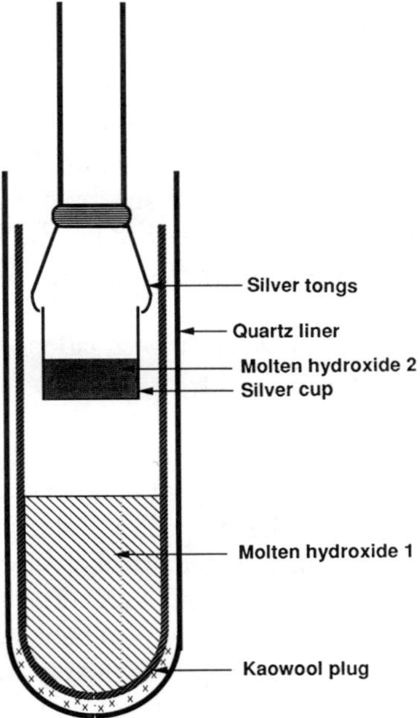

FIGURE 6. Suspended cup method.

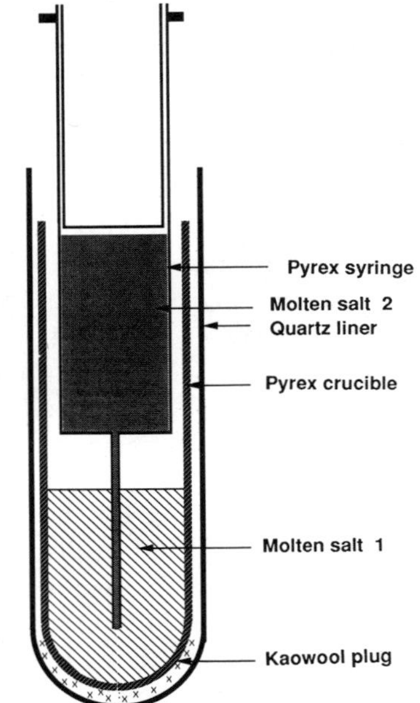

FIGURE 7. Method of continuous injection.

suitable technique of mixing for a particular system should always be made in accordance with the physicochemical properties of the melt under investigation.

The two drop methods described above allow several successive additions of B to be made during the same experiment, for example, to measure the composition dependence of the enthalpy of mixing. The main advantages of the drop methods are therefore their simplicity and rapidity; this contrasts with those more direct and accurate techniques, described below, which make it possible to measure the mixing of liquid components directly and, in principle, with a greater accuracy.

Some examples are given below of investigations of very reactive melts, in which great care must be taken in order to obtain reliable enthalpy of mixing measurements.

For the investigation of zinc halide-containing melts,[24] a novel experimental arrangement and procedure was adopted in order to control the weight losses due to evaporation of the zinc halides at 665°C. This is shown schematically in Fig. 8. Inside the fused-silica liner, a cylindrical crucible contains the low-vapor-pressure salt in the liquid state (LiX, CsX, AgX). The zinc halides are contained in an evacuated "double break-off" fused-silica bubble. The pressure of the zinc halide liquids inside this bulb was approximatively 0.5 to 0.7 atm, while the argon pressure in the liner was maintained at 1 atm.

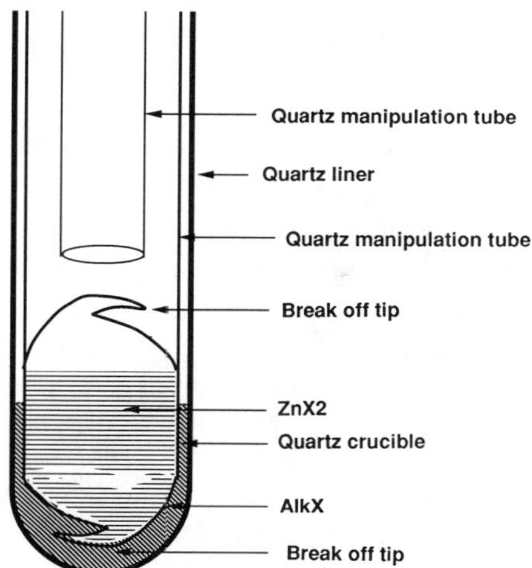

FIGURE 8. Double break-off method.

The double break-off tube eliminated the problem of volatilization of the zinc halide from the time when the liner was introduced into the calorimeter until the mixing was initiated. This period ranged from 45 min to 2 h.

The mixing of the two liquids was started by crushing the lower break-off tip against the bottom of the crucible. As the pressure in the bulb was lower than 1 atm, a portion of the melt contained in the crucible was sucked up into the bulb and was mixed there with the zinc halide. The mixing process was then completed by crushing the upper break-off tip. This allowed the zinc halide to be released into the crucible and mixed with the remainder of the melt. However, after the mixing operation was started, zinc halide was lost by evaporation and an endothermic baseline shift was observed. In order to correct for this baseline shift, a series of blank experiments was performed.

For the investigation of lead oxide-containing melts, the following experimental procedure was adopted[25,26] (Fig. 9). About 60 g of oxide melt was contained in an 80% gold–20% palladium crucible of about 17-mm diameter and 75-mm height. The lead oxide to be dissolved in the melt was kept in a very shallow platinum cup of about 10-mm diameter. This cup was attached, by means of three platinum wires, to a fused-silica tube which could be manipulated from outside the furnace system. The solution reaction was initiated by lowering the platinum cup into the melt. Stirring was accomplished by means of a platinum-covered graphite plunger. The liquid PbO was displaced and brought into reaction by inserting the plunger into the platinum cup. The other oxide component was added to the solvent in a similar manner, but in this case adequate stirring was achieved simply by moving the platinum cup up and down in the gold–palladium crucible. Corrections were made for the heat change associated

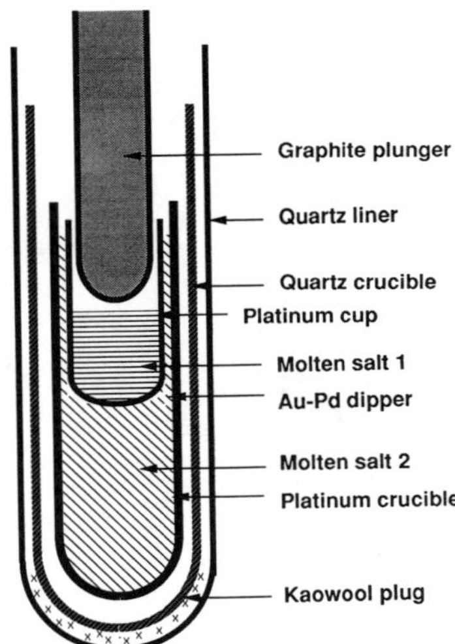

FIGURE 9. Experimental arrangement for the investigation of lead oxide melts.

with each stage. This heat change is endothermic in character and is largely due to mass displacement in the vertical temperature gradient of the calorimeter. When a plunger was used, the stirring correction was 10–50% of the total heat of reaction; without a plunger, the correction was 10–20%.

Enthalpies of mixing in binary liquid carbonate mixtures have been measured.[27] Due to the corrosive nature of molten alkali carbonates, these cannot, unlike most other salts, be contained in fused-silica containers and be introduced by the usual "break-off" technique. Experience has shown that attack on Palau (20% Pd–80% Au) in "acid" alkali carbonate melts, that is, in melts kept under a relatively high CO_2 pressure, may be considered negligible.

The experimental arrangements shown schematically in Fig. 10 were used to perform these calorimetric measurements. The plunger as well as the dipper crucible could be manipulated from outside the calorimeter proper.

Immediately before insertion into the calorimeter, the fused-silica liner with its contents was preheated for a period of 15 min at about 50 K above the operating temperature of the calorimeter ($\sim$1150 K). The weight loss of the most volatile carbonate (Rb_2CO_3) was about 0.3%; in spite of the relatively small vaporization losses, attack by the vapors on the fused-silica liner and shield was considerable, and the lower parts of the device had to be rebuilt after 10 to 15 experiments.

The mixing of the two salts was achieved by vertical manipulation of the plunger and dipper. After the initial mixing process, three additional stirring operations were carried out at 1-min intervals to ensure complete mixing.

Mixtures of $K_2S_2O_7$, V_2O_5, and K_2SO_4 were investigated using the follow-

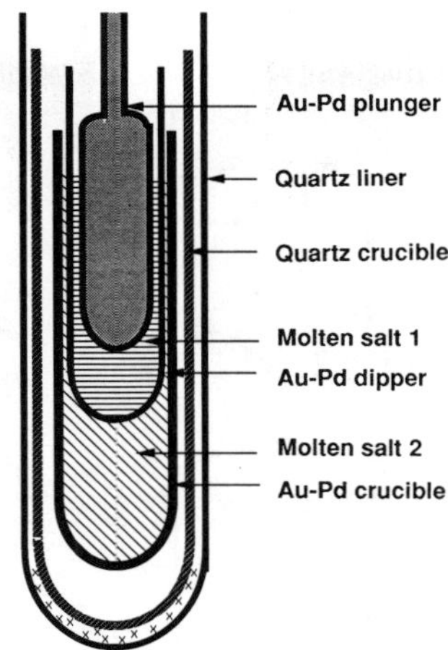

FIGURE 10. Experimental arrangement for the investigation of carbonate melts.

ing experimental technique.[28] All experiments concerning measurements of the heat of mixing were performed by addition of a solid (V_2O_5 or K_2SO_4) to a liquid ($K_2S_2O_7$ or $K_2S_2O_7$–V_2O_5 mixtures). This addition was made by the "indirect drop method," one of several addition methods that have been described earlier.[19] However, in these experiments, a modified experimental setup was constructed as shown in Fig. 11. The outer quartz tube (B) has an outer diameter of 18 mm and is 600 mm long; the lower 100-mm part fits very closely to the alumina walls of the calorimetric cavity in order to secure an optimized horizontal heat conductivity. A standard-taper joint of borosilicate glass at the top is connected to the quartz tube outside the calorimeter by a taper joint. A slow gas flow is maintained through the system before and during the experiment. Through the taper joint, two Pyrex or quartz tubes lead into the calorimetric cavity. The inner tube (C) (outer diameter, 6 mm) is equipped with a small funnel at the top outside the calorimeter and a hole in the wall just above the smoothly ground end of the glass rod inside the calorimeter. The end of the glass rod also acts as a stopper (E) in the reservoir of the outer tube, which is well tightened to the taper joint by conventional fittings consisting of a screw and a silicon packing. Just below the reservoir and in close contact with it, the crucible (F), containing the melt, is located. This quartz or borosilicate crucible has outer dimensions close to the inner dimensions of the tube B to optimize the horizontal heat flow. The experiments concerning the heat of mixing were performed as follows. The crystalline V_2O_5 or K_2SO_4 was weighed with an accuracy of 10^{-2} mg. The final weight of the crystals was measured after the dust had been removed from their surface by several drops from about 50 cm. The crystals were accepted as dust-free when the deviation between the last two preliminary weigh-

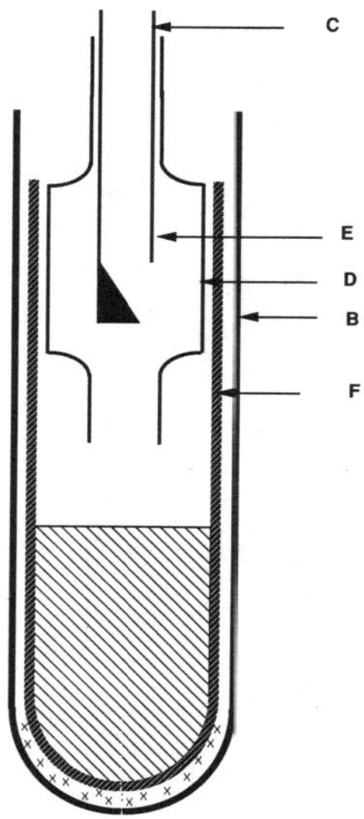

FIGURE 11. Experimental arrangement for the investigation of vanadium oxide melts. See text for description.

ings was less than a few hundredths of a milligram. A crystal, typically with a mass of 10–30 mg, was dropped through the funnel of the tube C into the reservoir D, and the thermal disturbance was registered on a recorder and fed to a system for automatic integration. After the baseline was reestablished—usually within 20 min—the tube C was raised about 2 cm by manipulation from the outside, allowing the crystal to drop from the reservoir into the melt below. The possible thermal change could be integrated manually from the recorder or—usually—indirectly by the connected integrator. In both cases, calibration was performed by dropping five small gold boules, with known weights and thermal capacities similar to the ones in the experiments, directly into the melt. In order to check the indirect drop method, experiments with a two-step drop of gold boules were also performed. No thermal change could be recorded in any case by the final drop of the chemically inert gold boules from the reservoir to the melt. However, different thermal effects were found by dropping gold boules into the reservoir instead of directly into the melt, a consequence of the inhomogeneity of the thermopile system. Values corresponding to the direct drop were therefore applied.

Up to 15 additions were made to the same melt in experiments lasting up to 24 h. Each series was usually terminated by weighing the solidified melt in the crucible, in order to check that all the added crystals had reached the melt. In this way, the possible escape of volatile components (SO_3 or H_2SO_4) during the experiment could also be checked. For the V_2O_5–$K_2S_2O_7$, system an average loss of 0.6% by weight was found. For the series involving the ternary system V_2O_5–$K_2S_2O_7$–K_2SO_4, the average weight loss was found to be 0.05%. An experimental series was usually terminated when the point of saturation (indicated by no thermal change) of the melt was reached. The solidified melt was usually crushed in order to recover the final crystal and thus confirm that saturation was in fact reached.

Sometimes, the crystal in the reservoir stuck to the wall, presumably due to a reaction with tiny amounts of volatile compounds from the crucible. In these cases—which occurred frequently when K_2SO_4 crystals were added—the tubes (C and D) were removed, the reservoir was cleaned, and the tubes were remounted in the calorimeter. The experimental series could thereafter be continued when a stable baseline was again obtained. In the present case, where a solid (B) is dropped into a liquid (A), both at the same temperature, T, the molar enthalpy of liquid–liquid mixing, ΔH_{mix}, at the temperature T is given by

$$\Delta H_{\mathrm{mix}} = \Delta H_{\mathrm{exp}} - x_B[H_B(\mathrm{l},T) - H_B(\mathrm{s},T)] \tag{2}$$

Here ΔH_{exp} is the molar heat evolved at constant pressure during the experiment, x_B is the mole fraction of the solid, and the term $H_B(\mathrm{l},T) - H_B(\mathrm{s},T)$ is the molar heat of fusion of the solid at the temperature T. This molar heat of fusion at 430°C was found by extrapolation from tabulated data[29] to be 61.84 kJ/mol for V_2O_5 (mp 670°C) and 49.18 kJ/mol for K_2SO_4 (mp 1069°C).

In the case of the ternary system V_2O_5–$K_2S_2O_7$–K_2SO_4 where the liquid, A, has the composition $V_2O_5 \cdot nK_2S_2O_7$ (n = 1, 2, or 3), the molar enthalpy of liquid–liquid mixing, ΔH_{mix}, at the temperature T is given by

$$\Delta H_{\mathrm{mix}} = \Delta H_{\mathrm{exp}} + \Delta H_{\mathrm{premix}} - x_B[H_B(\mathrm{l},T) - H_B(\mathrm{s},T)] \tag{3}$$

where the term $\Delta H_{\mathrm{premix}}$ is the enthalpy of mixing found for n = 1, 2, or 3 in the binary system V_2O_5–$K_2S_2O_7$. The spread (uncertainty) in the value of the heat of mixing for a particular drop was around ±2%. However, the deviation between the measured value and the true value is probably ±5%. The uncertainty in the temperature is ±1°C. When the vapor pressures of the salts to be mixed are too high, however, it is impossible to use a classical mixing device. Suitable cells had to be constructed for this purpose.

Recently, mixtures of $BiCl_3$ and KCl were investigated.[30] Measurements were carried out in a high-temperature Calvet-type microcalorimeter at 690 K, 22 K below the normal boiling point of $BiCl_3$.[31] At this temperature, the vapor pressure of $BiCl_3$ is 69.4 kPa. Until now, measurements of the enthalpy of mixing of salts have been carried out at temperatures much lower than the boiling points of the components. Only $[(1-x)ZnCl_2 + xMCl)]$ and $[(1-x)ZnBr_2 + xMBr]$ (M = Li, Cs) had been examined by Papatheodorou and Kleppa[24] under similar conditions; the difference between the boiling point of $ZnBr_2$ and the measuring temperature was 27 K. We have improved the technique described by Papatheodorou and Kleppa so that it was possible to avoid evaporation of the more volatile component, not only before measurement, but also after breaking the ampoule.

Pyrex glass was used to prepare the ampoules (Fig. 12). The spherical part was a thin-walled membrane of strictly controlled mechanical properties. The thickness of the membrane was selected so that it was possible to break the ampoule in the calorimeter without any measurable thermal effect but, on the other hand, so that the membrane was strong enough not to be destroyed by atmospheric pressure after evacuation of the ampoule. The ampoules were filled in a glove box with weighed portions of $BiCl_3$. After evacuation by means of a rotary pump, the ampoules were sealed and stored in vacuum to avoid diffusion of air through the membrane.

Other parts of the calorimetric cell were also made of Pyrex glass. The crucible, to which a weighed portion of KCl was introduced prior to measurement, was provided at the top with a ground-glass ball joint to close it tightly when the ampoule was broken. The crucible cover, provided with the male part of the ball joint, formed one body with a follower which moved the ampoule down during the crushing. The distance between the lid joint and crucible joint was adjusted before measurement by means of spacers so that it would be from 0.4 to 0.6 mm. The crucible together with its lid and follower was placed in a protective quartz tube filled with argon.

The time elapsed from introduction of the arrangement into the calorimeter to the beginning of measurement was 4 h. By breaking the ampoules, half of the

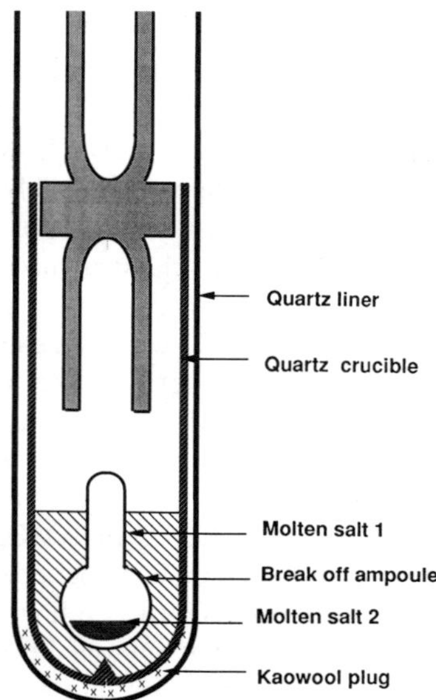

FIGURE 12. Experimental arrangement for the investigation of bismuth chloride melts.

spherical part was destroyed so that it was possible to mix thoroughly liquid bismuth chloride with solid potassium chloride. Measurements in which the ampoules were not completely broken were disregarded.

No endothermic effects related to evaporation were found. Any baseline shift after mixing was very slight; it was never larger than 1 mm, which is less than 1% of the peak height. In most cases, no measurable change in the baseline position was found.

The calibration constant k of the calorimeter was determined on the basis of the thermal change for zinc solidification. Calibration measurements were carried out in the same cell in which enthalpy of mixing was measured. The molar enthalpy of fusion of zinc, $\Delta_{fus}H_m = 7.343 \pm 0.083$ kJ/mol was taken from Chiotti *et al.*[32] The standard deviation $s(k)$ of the calibration constant of the calorimeter was $0.021k$.

The same method was used for the investigation of mixtures of $BiCl_3$ with the other alkali metal chlorides.[33]

The experimental constraints are most severe when the mixture under consideration has a sublimable component. Very recently, ternary mixtures of $AlCl_3$, KCl, and $AlCl_3NH_3$ were investigated.[34] The cell shown in Fig. 13 had therefore to be constructed; both reactants, the $AlCl_3$ + KCl liquid mixture and

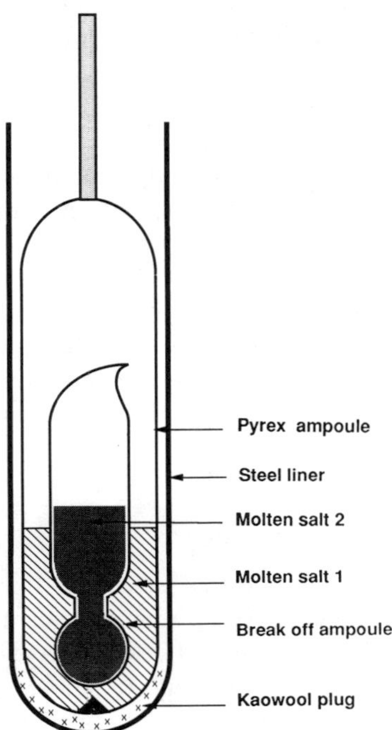

FIGURE 13. Experimental arrangement for the investigation of aluminum chloride melts.

$AlCl_3NH_3(l)$ were contained in closed Pyrex ampoules, both kept inside the calorimeter proper during measurements. The inner ampoule containing $AlCl_3NH_3$ had a very thin base which could easily be broken on the sharp edge of the outer ampoule with a rapid movement of the manipulation rod sticking out of the calorimeter.

Calibration was performed using $\alpha-Al_2O_3$ from the National Bureau of Standards. Because the experimental cell was closed, the calibration could not be carried out in the cell itself, but had to be performed directly in the steel protection tube situated in the calorimeter proper. The calibration constant obtained in this manner is somewhat uncertain. Tests have been made[35] by varying

- the rest position of the $\alpha-Al_2O_3$ piece relative to the calorimeter proper
- the cell arrangement by letting the $\alpha-Al_2O_3$ piece fall into a Pyrex tube situated inside the steel protection tube.

Variations of the order of $\pm 5\%$ were observed.

The procedure employed in the experiments was as follows. The inner cell was first filled with the desired amount of $AlCl_3NH_3$ in a glovebox, and transferred to a vacuum line evacuated to 10^{-5} torr. The cell was then back-filled with Ar to adjust the pressure so that the pressure in the inner and the outer cell were approximately equal at the working temperature of the calorimeter. This was done to ensure stable cell behavior when the inner cell wall was broken during the mixing experiment. The cell was then sealed and transferred back to the glovebox, where it was mounted into the outer cell after this cell had been filled with the necessary amounts of $AlCl_3$ and KCl. The outer cell with its contents was then transferred to the vacuum line, evacuated to 10^{-5} torr, sealed, and fused onto the Pyrex manipulation rod.

Before the cell was introduced into the calorimeter, it was preheated at a somewhat higher temperature than the calorimeter temperature to speed up the melting of the $AlCl_3-KCl$ mixture. Complete dissolution of KCl(s) was observed visually.

Complete mixing of the three salts after breaking the inner ampoule was ensured by rapid stirring of the cell. These manipulations had a minor influence on the measured heat change and no baseline shifts were observed.

3.2. Methods for Solids

3.2.1. Direct Methods for Solids

A differential calorimeter used by Fischer and Lorenz[36] to determine the heat of formation of the spinel Al_2FeO_4 from its component oxides between 800 and 1000°C is shown in Fig. 14. A nickel cylinder (A) was bored out and carried two alumina tubes (B and C; 100 mm long, 6 mm wide, 2 mm wall thickness); the space between the tubes and the block was filled with alumina powder. One

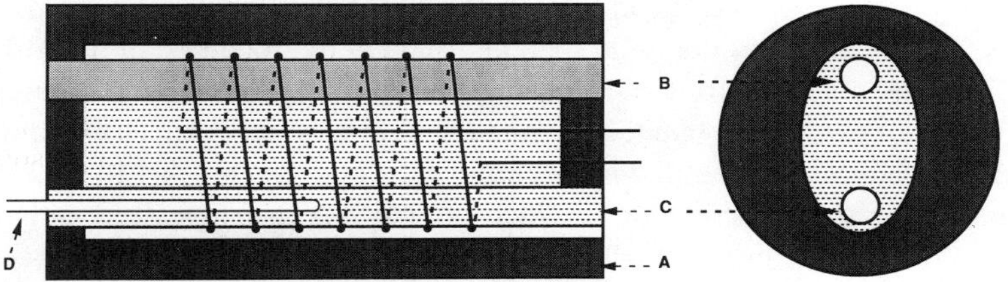

FIGURE 14. Differential calorimeter to determine the enthalpy of formation of the spinel Al_2FeO_4 from its component oxides. A, Nickel cylinder; B and C, alumina tubes; D, thermocouple.

tube was packed with alumina powder in which was embedded a thermocouple (D). The other tube was packed with the mixture of FeO and Al_2O_3 powders, in which was embedded a Pt wire, used as a resistance heater for calibration, 0.2 mm diameter and 5 cm long attached to 0.8 mm diameter Pt leads. Twenty Pt/Pt–Rh thermocouples connected in series were wound round the two tubes so that their even junctions lay against one tube and their odd junctions against the other. The block was placed in a sillimanite tube in a platinum-wound furnace. The tube could be evacuated and filled with argon. The emf produced by the differential thermocouples was measured on a galvanometer (sensitivity, 6×10^{-7} V/mm). Heat changes of 70–200 cal, extending over several hours, were measured by this method with an accuracy of $\pm 15\%$.

3.2.2. Indirect Methods for Solids

Direct reaction calorimetry can be used to measure the enthalpy of formation of solid compounds, but there are some limitations to this method, as stressed previously.

Dissolution calorimetry has often been employed to measure the enthalpy of formation, ΔH_f, of solid compounds as the difference between the heat of dissolution of the compound A_xB_{1-x} and that of a solid mechanical mixture $[xA + (1-x)B]$ into the same solvent S:

$$A_xB_{1-x} + S \quad\quad \rightarrow [xA + (1-x)B]_S \quad\quad \Delta H_1$$
$$xA + (1-x)B + S \rightarrow [xA + (1-x)B]_S \quad\quad \Delta H_2$$

The enthalpy of formation of the solid compound A_xB_{1-x} is $\Delta H_f = \Delta H_2 - \Delta H_1$. For practical reasons, it is often more convenient to use this method with a large amount of solvent, for example, to measure the heats at infinite dilution of A_xB_{1-x} and of $[xA + (1-x)B]$. Then:

$$\Delta H_f = x\Delta H_A^\infty + (1-x)\Delta H_B^\infty - \Delta H_{A_xB_{1-x}}^\infty$$

When possible, it is advisable to use one of the two components as the solvent in order to obtain a better accuracy.

Many years ago, the solvents used for dissolution were room temperature liquids, mainly aqueous inorganic solvents, such as hydrochloric or hydrofluoric acid. The enthalpies of formation obtained in this way were often of poor accuracy since they were obtained as small differences between large numbers. This method, however, is still widely employed when the enthalpy of dissolution is needed at room temperature or when no other high-temperature solvent is available. Of course, there is no universal solvent, and the choice of a dissolution medium should be made in accordance with some compromise between several requirements, such as reasonable duration of dissolution experiments and sufficiently low vapor pressure of the melt in order to avoid losses by evaporation.

Solution calorimetry has been used, in particular, in the field of geology,[37] and several oxide melts have been found to be good solvents: $2PbO \cdot B_2O_3$, $3Na_2O \cdot 4MoO_3$, $9PbO \cdot 3CdO \cdot 4B_2O_3$, $0.35Na_2O \cdot 0.35B_2O_3 \cdot 0.30SiO_2$, and some mixtures of $Na_2O \cdot P_2O_5$. In Table I, the main characteristics of these solvents are listed.

TABLE I. Solvents for High-Temperature molten salt calorimetry[a]

Solvent	Temp. range(°C)	Uses	Limitations
$2PbO \cdot B_2O_3$[b]	650–900	Excellent for SiO_2, Al_2O_3, GeO_2, Fe_2O_3, Cr_2O_3, Ga_2O_3, MgO, coO, NiO,CuO, ZnO, CdO, CaO, Cu_2O	Forms insoluble precipitates with TiO_2, SnO_2, Mn_2O_3. Cannot be used under reducing atmosphere. Can be used for Mn^{2+-} (and Fe^{2+-}) containing oxides under inert atmosphere when oxides can be maintained
$9PbO \cdot 3CdO \cdot 4B_2O_3$[c]	650–900	Same as above	Same as above
$3Na_2O \cdot 4MoO_3$[d]	650–700	Good for TiO_2, SnO_2, Mn_2O_3, MgO, CoO, NiO, ZnO, CdO	SiO_2 and other acid oxides dissolve slowly. Not used under reducing atmosphere. Vaporizes significantly at higher temperature
$0.35Na_2O0.35B_2O_30.30SiO_2$[e]	800–900	Good for SiO_2, Al_2O_3, MgO, CaO, FeO, MnO under reducing atmosphere	

[a]From Ref. 37.
[b]Solvent of choice when applicable.
[c]$2PbO \cdot B_2O_3$ generally preferable.
[d]Does not form a glass on cooling.
[e]Hygroscopic; carefully prepared and stored.

For instance, a series of aluminum silicates were investigated by solution calorimetry at 968 K.[38] Several studies were conducted on $NaAlSi_3O_8$ (orthose)[39] and its mixtures with, for instance, the analogous compound $KAlSi_3O_8$.[40] The same technique was also used to investigate chromite spinels.[41]

The enthalpies of formation of several alkali metal titanates were measured by Mitsuhashi and Fujiki.[42] These compounds, which are described by the general formula $mM_2O \cdot nTiO_2$, where $(m,n) = (1,1),(4,5),(1,2),(1,3)$, and $(1,6)$ with $M = Na$ and $(m,n) = (1,1),(2,3),(1,2),(1,4)$, and $(1,6)$ with $M = K$, are of increasing importance because of possible applications such as in the production of thermal insulators and ion exchangers. The solvent melt used as a dissolution medium at 970 K was $3Na_2O \cdot 4MoO_3$.

4. Reaction Calorimetry at Very High Temperature $(T > 1000°C)$

Calorimetry at very high temperature $(T > 1000°C)$ was made possible by the development of new equipment.

The formation of a high-temperature liquid mixture, metallic (alloy) or ionic (molten salt system), does not generally result in significant thermal effects; this makes it necessary to obtain the enthalpies of formation from direct measurements using a high-temperature calorimeter with a high sensitivity. Up to 1000°C, such measurements can be carried out successfully using apparatuses similar to those described above. Beyond this temperature, however, the technical difficulties are such that a new apparatus must be designed.

For investigations in the range 1000–1500°C, such an apparatus exists[43,44]; for the sake of clarity, its principle will be reiterated here and the improvements of the calorimetric detector and the mixing and stirring systems will be described in detail together with the method used for the acquisition and treatment of calorimetric data.

The enthalpy of formation ΔH_M of a binary liquid mixture at the temperature T_E is obtained from the measurement of the energy related to the introduction of a solid or liquid sample A into a liquid bath B, at the experimental temperature T_E:

$$A_{liq(or\ sol)} + B_{liq} \rightarrow AB_{liq}$$

In this kind of apparatus, as in many other calorimeters, the detector consists of platinum- and rhodium-based thermocouples. One set of their junctions is in contact with the crucible where the experiment takes place and the other one with a reference enclosure where no thermal change arises.

The magnitude of the thermal changes to be measured and the volatility and considerable chemical reactivity of the components (A or B or both) imply many constraints that should be taken into account in the development of such an apparatus (e.g., inert refractory materials, purified atmosphere).

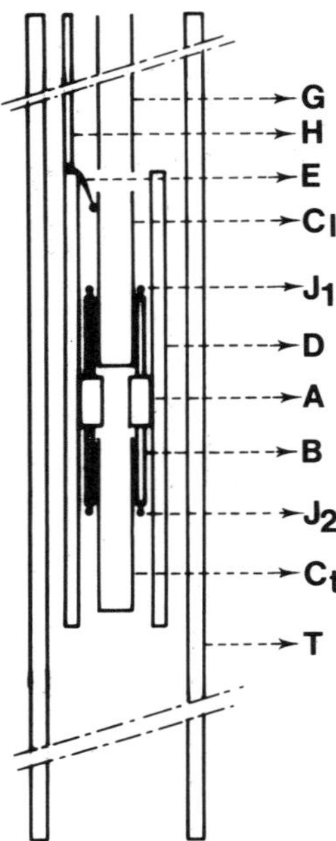

FIGURE 15. Vertical section of the very high temperature calorimeter. See text for description.

The whole calorimetric assembly consists of the following parts. (1) The vertical cylindrical furnace consists of a graphite resistor surrounding a gastight alumina tube, 23 mm i.d., 600 mm long, in which the calorimetric detector is located and localizing the experimental chamber. The geometry of the resistor is such as to provide a constant-temperature zone, 140 mm long, in the central part of the tube. This furnace has an external water-cooled jacket and can be heated up to about 2000°C. (2) The furnace is supplied with low-voltage current. Its temperature is controlled by an electronic system monitored by a Pt/Pt–10% Rh thermocouple within the central part of the furnace. (3) A set of valves and flowmeters enables evacuation of the furnace and the experimental chamber, and the flow or maintenance under pressure of a purified gas. (4) The samples are introduced into the calorimeter and maintained under experimental conditions, at ambient temperature and after a preliminary evacuation, with a very simple charging device very similar to that previously designed for Calvet calorimeters. (5) The features of the calorimetric detector, and of the mixing and stirring systems, located within the experimental chamber, are strongly dependent on the nature of the experiment. (6) The data treatment is made easier by a data acquisition system coupled to the calorimeter.

The equipment mentioned in parts (1)–(4) of the above description, purchased from Setaram (Lyon, France), was used without any further modification.

On the other hand, in order to obtain significant values of the enthalpies of formation for several mixtures, it was necessary to modify the calorimetric detector and to design original mixing and stirring systems.

Figure 15 is a vertical section of the sensing device used for the detection of thermal changes: C_1 and C_t are the laboratory and reference crucibles, respectively. These alumina crucibles are placed vertically and coaxially in the center of the experimental chamber. The gastight alumina tube T constitutes the external shield of the experimental chamber.

The middle part of each crucible is in contact with the junctions of the sensing thermocouples, J_1 and J_2. The tube D acts both to maintain in the proper place the ring supporting the thermocouples and as a thermal screen to the assembly.

The whole assembly is suspended in the central part of the furnace by three hollow alumina tubes (H) which protect the junction wires of the thermocouples.

A systematic study showed that, with this kind of calorimeter, the two following conditions are very important in order to obtain reproducible and accurate values of the thermal changes upon mixing: (1) a high sensitivity of the detector, and (2) an appropriate integration of the thermal flux produced within the laboratory cell by the heat change upon mixing. The former condition is best met by using a sufficient number of thermocouples, and the latter by having a regular arrangement of the thermopile junctions around the laboratory crucible.

Several calorimetric detectors were built in accordance with these conditions. In one detector, shown in Fig. 16, the 16 upper junctions of the thermocouples were located alternately on two horizontal levels about 10 mm apart. For

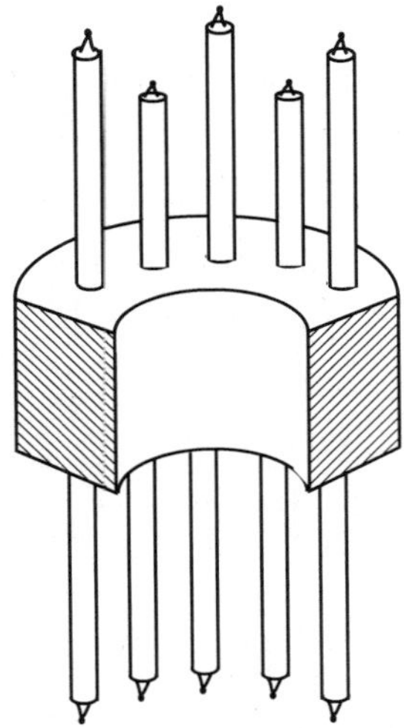

FIGURE 16. Vertical section of the perspective of a calorimetric detector with junctions evenly distributed on two levels.

technical reasons, and since no thermal excursion arises within the reference crucible (C_t), the lower junctions were on the same horizontal level. The other detector assembly (Fig. 17) has more thermocouples (25–66). These arrangements were obtained using alumina insulating tubes with six channels. The construction is somewhat different from that of the detector in Fig. 16. The alumina tubes are cemented directly on the ring instead of within holes bored into the ring.

The advantages of these detectors with respect to the commercial version, shown in Fig. 18, with its 16 thermocouples located on the same level, are obvious. First, the detector shown in Fig. 16 was calibrated by dropping NBS α-alumina into the crucible C_1 filled with increasing amounts of liquid gold. This procedure ensured a smaller dependence on the liquid level, and hence on the position of the heat source. Second, the sensitivity of the detector shown in Fig. 17 can be four times higher than that of the commercial version.

The choice of a detector is made in accordance with the requirements of the experiment to be carried out. For instance, the determination of a limiting partial enthalpy, obtained from a small amount of A added to a liquid bath B, does not require a detector similar to that shown in Fig. 16 since the bath level is almost unchanged during the course of the experiment.

Owing to its simple configuration, the so-called "drop method," already described, is usually used to obtain the enthalpy of formation of a liquid mixture AB. However, it should be stressed once more that the enthalpy change during liquid–liquid mixing is calculated as the difference between two rather large quantities, thus introducing large errors limits. For instance, numerically, the

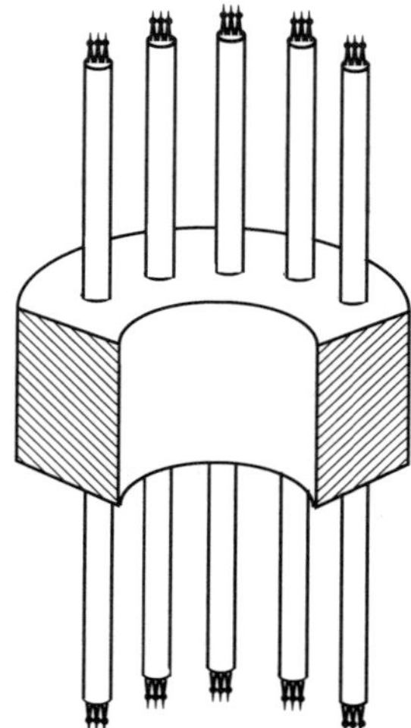

FIGURE 17. Vertical section of the perspective of a calorimetric detector with 66 junctions.

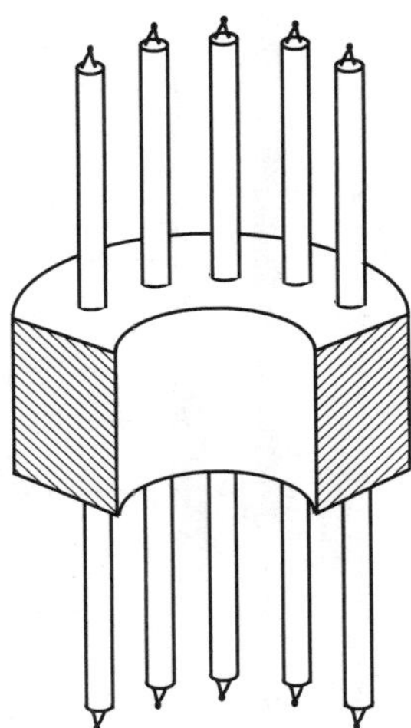

FIGURE 18. Vertical section of the perspective of a commercially available calorimetric detector with only 16 junctions distributed on the same level.

enthalpy of fusion of silicon ($\Delta H_{\text{fus,Si}} = 50.54$ kJ/mol) is significantly higher than the maximum value of the enthalpy of mixing of most silicon-based alloys. The same situation pertains for molten ionic mixtures such as K_2SO_4–NaF,[45] for which the enthalpy of mixing is found to be 6.7 kJ/mol at $x_{\text{NaF}} = 0.6$ and the correction term about 250 kJ/mol.

In order to improve the results, a necessary procedure is therefore to eliminate or severely limit the correction term ΔH_c. This can be achieved if the samples A are introduced into the liquid bath B at a temperature equal or close to the experimental temperature T_E.

The "indirect drop method" was adapted for use with this kind of calorimeter, as shown schematically in Fig. 19. The sample A, contained in the charging system at temperature T_O, is dropped into the calorimeter and remains in the funnel (F) thermostated at temperature T_F. This temperature T_F is measured with a Pt–6% Rh /Pt–30% Rh thermocouple located within the small alumina sphere (S) (6 mm diameter) which closes the aperture of F. When the sample A reaches thermal equilibrium, the funnel (F) is opened by lifting the thin alumina rod (M). The mixing process takes place in C_1. The temperature T_F, and therefore the magnitude of the correction term, depends on the position of the funnel in the tube (T). The funnel (F) may be attached with a small alumina wedge at a distance from crucible C_1 which can be adjusted, for example, at temperatures equal to or less than T_E.

When the temperature of the funnel, T_F, is less than the fusion temperature of sample A, $T_{\text{fus,A}}$, the experiment is generally quite facile. On the other hand, many factors have to be taken into account to obtain reliable results when $T_F >$

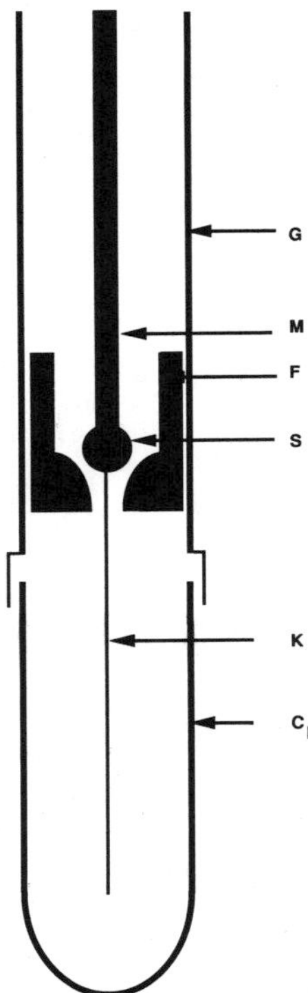

FIGURE 19. Adaptation of the indirect drop method. G, Alumina drop tube; M, alumina rod; F, funnel; S, alumina sphere; K, stirrer; C_1, laboratory crucible.

$T_{fus,A}$. The liquid A should have suitable physicochemical properties (vapor pressure, viscosity, chemical reactivity with respect to the container, surface tension, etc.) to remain in the funnel F during thermal stabilization and to completely flow out when F is open. This can be achieved with molten metals, but these conditions are very difficult to fulfill with molten salts since they generally completely creep out before F is open.

For some mixing experiments, a large difference between the densities of the components A and B restricts obtaining a homogeneous mixture AB. In these cases, a stirring device completes the previous system of mixing. A thin alumina rod (K), sufficiently long to dip into the liquid bath, is added to S and acts as a stirrer. The vertical movement of M homogenizes the liquid AB. Such a device was used to investigate the ternary alloys Al–Ge–Si[46] (Si is added to the liquid Al + Ge) and the ionic mixture $LiF–K_2SO_4$.[47]

In order to improve the efficiency of this calorimeter while facilitating the processing of experimental data, a system of data acquisition was assembled for a more complete automation of the apparatus. The data obtained are directly processed so as to yield directly the enthalpy of mixing.

The energy Q_p, evolved at constant pressure into the calorimeter cell, can be measured according to Tian[8]: the increase of the crucible temperature gives rise, through the thermocouples, to an emf E, which is recorded against time [$E = f(t)$]. The area S contained between this curve (thermogram) and the baseline (experimental zero) is proportional to the energy amount Q_p. This area S is found by the integration of the experimental signal $E = f(t)$ of the thermopile from $t = t_0$ to $t = t_1$, where t_1 is the time when the electrical signal E has attained the constant value E_0 equal to the electrical signal before the specimen has been added ($t = t_0$):

$$Qp = KS = K \int_{t_0}^{t_1} |\, E(t) - E_0 \,| \; dt$$

where K is a constant which has to be determined by the calibration of the system at that particular temperature T_E.

The classical method for evaluation of the integral is the measurement of the area of the thermogram by integration with a manual planimeter (e.g., OTT planimeter). Another way to determine the integral is by the trapezoid method: the value of the function to be measured is determined at given short time intervals. The area is given by the sum of the successive values of ($E - E_0$), if $E(t_1) = E_0$ and the time intervals are equal during calibration and experiment and if the time interval is short compared with the time of one mixing reaction.

An automatic data acquisition system must be able to measure the thermo-emf values with constant frequency, to store the data, and to make simple calculations for baseline corrections described later on. For this purpose, we have used an Apple II (or any PC-compatible) microcomputer combined with a timing unit and a 12-bit A/D converter, with five input channels. A Preston preamplifier is used as a linear amplifier of the thermo emf's for each channel. A program, directly in assembler language, has been established for the simultaneous operation of four calorimeters (two high-temperature Calvet-type and two very high temperature calorimeters).

The signal sources, that is, the thermocouples, were connected to the preamplifiers, and the output of the preamplifiers was fed by the scanner to the A/D converter. The time base and the A/D converter were programmed to scan the five channels at a specified rate and to take a mean value over a period of 3 s. To the fifth channel, a coded signal was sent to identify the calorimeter in which a reaction was initiated or was proceeding.

The voltage U at that channel is generated by an operational amplifier and four switches in the following manner:

$$U = U_1 + U_2 + U_3 + U_4 + U_0$$

where U_1 is a constant voltage yielding a signal a little greater than necessary for a digital reading of 100, U_0 is a small offset voltage to avoid negative input voltages, and the other voltages U_j are adjusted to $U_j = 2(j - s)U_1$. Immediately before a reaction in a calorimeter is started, the corresponding switch is turned to the "ON" position, adding the voltage U_j to U. By the first two digits of the voltage input signal measured on channel 5, all calorimeters with reactions running are easily identified. The program developed provides a definition of the baseline by the measurement of the dwell period value of the emf of the thermopile. Since the signal switch is turned on after the emf has become constant, a period of 3 s after switching is sufficient for the computation of the baseline. After that, the integration procedure is started. There is also built in a correction method for the integral values by means of triangles or rectangles, if the value of the thermo emf after reaction is different from the pre-experimental baseline value.

At fixed time intervals, the measured values (or the difference from the baseline), the time passed since the start, the integral value $[=\Sigma_i (E_i - E_0)_i]$, and the integral values corrected by the two classical integration correction methods are printed.

All improvements described above enabled us to measure the enthalpies of formation of many metallic and ionic mixtures such as Au–Sn,[48] Ag–Au–Pd,[49] Al–Ge–Si,[46] NaF–Na_2SO_4 and NaF–K_2SO_4,[50] NaF–Rb_2SO_4,[51] ZrF_4–MF (M = Li, Na, K, Rb),[52] and ternary ZrF_4-based melts[53] over a temperature range between 1000 and 1500°C. Whenever possible, we compared our results with those obtained from other calorimetric methods.

In the course of a critical analysis[54] of all the calorimetric data published on the enthalpy of mixing of the Au–Sn system, we showed that the difference is less than 2% between two sets of measurements at 1000 K obtained either with a Calvet microcalorimeter or with the apparatus described above. Although at higher temperatures the few data available for other binary alloys do not make possible such a comparison, this good agreement evidenced the reliability of the method.

Very recently, a new automatic sample adder was designed and developed which enables a completely automated operation of the very high temperature enthalpimeter.[55] It allows a complete experimental run with successive addition of 30 samples; each individual mixing experiment is computer operated and, as described above, calorimetric data are also automatically integrated.

It should be noted, however, that no apparatus or method can be considered universal in the domain of high-temperature calorimetry, and adaptations should always be made in accordance with the particular requirements of the system under investigation.

5. Non-Reaction Calorimetry

5.1. Enthalpy and Heat Capacity Determinations

Heat content measurements at high temperature are generally easier than enthalpy of formation determinations. The enthalpy increment of a substance between the temperatures T_1 and T_2 (with $T_1 < T_2$), $[H(T_2) - H(T_1)]$, is generally measured by drop calorimetry. Two general techniques are employed, depending on whether measurements are carried out at high temperature (T_2) or at low temperature (T_2):

1. The sample is heated at high temperature (T_2), and the actual heat content measurement is performed in a calorimeter at the experimental temperature T_1.
2. The sample is at low temperature (T_1), and the actual heat content measurement is performed in a high-temperature calorimeter at the experimental temperature T_2; this is the so-called "reverse drop method."

This method should be preferred in principle for melts since nonequilibrium final states can be obtained on cooling.

Figure 20 shows the enthalpy increment $[H(T) - H(298)]$ measured for a hypothetical solid and liquid sample. The enthalpy of fusion can be deduced as the difference between the enthalpy of the liquid and of the solid at the fusion temperature T_F. The heat capacities of the solid and the liquid, $C_{p,s}$ and $C_{p,l}$ can also be obtained from the temperature dependence of $[H(T) - H(298)]$.

Although there is no fundamental difference between the calorimeters used for the investigation of chemical reactions and those for transitions such as transformations or fusion of compounds or precipitation from a supersaturated solution, the construction is usually adapted to a particular type of reaction, and heats of transition may therefore be treated separately. It should be added that enthalpies of transformation, in particular, of second-order transformations, may also be obtained with specific heat methods.

Using a high-precision adiabatic-shield drop calorimeter, Holm *et al.*[56] measured the enthalpy increments $[H(T) - H(298.15 \text{ K})]$ of the congruently melting 2:1 and 1:1 compounds of alkali chlorides and magnesium chloride; from the results obtained at several experimental temperatures corresponding to solid and liquid samples, they determined the enthalpies of fusion of these compounds. Values of the heat capacities for the molten salt mixtures were also derived.

Two samples of each of the compounds were loaded into platinum containers of known mass. The containers were evacuated carefully inside a

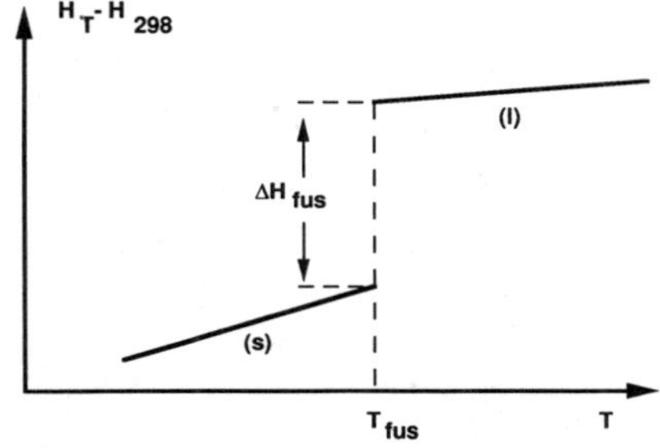

FIGURE 20. Determination of the enthalpy increment for a hypothetical liquid and solid sample.

glovebox to remove the air. The glovebox was filled with purified nitrogen. After evacuation, the containers were filled with purified argon. They were then sealed by arc-welding a cup-shaped platinum lid to the rim of the container.

The sample was equilibrated in a vertical laboratory furnace and lifted into the silver calorimeter, which was placed above the furnace. The calorimeter was surrounded by silver shields, electrically heated to maintain quasi-adiabatic conditions. The furnace temperature was measured by a Pt/Pt–10% Rh thermocouple and the calorimeter temperature by a quartz thermometer.

Steady-state conditions were usually obtained after 10 to 20 min, depending on the furnace temperature. The calorimeter temperature during the period of experiments ranged from 300 to 330 K with a mean of 315 K. The heat capacity values for the compounds were estimated from those of the binary compounds by the relation:

$$C_p = nC_p(\text{AlkCl}) + C_p(\text{MgCl}_2), \qquad n = 1 \text{ or } 2$$

For enthalpy determinations, different techniques such as differential thermal analysis (DTA) and differential scanning calorimetry (DSC) can be used. The purpose of these differential techniques is to record the difference between the enthalpy change which occurs in a sample and that in some inert reference material when they are both heated. These systems may be classified[57] into three types as follows: (1) classical DTA, (2) Boersma DTA, and (3) DSC. The most important differences between these three types are illustrated in Figs. 21 and 22.

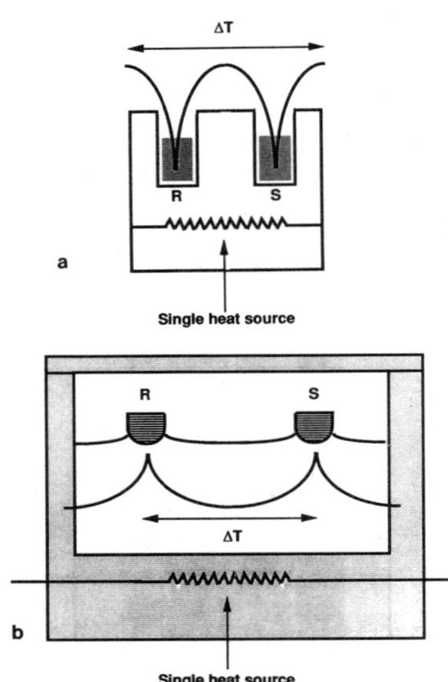

FIGURE 21. Schematic representation of two principal differential thermal analysis systems: (a) classical DTA; (b) Boersma DTA.

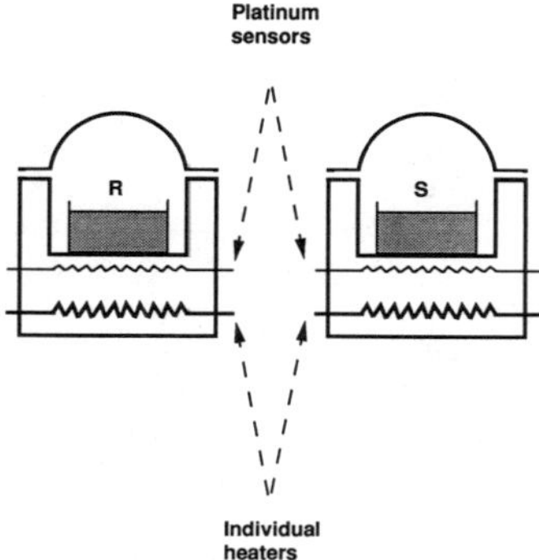

FIGURE 22. Schematic representation of a differential scanning calorimeter.

In the classical and Boersma DTA systems (Fig. 21), both sample and reference are heated by a single heat source. Temperatures are measured by sensors embedded in the sample and reference material (conventional DTA) or attached to the cells which contain the material (Boersma DTA). The thermal signal obtained is the temperature difference, $\Delta T = T_s - T_r$, between sample and reference, and it is usually graphically recorded against time (thermogram). The magnitude of ΔT, at a given time, is proportional to (1) the enthalpy change, (2) the heat capacities, and (3) the total thermal resistance to heat flow, R. High sensitivity requires a large value of R, but unfortunately the value of R depends on the nature of the sample and the extent of thermal contact both between the sample and the cell, on the one hand, and between this cell and the holder, on the other. The effects of variations in the thermal resistance caused by the sample itself are significantly reduced in the Boersma method, owing to attachment of the temperature sensors to the cells (Fig. 21).

With neither of these DTA systems, however, is it possible to make a simple conversion of the peak area, from a plot of ΔT against time, into energy units. This is because of (1) the need to know the heat capacities and (2) the variation of R, and hence the calibration constant, with temperature. Therefore, results obtained from DTA techniques should only be considered as semiquantitative (see, for instance, Fig. 23a), and heat flow differential enthalpic analysis (DEA; see, for instance, Fig. 23b) or DSC should be preferred.

The important difference between the DTA and DSC systems is that in the latter the sample and reference are each provided with individual heaters (Fig. 22). By means of electronic devices, the temperature of the sample holder is always kept the same as that of the reference holder by continuous adjustment of the heater power, whatever the temperature difference between the sample and reference (because of exothermic or endothermic reaction in the sample). The upper limit of DSC apparatus is generally 750°C.

Details of the theory and design of differential scanning calorimeters have been given by Watson *et al.*[58] and by O'Neil.[59]

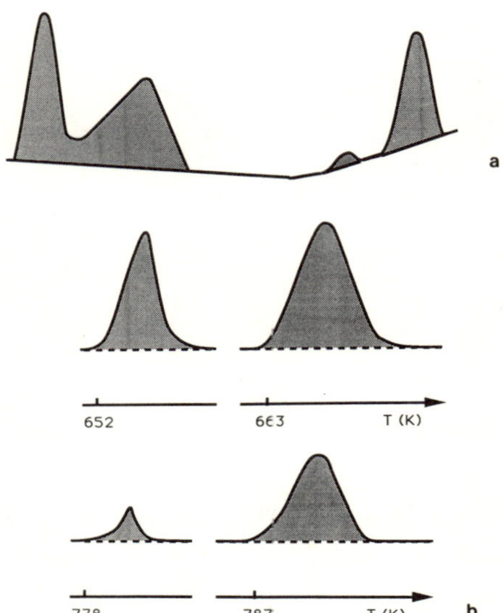

FIGURE 23. Phase transitions in Na_2UBr_6: (a) DTA thermogram; (b) DEA thermograms.

In addition to the determination of enthalpies, DSC gives access to heat capacities. When a sample material is subjected to a linear temperature increase, the rate of heat flow into the sample is proportional to its instantaneous heat capacity. By regarding this rate of heat flow as a function of temperature and comparing it with that for a standard material under the same conditions, we can obtain the heat capacity, C_p, as a function of temperature. The procedure has been described in detail and a precision for specific heat determinations of 0.3% claimed in some cases.[60]

Figure 24 shows the principal method of heat capacity determination. Empty cells are placed in the sample and reference holders. An isothermal base-line is recorded at the lower temperature, and the temperature is then programmed to increase over a range. An isothermal baseline is then recorded at the higher temperature as indicated in the lower part of Fig. 24. The two baselines are used to interpolate a baseline over the scanning section, as shown in the upper part of Fig. 24. The procedure is repeated with a known mass of sample in the sample cell, and a trace of dH/dT versus time is recorded. The deviation from the base line is due to the absorption of heat by the sample, and we may write:

$$dH/dT = mC_p(dT_p/dt)$$

where m is the mass of the sample, C_p the heat capacity, and (dT_p/dt) the programmed rate of temperature increase.

While the above equation would yield values of C_p directly, in order to minimize experimental errors, the procedure is repeated with a known mass of

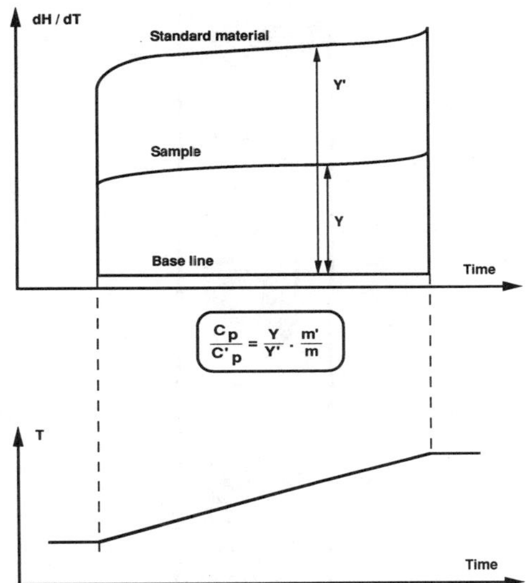

FIGURE 24. Heat capacity determinations by ratio method.

standard sample, the heat capacity of which is well established. Thus, only two ordinate deflections at the same temperature (Y and Y' in Fig. 24) are required to yield a ratio of the C_p values of the sample and standard.

A Calvet microcalorimeter can also be operated as a differential enthalpic analyzer if a linear temperature programmer is added to the temperature regulator; although its thermal inertia is such that the heating speeds do not exceed some degrees an hour, this inconvenience is largely compensated by the great sensitivity and the large experimental volume. The cell used for DEA is shown in Fig. 25. This microcalorimetric technique is very reliable and most efficient when the sample under investigation undergoes several phase transitions at temperatures very close together: it is then possible to separate the resulting thermal effects (Fig. 23b), while conventional DTA would only give overlapping peaks (Fig. 23a). Several uranium(IV) halides and compounds of alkali metals and uranium(IV) halides were successfully investigated[35,61—64] using such a technique.

A detailed exposition of such thermal methods has been given by Franzosini and Sanesi in Volume 1 of this series.

5.2. Calorimetric Determination of Phase Diagrams

Several physicochemical methods make it possible to determine the equilibrium lines of a phase diagram. The most extensively used so far is thermal analysis.[18]

Isobaric determination of phase equilibria in binary or multicomponent

FIGURE 25. Microcalorimetric cell used for DEA.

systems is based on either the temperature dependence or the concentration dependence of a physicochemical parameter. Figure 26a shows a very simple type of condensed binary system; the characteristic shape[65] of the thermal analysis thermogram (Fig. 26b) is obtained by following the thermal effects arising from a linear temperature variation against time (isopleth *XX'*); points *a'* and *b'* correspond to the intersections of the equilibrium lines *AE* and *ES* of the

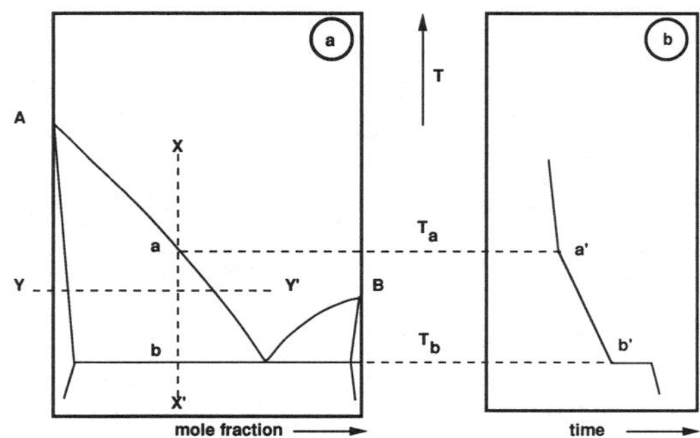

FIGURE 26. Phase diagram determination by thermal analysis: (a) hypothetical binary phase diagram; *XX'*, isopleth; *YY'*, isotherm; (b) corresponding thermogram.

phase diagram. In the same way, the equilibrium points c and c' can be detected by following the change in a physical parameter such as electrical conductivity or viscosity.

The choice of a suitable method for the determination of equilibrium lines depends on many factors such as the physical and chemical properties of components (reactivity, volatility, etc.) and the nature of the equilibrium phase diagram. The determination by thermal analysis, for instance, of a liquid miscibility gap or of a steep liquidus line is rather critical, and other methods seem more suitable.

In many cases, however, calorimetry, and microcalorimetry particularly, is able to provide valuable information; for example, it can be used either to obtain directly equilibrium points or to calculate indirectly some equilibrium lines of a phase diagram. This will be shown in the following.

The principle of the method is as follows. For the sake of clarity, the present work has been limited to binary and ternary mixtures, but generalization to multicomponent systems can be readily accomplished; at constant temperature and pressure, it is relatively easy to measure the enthalpy of formation ΔH_M of a liquid, single-phase mixture A–B.[19–22,66–70] Measurements are performed for mixtures of different compositions, and the plot of ΔH_M versus the mole fraction x_B of component B has a nearly parabolic shape. When, at the experimental temperature, more than one phase exists at the mole fraction x_B, then the enthalpy curve exhibits a characteristic shape which can easily be explained by the "lever rule."

Figure 27 gives an idea of the shape of the $\Delta H_M = f(x_B)$ curve obtained at several temperatures and for different kinds of liquidus. Only three kinds of fairly simple equilibrium diagrams have been chosen for this theoretical example (diagrams a–c in Fig. 27). The dashed lines on the figures correspond to the experimental temperatures T_1 and T_2 for which, in the whole concentration range, the mixtures are either single-phase (T_1) or two-phase (T_2). Diagrams a′–c′ in Fig. 27 show the shape of $\Delta H_M = f(x_A)$ curves obtained at T_1 and at T_2. We used the simplifying hypothesis, which is very often accepted, that enthalpy of mixing does not change with temperature. Each liquidus crossing corresponds to an angular point (g-g', h-h', etc.) on the $\Delta H_M = f(x_A)$ curve. The break points h' and i' refer to the intersection of $\Delta H_M = f(x_A)$—a quasiparabolic curve for a liquid single-phase region—with the straight line $h'i'$. The segment $h'i'$ arises from the linear variation (lever rule) of the amounts of the conjugated liquids L_1 and L_2, the enthalpies of formation of which are ΔH_M^{L1} and ΔH_M^{L2}. For the second diagram, the linear parts correspond to the existence of the compound A_xB_y. This microcalorimetric determination can therefore provide not only the coordinates of equilibrium points but also the thermodynamic quantities of formation of these mixtures, quantities which will be useful in the calculation of equilibrium phase diagrams. This technique is generally completed by differential thermal analysis performed with the same apparatus and with the same sample.

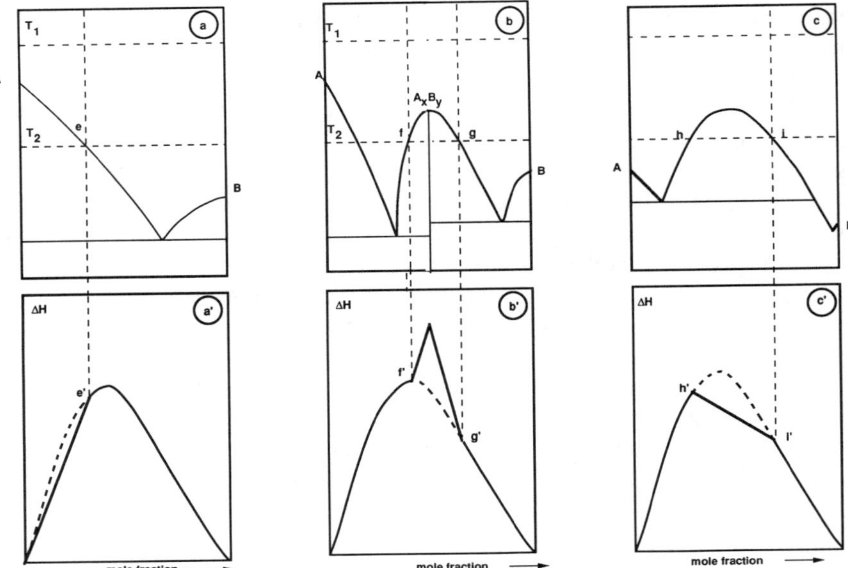

FIGURE 27. Schematic representation of the composition dependence of the enthalpy of formation of multiphase mixtures: (a) simple binary system; (b) binary system with a definite compound; (c) binary system with a liquid miscibility gap.

Isothermal calorimetry allows the enthalpy variations corresponding to the formation of single-phase liquid mixtures of many-phase liquid–liquid and sol-id–liquid mixtures at the same time to be measured directly and also, in some cases, the equilibrium temperatures of the liquidus to be determined. We showed that these measurements are obtainable over a wide temperature range for two- or three-component ionic and metallic mixtures; obviously, this process can be extended to *n*-component mixtures, with the usual difficulties of graphical representation.

Moreover, in the absence of a good knowledge of Gibbs enthalpies of formation and postulating some simplifying hypothesis, the liquidus lines or surfaces of the equilibrium phase diagram can often be calculated from such experimental results. If the temperatures calculated in this way agree with those obtained experimentally, it is then possible to propose a set of consistent values for the mixture considered.

Differential scanning calorimetry can also be used to determine phase diagrams. This is illustrated in Fig. 28. The A–B binary system is a hypothetical mixture exhibiting the following features:

- Component B exists in two allotropic forms.
- The compound A_xB_y melts incongruently.
- Solid solubility occurs in the A-rich region.

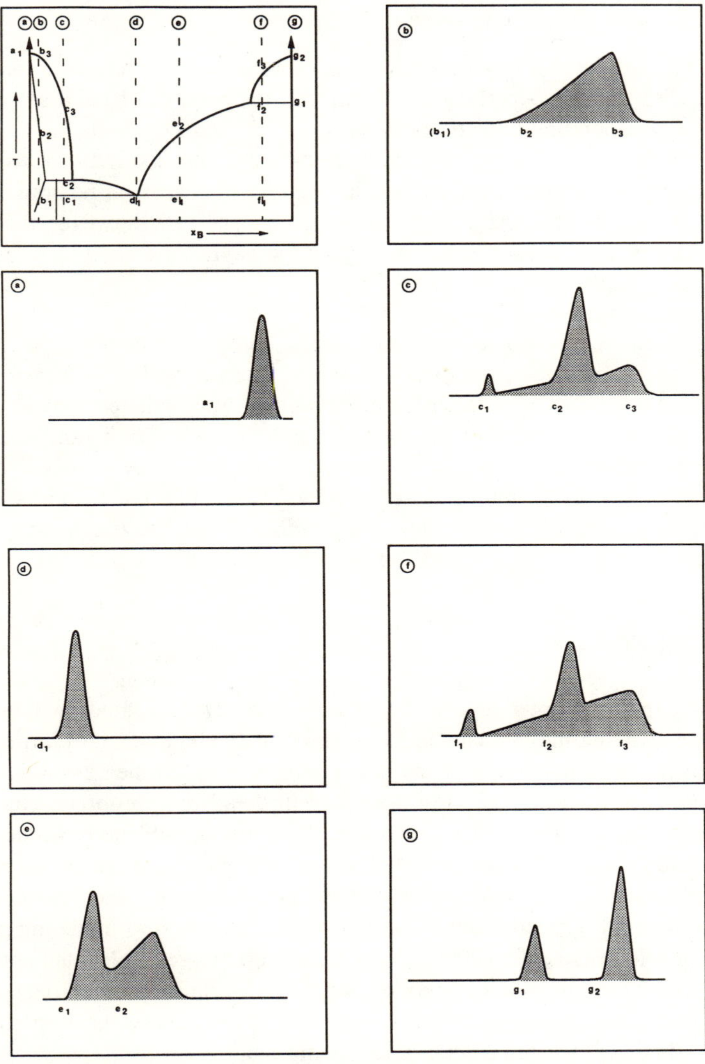

FIGURE 28. Phase diagram determination by differential scanning calorimetry. Thermograms (a) to (g) pertain to several mixtures corresponding to different compositions of the hypothetical A–B phase diagram.

Seven thermograms (a–g) are given as examples. They correspond to:

- fusion of pure A (a_1)
- solid-state transition and fusion of pure B $(g_1$ and $g_2)$
- fusion of the eutectic mixture E (d_1)

In the noneutectic mixtures corresponding to thermograms c, e, and f, the peaks c_1, e_1, and f_1, occurring at the eutectic temperature, have a smaller area than the corresponding eutectic peak d_1.

In the mixture corresponding to thermogram f, the first melt forms isothermally at the eutectic temperature and its amount increases on heating until the liquidus is reached at f_3. At the transformation temperature, however, the amount of pure B still present as a solid undergoes the isothermal solid-state transition indicated by peak f_2.

Thermogram e is self-explanatory.

The features of thermogram c are similar to those of thermogram f; in this case, however, peak c_2 corresponds to the isothermal peritectic transformation into liquid and mixed crystals of the amount of A_xB_y still present as a solid.

Finally, in thermogram b, the melting process begins at b_2 and ends at b_3, while the solid-state reaction denoted by b_1 can hardly be detected by DSC or, generally speaking, by a thermal method (other physical methods such as, for instance, electrical conductivity measurements are more sensitive for this purpose).

6. Conclusion

The present chapter reports the main calorimetric methods which can be used to investigate molten salts and their mixtures. The essential purification of melts has not been discussed in any detail here, since each melt system demands individual consideration; this was provided in detail in appropriate chapters of previous volumes of this series. Of course, it was not possible to cover all aspects of high-temperature calorimetry and all fields of applications. As in any research field, calorimetric experimentation with molten salts can yield results of both fundamental and applied interest. The former aspect covers modeling and theoretical developments of molten salts as high-temperature liquids; the latter deals with the many applications of molten salts as materials. These aspects are not independent, and molten salt technology should not ignore the important theoretical advances that have been or are being made.

References

1. F. D. Rossini, *Experimental Thermochemistry*, Interscience, New York (1956).
2. H. Von Wartenberg, *Z. Anorg. Chem.* **269,** 76 (1952).
3. H. Siemonsen, *Z. Elektrochem.* **55,** 327 (1951).

4. H. Zeumer and W. A. Roth, *Z. Phys. Chem.* **A173**, 365 (1935).
5. O. Kubaschewski and E. L. L. Evans, *Metallurgical Thermochemistry,* 3rd ed., Pergamon Press, London (1958).
6. F. Sauerwald and F. Fleischer, *Z. Elektrochem.* **39**, 686 (1933).
7. O. J. Kleppa, *J. Phys. Chem.* **59**, 175 (1955).
8. A. Tian, *J. Chim. Phys.* **20**, 132 (1923).
9. M. E. Melnichak and O. J. Kleppa, *J. Chem. Phys.* **52**, 1790 (1970).
10. R. Baboian, D. Laing, and S. N. Flengas, *Can. J. Chem.* **45**, 383 (1967).
11. L. S. Hersh and O. J. Kleppa, *J. Chem. Phys.* **42**, 1309 (1965).
12. S. N. Flengas and T. R. Ingraham, *J. Electrochem. Soc.* **106**, 714 (1959).
13. H. Bloom and S. B. Tricklebank, *Aust. J. Chem.* **19**, 187 (1966).
14. E. Calvet and H. Prat, *Microcalorimétrie,* Masson, Paris (1955); E. Calvet and H. Prat, *Récents progrès en microcalorimétrie,* Dunod, Paris (1958).
15. F. E. Wittig and F. Huber, *Z. Electrochem.* **60**, 1181 (1956).
16. P. Picker, C. Jolicoeur, and J. E. Desnoyers, *Rev. Sci. Instrum.* **35**, 676 (1968).
17. M. Laffitte, F. Camia, and M. Coten, *Bull. Soc. Chim. Fr.* **8B**, 43 (1968).
18. M. Gaune-Escard and J. P. Bros, *Thermochim. Acta* **31**, 323 (1979).
19. O. Kubaschewski and E. L. L. Evans, *La thermochimie en métallurgie,* Gauthier-Villard, Paris (1964).
20. H. Eslami, Thesis, Université de Provence, Marseille (1976).
21. M. Gaune-Escard, Thesis, Université de Provence, Marseille (1972).
22. H. Aghai-Khafri, Thesis, Université de Provence, Marseille (1974).
23. P. Leydet, Thesis, Université de Provence, Marseille (1962).
24. G. N. Papatheodorou and O. J. Kleppa, *Z. Anorg. Allg. Chem.* **401**, 132 (1973).
25. J. L. Holm and O. J. Kleppa, *Inorg. Chem.* **6**, 645 (1967).
26. T. Østvold and O. J. Kleppa, *Inorg. Chem.* **8**, 78 (1969).
27. B. K. Andersen and O. J. Kleppa, *Acta Chem. Scand. Ser. A* **30**, 751 (1976).
28. R. Fehrmann, M. Gaune-Escard, and N. J. Bjerrum, *Inorg. Chem.* **25**, 1132 (1986).
29. *JANAF Thermochemical Tables,* 2nd ed., National Bureau of Srandards, Washington, D.C., (1971); Suppl. (1974).
30. W. Lukas, M. Gaune-Escard, and J. P. Bros, *J. Chem. Thermodyn.* **19**, 717 (1987).
31. J. W. Johnson, W. J. Silva, and D. J. Cubicciotti, *J. Chem. Phys.* **69**, 3916 (1965).
32. D. Chiotti, G. Gartner, E. Stevens, and Y. Saito, *J. Chem. Eng. Data* **11**, 571 (1966).
33. Z. Benkhaldoun, Thesis, Université de Provence, Marseille (1985).
34. G. Hatem, M. Gaune-Escard, J. P. Bros, and T. Østvold, *Ber. Bunsenges. Phys. Chem.* **92**, 751 (1988).
35. Y. Fouque, M. Gaune-Escard, W. Szczepaniak, and A. Bogacz, *J. Chim. Phys.* **75**, 360 (1978).
36. W. A. Fischer and G. Lorenz, *Arch. Eisenhüttenw* **27**, 375 (1956).
37. A. Navrotsky, *Phys. Chem. Miner.* **2**, 89, (1977).
38. J. L. Holm and O. J. Kleppa, *Am. Mineral.* **51**, 1608 (1966).
39. J. L. Holm and O. J. Kleppa, *Am. Mineral.* **53**, 123 (1968).
40. D. G. Fraser and Y. Bottinga, *Geochim. Cosmochim. Acta* **49**, 1377 (1985).
41. F. Mueller and O. J. Kleppa, *J. Inorg. Nucl. Chem.* **35**, 2673 (1973).
42. T. Mitsuhashi and Y. Fujiki, *Thermochim. Acta* **88**, 177, (1985).
43. M. Gaune-Escard and J. P. Bros, *Can. Met. Q.* **13**(2), 335 (1974).
44. G. Hatem, P. Gaune, J. P. Bros, F. Gehringer, and E. Hayer, *Rev. Sci. Instrum.* **52**, 585 (1981).
45. G. Hatem and M. Gaune-Escard, *J. Chem. Thermodyn.* **11**, 927 (1979).
46. J. P. Bros, H. Eslami, and P. Gaune, *Ber. Bunsenges. Phys. Chem.* **85**, 333 (1981).
47. G. Hatem and M. Gaune Escard, 9th Experimental Thermodynamic Conference, London, April 16–18, 1980.
48. E. Hayer, K. L. Komarek, J. P. Bros, and M. Gaune-Escard, *Z. Metallkde,* **72**, 109 (1981).
49. J. M. Miane, M. Gaune-Escard, and J. P. Bros, *High Temp.-High Pressures* **9**, 465 (1977).
50. G. Hatem, M. Gaune-Escard, and A. Pelton, *J. Phys. Chem.* **86**, 3039 (1982).

51. G. Hatem and M. Gaune-Escard, *J. Chem. Thermodyn.* **16,** 897 (1984).

52. G. Hatem, F. Tabariès, and M. Gaune-Escard, *Thermochim. Acta* **149,** 15, (1989).

53. K. Mahmoud, Thesis, Université de Provence, Marseille (1989).

54. E. Hayer, K. Komarek, J. P. Bros, and M. Gaune-Escard, 4th International Conference on Liquid and Amorphous Metals, Grenoble, July 7–11, 1980.

55. D. El Allam, Thesis, Université de Provence, Marseille (1989).

56. J. L. Holm, B. J. Holm, B. Rinnan, and F. Grønvold, *J. Chem. Thermodyn.* **5,** 97 (1973).

57. Thermal Analysis Newsletter, No. 9, The Perkin-Elmer Corporation, Norwalk, Connecticut (1970).

58. E. S. Watson, M. J. O'Neil, J. Justin, and N. Brenner, *Anal. Chem.* **36,** 1233 (1964).

59. M. J. O'Neil, *Anal. Chem.* **36,** 1238 (1964).

60. M. J. O'Neil, *Anal. Chem.* **38,** 1331 (1966).

61. A. Bogacz, W. Wisniowski, Y. Fouque, J. P. Bros, and M. Gaune-Escard, *J. Cal. Anal. Therm.* **XIV,** 339 (1983).

62. A. Bogacz, J. P. Bros, Y. Fouque, M. Gaune-Escard, and W. Szczepaniak, *J. Chem. Soc., Faraday Trans. 1* 80, 2935 (1984).

63. Y. Fouque, J. P. Bros, M. Gaune-Escard, M. Wisniowski, and A. Bogacz, *Ber. Bunsenges. Phys. Chem.* **89,** 777 (1985).

64. J. P. Bros, M. Gaune-Escard, W. Szczepaniak, A. Bogacz, and A. W. Hewat, *Acta Crystallogr. Sect. B,* **43,** 113 (1987).

65. G. Tamman, *Z. Anorg. Chem.* **37,** 303 (1903); *Z. Anorg. Chem.* **45,** 24 (1905); *Z. Anorg. Chem.* **47,** 289 (1905); see also J. E. Ricci, *The Phase Rule and Heterogeneous Equilibrium,* Dover, New York (1966).

66. O. J. Kleppa, *J. Phys. Chem.* **61,** 1120 (1957).

67. O. J. Kleppa, *J. Phys. Chem.* **64,** 1937 (1960).

68. O. J. Kleppa, *J. Phys. Chem.* **65,** 843 (1961).

69. J. P. Bros, Thesis, Université de Provence, Marseille (1968).

70. G. Hatem, Thesis, Université de Provence, Marseille, (1980).

6

Ultra-High-Pressure Experimental Techniques

Q. Williams and R. Jeanloz

1. Introduction

Experiments on ionic solids and liquids under high compressions have been of continuing interest since the pioneering experiments of Bridgman in the early and middle part of this century.[1] High pressures are known to produce both structural and electronic (bonding) changes in ionic materials. For example, increases in coordination number, and thus atomic packing fraction, are known to occur within many crystalline alkali halides and alkaline earth chalcogenides at pressures from about 10^9 to nearly 10^{11} Pa.[2,3] Such structural changes also have been inferred to occur within molten salts at comparable pressures, with the evidence resting primarily on measurements of melt volumes and changes in the pressure dependence of the fusion temperature.[4–6] In addition to their effects on both liquid and crystal structures, ultrahigh pressures can induce dramatic changes in the bonding character of ionic materials. Alkali halides that are initially insulating at 1 bar, such as CsI, RbI, KI, and CsBr, undergo pressure-induced bandgap closure, with CsI ultimately metallizing at a pressure in excess of 100 GPa (1 Mbar).[7,8] Again, comparable changes in electronic transport properties have been inferred to occur in compressed liquid CsI.[9,10]

The purpose of this chapter is to summarize the experimental techniques that are available for ultra-high-pressure (P)–temperature (T) research. The "ultrahigh" regime we loosely define as temperatures of order 10^3 K being achieved in samples that are at pressures in the range 10^{10}–10^{11} Pa. Our focus on these conditions is motivated by the fact that the energy density achieved in condensed matter at pressures of 10^{11} Pa is comparable to, or even well in excess of, typical

Q. Williams • Institute of Tectonics, University of California, Santa Cruz, California 95064. *R. Jeanloz* • Department of Geology, University of California, Berkeley, California 94720.

193

bonding energies (~ 1 eV $\approx 10^2$ kJ/mol of atoms).[11] Thus, major changes in atomic packing and in interatomic bonding forces are expected, and indeed observed at ultra-high pressures. The simultaneous generation of high temperatures in samples while at high pressures is emphasized here in the context of research on molten materials.

Our intention is not to provide a detailed and exhaustive review of high-pressure techniques, but to give a summary overview that emphasizes recent developments and capabilities, especially at the highest P/T conditions now used in experiments. The lower-pressure range below ~ 1 GPa, including the use of gas-loading apparatuses and ultra-large-volume (1 liter) presses, is discussed elsewhere (Spain and Paauwe, 1977[12]; Ulmer, 1971[13]; Edgar, 1973[14]; Ulmer and Barnes, 1987[15]; Bridgman, 1949[1]; Holloway and Wood, 1988[16]). We note that further details can be found in numerous conference proceedings, including those of the International Association for Research and Advancement of High Pressure and Technology (AIRAPT Proceedings[17-20]), the European High Pressure Research Group (EHPRG[21,22]), the U.S.–Japan Seminars on High-Pressure Research,[23-25] and the American Physical Society's Topical Conferences on Shock Waves in Condensed Matter.[26-28]

2. Overview

There are two basic approaches to generating ultra-high pressures: dynamic and static. As illustrated in Fig. 1, the dynamic approach involves producing a high-pressure shock wave by impacting a projectile into a target that contains the sample. The duration of the experiment is short, typically less than 1 μs, and it is limited by the time required for the shock wave to pass through the entire target. In contrast, elevated pressures are sustained indefinitely in static experiments. These are based on squeezing a sample between anvils, either directly or indirectly via a pressure medium (e.g., a fluid or soft solid) that surrounds the sample. We further subdivide the static techniques between those that accommodate comparatively large sample volumes and those that do not. The former include the large-volume multi-anvil designs recently pioneered in Japan,[29-31] as well as the more conventional piston–cylinder apparatus,[32] whereas the small-scale experiments are exemplified by the diamond-cell technique.

The distinctions between the various experimental techniques, shock wave (dynamic), large-volume static, and diamond cell (small-volume static), are summarized in Fig. 2 and Table I. Clearly, there is a trade-off between the peak pressures that can be attained, the sample volume, and the experiment duration. This can be readily understood in that the force F required to achieve a peak pressure P in a sample of characteristic linear dimension L is given approximately by

$$P \sim FL^{-2} \tag{1}$$

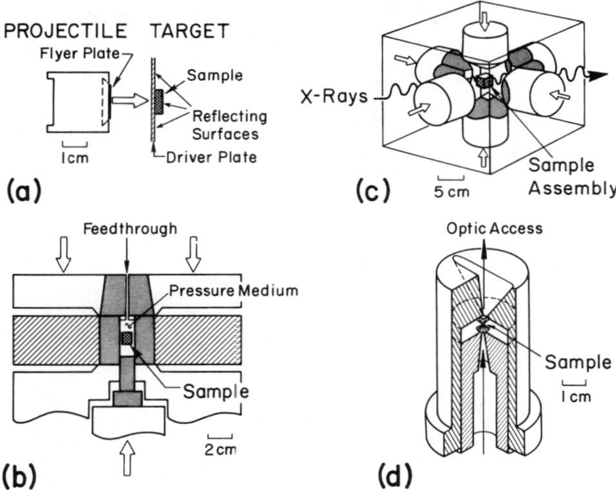

FIGURE 1. Schematic illustrations of three ultra-high-pressure techniques: dynamic (a), large-volume static (b, c), and diamond cell (d) (open arrows indicate force or impact). (a) Shock-wave (Hugoniot) measurements consist of a metal plate (flyer plate) embedded within a projectile impacting a stationary target at velocities of the order of kilometers per second. The target includes a sample attached to the back of a metal plate (driver plate). [36] (b) Piston–cylinder experiments involve compressing a pressure medium (either liquid or solid) by advancing the piston within the cylinder (shown here in cross section), thereby pressurizing the sample. Typically, hard metals (e.g., WC) are used for the piston and cylinder, although sintered diamond has also been used for the end of the piston. A narrow feedthrough hole, appropriately sealed within the cylinder, allows electrical leads to be brought into the sample area from the outside. The leads are used for monitoring temperature and pressure and for bringing current to a resistance heater around the sample (not shown).[32] (c) Cubic anvil design of large-volume press has been extensively used in Japan since the late 1970s.[32] The sample assembly often consists of further (pressure intensifying) anvils, gasketing material, pressure media, calibration standards, and the sample. In this design, X rays and electrical leads can pass through gasketing material between the six anvils, thus traversing the entire pressure vessel and sample assembly along a diagonal. The anvils are typically made of steel, with WC or sintered diamond tips. (d) Diamond cell of the Mao–Bell ("Megabar") design [85] consists of two gem-quality diamond crystals being pressed together in a piston–cylinder assembly (shown here in cutaway view). The sample is placed between the points of the two diamonds and can be directly viewed through the diamond anvils. X rays and other forms of radiation can be used to probe the sample along the same optical path. The piston and cylinder are typically made of metal (e.g., hardened steel with WC backing plates behind the diamonds) and are placed within a lever-arm assembly or an opposed-anvil press in order to apply force to the diamonds.

Although Eq. (1) neglects frictional effects and the pressure dependence of material strength, it does illustrate that increasing P requires either an increase in F or a decrease in L (proportional to $V^{1/3}$, where V is sample volume) or both. It is evident from Fig. 2 that the peak pressures currently achieved in large-volume presses require considerably higher applied forces than do the pressures generated in the diamond cell, and this is reflected in the large dimensions of these apparatuses in comparison to the diamond cell (cf. Refs. 29 and 149).

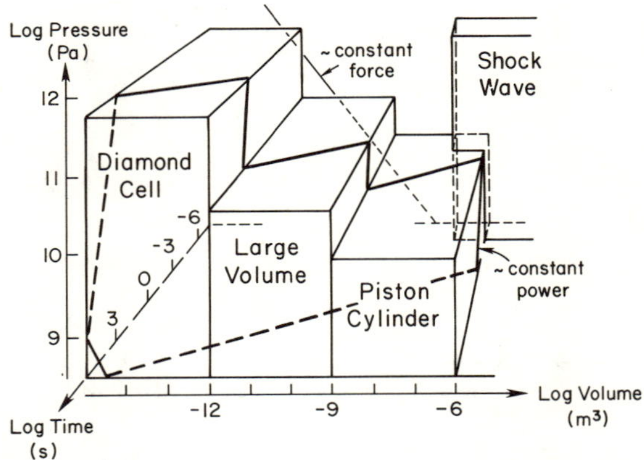

FIGURE 2. Relationship between peak pressures, sample volumes, and duration of experiments achieved in diamond-cell, large-volume, and piston–cylinder devices and in shock-wave experiments. The straight line on the back face indicates the increasing sample volume that can be accommodated with decreasing peak pressure, assuming constant force being applied to the sample (2.2 kN, neglecting frictional forces). The inclined plane indicates a power of 1 W per unit strain of the sample being dissipated in elastic compression. Although the time scale is, in principle, indefinite, the three static techniques are limited in time by the typical duration of laboratory experiments ($\sim 10^6$ s).

Similarly, the effect of time is illustrated by the power required for a given experiment. We take the power to be proportional to the elastic energy change in the sample (again placing frictional and strength effects within constants of proportionality), so that

$$\text{Power} \sim PL^3 t^{-1} \sim FLt^{-1} \tag{2}$$

Here, t is the duration of the experiment, and the volume change of the sample is taken to be proportional to L^3. Thus, for a given power dissipation or applied force, sample dimension and pressure attained trade off with experiment duration: for example, either small samples can be compressed indefinitely (diamond cell), or large samples can be compressed transiently (shock wave) to achieve the highest pressures (Fig. 2).

The main advantages of the shock-wave and diamond-cell techniques are that they span the largest P–T range that is currently accessible (Fig. 3). The particular drawbacks of each of these, short duration and small sample size, are mitigated by specific advantages. For example, the pressure–volume–internal energy (P–V–E) state achieved under shock loading can be directly measured, as opposed to being indirectly calibrated. Also, in both shock-wave and diamond-cell experiments, the sample can be observed *in situ* while at ultra-high pressures and temperatures. In comparison, the large-volume static techniques offer neither the advantages of direct observation of the sample nor those of generating the

TABLE I. Ultra-High-Pressure Techniques

Technique	Advantages	Disadvantages
Dynamic		
Shock waves	Ultra-high P–T achieved	Short duration:
	Absolute P,V,E measurement	Disequilibrium
	Large sample	Difficult measurement
		P,T not independently controlled
		Heterogeneous, nonhydrostatic at low P
		Expensive technique
Static		
Large volume	Large samples: synthesis	P–T range limited
	Diminished P–T gradients	No (poor) direct observation of sample
		Expensive installation, maintenance
		Indirect P–T calibration
Diamond cell	Ultra-high P–T achieved	Small sample volume
	Direct sample observation	Indirect P calibration
	Inexpensive technique	

highest P–T conditions now possible, yet they also do not suffer from the disadvantages of either small samples or brief experiments (Table I). Finally, both diamond-cell and large-volume static techniques require that the sample pressure be independently calibrated. Thus, because the sample pressure is obtained directly in shock-wave measurements, these play a special role in calibrating ultra-high-pressure static experiments.

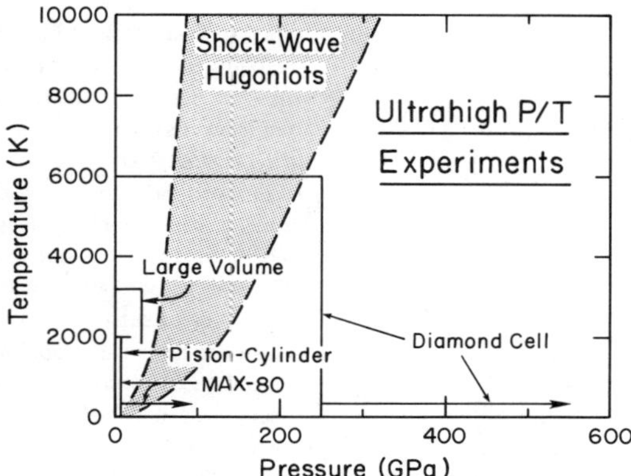

FIGURE 3. Pressures and temperatures that can be achieved with ultra-high P–T techniques. The Hugoniot (shock-wave) temperatures depend on the nature of the initial sample material: only an approximate range of P–T conditions is indicated by the shading. Typically, much higher pressures have been achieved in static experiments at 300 K (arrows) than at high temperatures; MAX-80 is a specific, cubic-anvil-type large-volume press (Fig. 1c) that has been developed in Japan.[31]

3. Shock-Wave Techniques

3.1. Hugoniot Relations

The principle behind shock-wave experiments is illustrated in Fig. 4. Upon impact of a projectile into a stationary target, large-amplitude stress waves are sent into both the target (forward-traveling wave) and the projectile (backward-traveling wave) at velocities exceeding the speed of sound in the uncompressed material. At sufficiently high impact velocities, typically over 1 km/s, these shock waves bring the compressed region to a stress well above the dynamic yield strength of the material. Thus, the shock-compressed material effectively acts as a fluid under one-dimensional loading (uniaxial strain, prior to unloading), and the relevant conservation relations are hydrodynamic in nature.[33-35] In the following, we describe in sequence the mass-, momentum-, and energy-conservation relations.

A volume of unit cross-sectional area and of length U_s, the shock velocity in the target, is encompassed in unit time by the forward-traveling shock (Fig. 4,

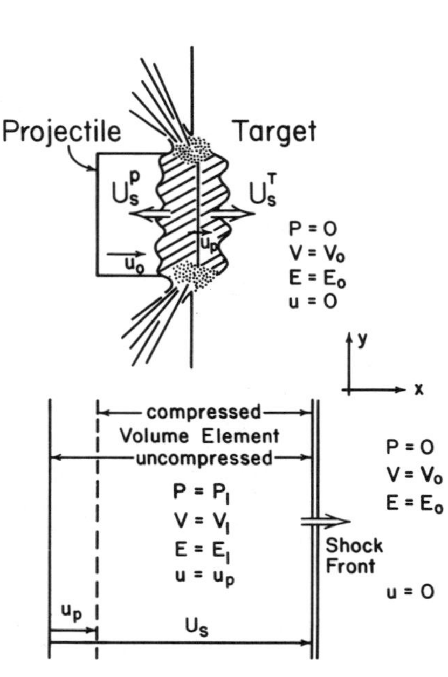

FIGURE 4. Illustration of a shock experiment. In two dimensions (*top*), the impact of a projectile traveling at velocity u_0 into a stationary target generates shock waves traveling forward into the target (at velocity U_s^T) and backward into the projectile (shock velocity U_s^P). The shock fronts are schematically shown as wavy lines, but are in fact very nearly planar. In the compressed region (hatched), the projectile is decelerated to the particle velocity u_p, which is also the velocity to which the target material is accelerated: constant particle velocity ensures continuity across the projectile–target interface (the driver plate has been left out for simplicity). Wherever the shocked region encounters vacuum, a rarefaction or release wave is generated which tears the material apart (stippled region). The rarefaction front eats into the shocked region and creates the outward spray of material. In one dimension (*bottom*), a volume element of unit cross-sectional area is shown as it is encompassed by a shock wave (double line) traveling at velocity U_s to the right. The material is initially at rest (velocity $u = 0$) and zero pressure in front of the shock wave, and it is accelerated to velocity u_p and taken to pressure P_1 behind the shock. The initial volume (V_0) and energy (E_0) are decreased and increased, respectively, to V_1 and E_1 across the shock front (from Jeanloz[11]).

bottom). Therefore, the mass in this volume element prior to compression is $\rho_0 U_s$, if $\rho_0 = 1/V_0$ is the uncompressed density. After passage of the shock front, the material is accelerated from its initially stationary state ($u = 0$) to a finite particle velocity u_p, and this causes the initial volume element to be compressed to a length $U_s - u_p$. With a compressed density $\rho_1 = 1/V_1$, the mass in the volume element after passage of the shock front must equal that prior to compression, by conservation of mass:

$$\rho_0 U_s = \rho_1 (U_s - u_p) \tag{3}$$

Setting the initial pressure, P, equal to 0, the pressure in the shock-compressed region (P_1) is simply given by the force per unit area or, equivalently, the mass in the volume element ($\rho_0 U_s$) times its acceleration (u_p) per unit time. Therefore,

$$P_1 = \rho_0 U_s u_p \tag{4}$$

and the compressed material now has a kinetic energy relative to the material in front of the shock equal to one-half its mass times its velocity squared, or

$$\text{KE} = \frac{\rho_0 U_s u_p^2}{2} \tag{5}$$

The internal energy change on shock loading is thus given by the $P-V$ work done on the material minus the kinetic energy acquired during the shock process: the fact that the particles are now moving at velocity u_p does not affect the internal thermodynamic state of the compressed material. Note that because there is essentially no heat flow on the time scale of the shock compression (rise times or "widths" of shock fronts are of order 10^{-9} s), the process is truly adiabatic. However, because of the irreversible nature of shock compaction, it is not isentropic. As the volume change corresponds to the particle velocity (u_p) per unit area (Fig. 4, bottom), the $-P\,dV$ work is simply $\rho_0 U_s u_p^2$ from Eq. (4). Combining with Eq. (5), the internal evergy change is therefore

$$E_1 - E_0 = -P\,dV - \text{KE} = \frac{\rho_0 U_s u_p^2}{2} \tag{6}$$

This means that half the work done by the impact goes into increasing the internal energy of the target, and the other half is lost to the kinetic energy associated with accelerating the target material to velocity u_p.

Equations (4)–(6) are known as the Hugoniot relations and describe the $P-V-E$ state achieved upon shock compression produced by a given impact velocity and given projectile and target materials. The set of such high-pressure states that are produced in a given target material (e.g., with increasing impact velocity) is labeled the Hugoniot. This is not a thermodynamic path, but it is a well-defined equation of state that is obtained under dynamic loading. In order to measure the Hugoniot, three of the six variables in Eqs. (4)–(6) must be measured ($E_1 - E_0$ is treated as a single variable).

A typical experimental configuration, illustrated in Fig. 1a, involves a metal flyer plate impacting a metal driver plate on the back of which is positioned the sample.[36] If light is reflected off the back surfaces of the driver plate and sample, it can be directed to a streak camera that records the destruction of these surfaces as the shock enters and exits the sample. From this observed travel time of the shock wave through the sample and the previously measured thickness of the sample, U_s may be determined with better than 1% precision. Additionally, the initial density (ρ_0) of the sample is determined prior to the experiment.

The third variable that is usually measured is the particle velocity. If the sample and flyer plate are made of the same material, and if the driver plate is ignored, then symmetry dictates that $u_p = u_0/2$. The impact velocity u_0 is readily measured, so u_p can also be obtained to better than 1%. Therefore, the Hugoniot equation of state is entirely determined by the measurement of ρ_0, U_s, and u_p (via u_0). Using Eqs. (3) and (4), it can be expressed either as a $P–V$ relation or as a relationship between the directly measured variables U_s and u_p.

The purpose of the driver plate is not only to hold the sample in place but, more significantly, to provide a finite thickness of material within which the shock wave can reach a steady profile after its generation at the impact surface. As the flyer and driver plates are typically made of the same material, the impact velocity directly yields u_p in the driver, as just described. To obtain the particle velocity in the sample, the Hugoniot of the driver-plate metal must be known; that is, the form of $U_s^D(u_p)$ must have previously been determined by means of symmetric-impact experiments with identical flyer-plate and target materials. Often, the resulting Hugoniot equation of state is expressed as a linear relation between shock and particle velocities, with a and b being material properties:

$$U_s^D = a + bu_p^D \tag{7}$$

Then, conservation of momentum requires that the pressure be the same in the shock-compressed sample (superscript s) and driver plate (superscript D), or

$$\rho_0^s U_s^s u_p^s = \rho_0^D U_s^D u_p^D = P_1 \tag{8}$$

Finally, the particle velocity in the sample must equal the impact velocity minus the particle velocity in the driver so that continuity is maintained at the sample–driver interface:

$$u_p^s + u_p^D = u_0 \tag{9}$$

Solution of Eqs. (7)–(9), with four variables being measured (u_0, ρ_0^D, ρ_0^s, and U_s^s), yields the particle velocity in the sample, u_p^s, as well as u_p^D and U_s^D in the driver. This approach, called impedance matching, yields the Hugoniot equation of state for an arbitrary sample material.

3.2. Hugoniot Temperatures and Sound Velocities

The temperature of a sample increases under shock loading both because the compression is adiabatic and because it is irreversible. In fact, the dominant contribution to the Hugoniot temperature is the entropy increase across the shock front. Therefore, there is considerable interest in measuring the temperature along the Hugoniot because the irreversible temperature increase is not easy to calculate with reliability. As the internal energy is known from Eq. (6), determining the Hugoniot temperature effectively yields a measurement of the specific heat, $C_v = (\partial E/\partial T)_V$. Similarly, the sound velocity

$$c = \sqrt{(M/\rho)} \tag{10}$$

can be related to the Hugoniot P–V curve: the elastic modulus M is simply the adiabatic bulk modulus $K_s = \rho\,(\partial P/\partial\rho)_s$ if the sample is a liquid, and $M = K_s + 4\mu/3$ for a solid, with μ being the rigidity.

Measurements of temperature and sound velocity have been especially important for determining the pressures at which samples melt under shock loading. As the Hugoniot crosses the melting curve, the latent heat of fusion, ΔH_f, causes the temperature to drop. Thermal energy is used to structurally disorder the system, with the temperature decrease being of the order $\Delta H_f/C_v \approx 10^3$ K (the temperature change associated with the latent heat of fusion at constant volume). Similarly, the sound velocity drops from $c = \sqrt{[(K_s + 4\mu/3)/\rho]}$ in the solid to $c = \sqrt{(K_s/\rho)}$ in the melt. Thus, the combined changes in temperature and sound velocity are diagnostic of melting along the Hugoniot, and these measurements have been of central importance in determining melting temperatures and melt properties at ultra-high pressures.

Hugoniot temperatures are measured by spectroradiometry (Fig. 5). That is, the thermal emission generated at or just behind the shock front is observed through either the unshocked sample itself (if the sample is transparent) or, for opaque materials, through a window placed on the back of the sample. This is facilitated by using a mirror that is destroyed once the shock emerges from the rear surface of the target. The mirror directs the thermal radiation to a spectrometer, which might consist of a diode array on a monochromator. Alternatively, 3 to 8 individual wavelengths may be recorded through separate filters and detectors as a function of time. In either case, the blackbody-like radiation is generally measured at visible to near-infrared wavelengths.

Two factors are of critical concern in shock temperature measurements on opaque materials[37,38]: first, that the interface between sample and window be without any gap or topography; and second, that the material properties of the sample and window are such that their shock impedances ($\rho_0 U_s$) are similar. The presence of any space or locally poor contact between the target and window can result in anomalously high temperatures being generated at their interface. Such temperatures are produced because residual gas and surface contaminants within

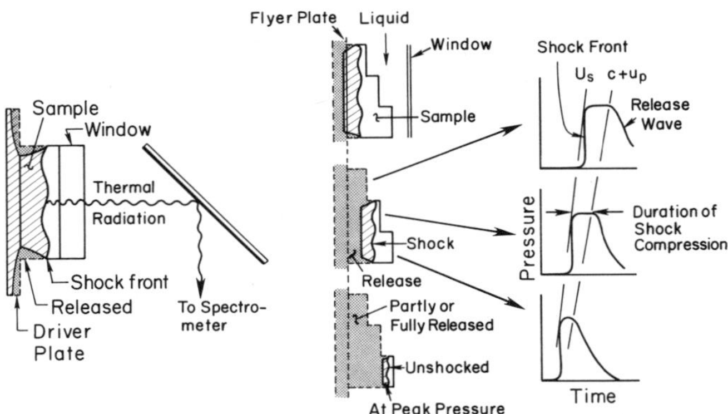

FIGURE 5. Illustrations of shock temperature and sound velocity measurements.[36] *Left:* To measure temperature along the Hugoniot, the thermal radiation (wavy arrow) from the shock-compressed portion of the sample is directed toward a spectrometer by means of a disposable mirror. The temperature is obtained by fitting the thermal radiation at visible wavelengths to a blackbody (Planck) function. Hatching (peak pressure) and stippling (decompressed region) have the same meaning as in Fig. 4, and release interfaces are shown as dashed lines for clarity. The back of the sample is viewed through a transparent window (typically either Al_2O_3 or LiF). If the sample is opaque, only an "interface temperature" is obtained at the moment the shock front reaches the sample–window interface (and begins to be modified by the presence of the interface). *Right:* Sound velocity (c) at Hugoniot P–T conditions is determined by measuring the rate at which the release wave (traveling at velocity $u_p + c$) catches up with the shock front (traveling at velocity $U_s < u_p + c$). This can be done by observing a "stepped" sample, with three or more regions of differing thickness, through a liquid contained behind a window. By using a liquid that luminesces brightly under shock (e.g., bromoform), it is possible to accurately time the arrival of the shock front and forward-traveling release wave through each section of the sample. Alternatively, for transparent samples, several samples of differing thicknesses can be impacted simultaneously by a driver plate, and the decrease in the luminosity of the sample caused by the catch-up of the rarefaction wave with the initial shock wave observed. In order to ensure that this catch-up of the rarefaction wave can be observed, the sample is impacted with a thin flyer plate (for clarity, no driver plate is shown).

the gap can achieve very high temperatures as they are compressed between the back of the sample and the window. Impedance matching of the materials within the target assembly ensures that any rarefaction waves or reflected shock waves produced at the interface are of low amplitude; hence, these do not significantly contribute to the release or reshocking of the sample. Finally, we note that the effect of thermal conductivity between the sample and window may be important. If the shock generates different temperatures within the two materials, as would normally be expected, these begin to equilibrate over the duration of the temperature measurement. The resulting interface temperature that is observed may therefore yield a biased estimate of the Hugoniot temperature of the bulk of the sample.

Sound velocity measurements also are conducted radiometrically but, in these, one is concerned with the location and intensity of the emission rather than its spectral characteristics. In particular, when the flyer plate impacts the target, a shock wave propagates both forward into the sample and backward through the flyer. Wherever the shock emerges from a high-impedance material (e.g., the flyer plate) into a low-impedance material (e.g., the projectile behind the flyer plate), a rarefaction is produced. Thus, a rarefaction (or release) wave is formed that moves from the back surface of the flyer through the compressed flyer–target assembly at a velocity equal to the particle velocity (velocity of the frame of reference) plus the sound speed of the compressed material ($u_p + c$). The reason that the release travels at the sound velocity is that the decompression is usually isentropic (except under special circumstances involving phase transformations[39]).

Measuring the sound speed in the shock-compressed sample is critically dependent on locating the release wave within the sample. Typically, its location is most easily determined when it intersects (catches up with) the slower traveling initial shock wave. Figure 5 illustrates an experimental configuration for measuring sound velocities, with a sample (or samples) of varying thicknesses being impacted.[40] The back of the sample assembly is immersed in a transparent liquid, chosen because of its bright emission when shocked. Thus, the progress of the shock through the sample can be monitored as a function of time by measuring the sample luminosity as a function of position on the stepped sample. From the observed duration of the shock state as a function of sample thickness (Fig. 5), the time at which the release wave overtakes the shock wave, or would overtake it, can be determined. For transparent samples, light from the shock front can be monitored continuously through the unshocked sample material. Hence, no liquid is required behind a transparent sample and the overtake of the shock by the release wave is directly observed as a decrease in emission from the sample.

In such measurements, the sound velocity within the shocked flyer plate must be accurately known in order to get reliable values for the sample. Therefore, many of the measurements of sound velocities under shock have been conducted on typical flyer-plate materials such as Fe, Al, and Ta. For the case in which both the flyer plate and the target are the same material, the sound velocity is calculated from[41]:

$$U_s t_{\mathrm{ovt}}/T_F = (c\rho + U_s\rho_0)/(c\rho - U_s\rho_0) \tag{11}$$

where T_F is the flyer-plate thickness and T_{ovt} is the measured time that the rarefaction wave takes to catch up to the shock wave. Thus, $U_s t_{\mathrm{ovt}}$ is the distance traveled by the shock wave in the sample prior to being overtaken: Eq. (11) results from setting the sum of the travel times of the shock (backward) through the flyer plate and the rarefaction wave (forward) through the flyer plate and sample equal to that of the shock through the sample. Also, Eq. (3) in the form $u_p/U_s = 1 - (\rho_0/\rho)$ has been used to eliminate particle velocity from the final equation. Relations of similar form can be used to derive the sound velocity of the sample material when the flyer plate and sample are different but the flyer plate properties are known.

3.3 Other Shock Techniques

There have been limited applications of other probes to shock-loaded samples. Of particular interest are X-ray diffraction experiments,[42,43] absorption,[44] Raman,[45] and fluorescence[46,47] spectroscopies, and DC electrical conductivity.[48,49] The first three techniques each involve a pulse or flash of light incident on the sample. The geometry of the absorption experiment is similar to that shown on the left of Fig. 5, with the sample–driver interface being coated with a reflecting surface. Thus, in visible absorption experiments, a high-intensity flash (typically from a xenon lamp) is reflected from the rear surface of the shocked material, and the reflected light dispersed spectrographically and observed using a streak camera.[44] The X-ray experiments involve 40- to 50-ns duration pulses of unfiltered copper X rays striking the sample synchronously with the shock front. Using a film detection system, diffraction signals between Bragg angles of 33° and 50° have been recorded.[42] Raman spectra have been measured with a 10-ns, 248-nm (KrF) laser pulse, with the beam oriented along the axis of impact and scattered light collected at 45° from this axis.[45] Fluorescence measurements have involved placing an optical fiber against a window on the rear of a sample and using the fiber to transmit both excitation radiation from its source and the sample fluorescence to a remote detector.[46,47] Using this technique, a sequence of fluorescence spectra can be measured throughout the duration of a shock experiment. Electrical conductivity measurements are made with leads oriented such that current flows either in the direction of shock propagation or parallel to the shock front.[48,49] A circuit of known resistance is placed in parallel with the sample, and a continuous voltage placed across the sample. The DC electrical conductivity of the sample during the shock event is then derived by observing the voltage through the sample relative to the resistor.

4. Large-Volume Presses

4.1. Motivation

The main reason for turning to static high-pressure techniques is that these offer the possibility of carrying out long-term, equilibrium experiments with independently variable pressure and temperature, in contrast to dynamic experiments (Table I). Of the static techniques, the large-volume press has the special advantage of large sample dimensions. This is particularly important in two respects. First, it allows large quantities of a material to be manufactured at high pressures, which is advantageous if the sample quenched to ambient conditions is a material with interesting properties (e.g., superhard materials such as diamond or cubic boron nitride). In addition, calorimetry and other analytical techniques are often carried out more readily on large rather than small quantities of sample material.

The second advantage of large volumes is that they allow temperature gradients to be diminished in ultra-high P–T experiments. This can be seen by considering a sample of linear dimension L ($V \sim L^3$) being heated with a power Q. The condition for steady-state heat flow can then be written as the dimensional relation

$$Q/L^3 \sim kT/l^2 \tag{12}$$

where l is the characteristic length scale on which the temperature T changes significantly across the sample, and k is the average thermal conductivity of the sample assembly.[50] The magnitude of the temperature gradient relative to the sample dimension is therefore given by the nondimensional ratio

$$L/l \sim (Q/LkT)^{1/2} \tag{13}$$

This shows that increasing sample volume produces more uniform temperatures across the sample by an amount which scales as $L^{-1/2} \sim V^{-1/6}$. Thus, high-pressure studies requiring the most uniform thermal conditions possible are best carried out with samples of the largest dimension that can be accommodated in a given P–T regime.

4.2. Types of Large-Volume Presses

The primary differences between types of large-volume presses involve four basic factors. First, the number of anvils used to compress the sample varies between two and eight. Second, different large-volume presses often incorporate different anvil materials: although tungsten carbide has been the most commonly used material, anvils have also been made of high-strength steel, alumina, and sintered diamond. Third, the size of anvils varies dramatically, from blocks with 1-mm edges up to those with 100-mm edges. Finally, the number of pressure-intensifying stages, and the geometry of these stages, differs between large-volume cell designs, depending in detail on the intended pressure range and sample size (Figs. 1c, 6, and 7). Indeed, it is not uncommon to have an arrangement such as a pair of opposed anvils compressed by a six-(cubic) anvil press, with the space surrounding the smaller opposed anvils being filled with a soft, pressure-transmitting medium such as pyrophyllite.

We subdivide the designs into two basic types, as illustrated in Fig. 1. The first, the piston–cylinder device (Fig. 1b), is basically an opposed-anvil uniaxial press. The sample and pressure medium are contained by the cylinder around the pistons, but it is only the pistons that are active elements. The cylinder acts as a passive jacket or gasket. In contrast, the multi-anvil design (Fig. 1c) involves simultaneous compression from several directions: there is no passive retaining element involved. Many specific geometries are possible, and the particular design shown (cubic anvil press) has been taken to higher pressures than older designs such as the tetrahedral anvil press.[51]

The advantage of the piston–cylinder device is that it can be more easily sealed to contain a fluid pressure-transmitting medium than the multi-anvil press. Also, the construction and maintenance of the piston–cylinder press is simpler: the requirement that multiple anvils be aligned and synchronized to compress a sample uniformly is avoided. In contrast, the multi-anvil design is inherently better suited to provide an isotropic, uniform pressure distribution than the axially symmetric piston–cylinder device. The maximum pressure generated in multi-anvil presses (>40 GPa; Fig. 3) far exceeds that generated thus far in the piston–cylinder press (about 8 GPa; R. Boehler, private communication). As there is currently no known fluid medium for pressures above 14 GPa at 300 K[52,53] and as fluid media are less useful for high-temperature experiments than at room temperature (see discussion below), the advantages of the piston–cylinder device are becoming less of a factor in ultra-high-pressure research than is its limited pressure range.

Variations of the two basic designs of large-volume presses are illustrated in Fig. 6. The girdle apparatus is a modification of the piston–cylinder device that can generally achieve higher pressures than the piston–cylinder geometry. The tapered piston shape provides a pressure intensification and a self-sealing in the girdle geometry, but at the expense of large frictional losses (although not achievable in practice, the ideal piston–cylinder device would be friction free). The particular apparatus shown in Fig. 6 (left) has been specifically designed for manufacturing purposes.[54] With a sample volume of 1 liter, it possesses the largest volume yet achieved above 5 GPa. Its maximum $P–T$ range, 6 GPa and 2300 K (continuous heating) to 4300 K (flash heating), is sufficient to produce diamond, cubic BN, and other high-pressure phases that are of technological interest due to their hardness or electronic properties.

The split-sphere device (Fig. 6, right) generates exactly the same forces in

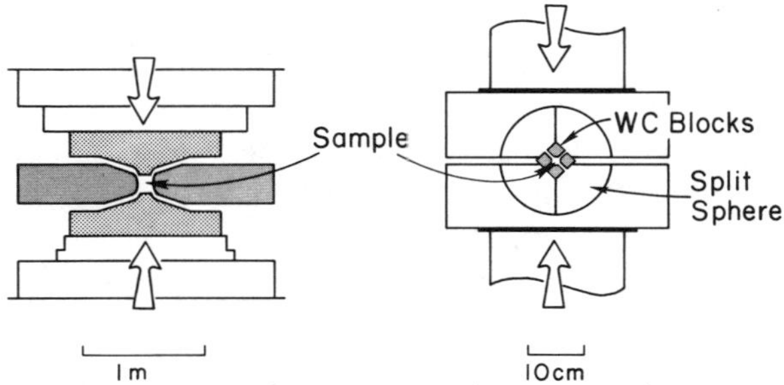

FIGURE 6. Two alternative designs of large-volume presses shown in cross section. *Left:* Girdle or belt-type press, which is a variant of the piston–cylinder design (Fig. 1b). This is an outline of the FB-120 design that accommodates a 1-liter sample inside a 30,000-ton (force) uniaxial press (after Fukunaga *et al.*[54]). *Right:* Split-sphere press can be considered a variant of the cubic anvil device (Fig. 1c). Force on the opposed piston compresses the hardened-steel sphere that has been split into six segments (three above, three below). These act like the anvils in the cubic design in compressing a second stage of anvils, shown here as WC blocks with the sample volume at the center (refer to Fig. 7). The pistons are part of a large (2000-ton force) uniaxial press (after Suito[55]).

the sample assembly as the cubic anvil press (Fig. 1c). It illustrates a geometry which ensures that the six opposed anvils compress the sample uniformly while only using two elements (the split sphere).[55] Its main advantage is that it can be installed inside a large conventional uniaxial opposed-piston press, and its limits, about 2500 K at pressures above 20 GPa, are comparable to the most extreme conditions that have been achieved in the cubic anvil device (about 25 GPa at high temperatures to 40 GPa at room temperature; Refs. 31 and 56–58).

The multi-anvil presses, whether of the split-sphere or cubic design, often feature multiple stages that lead to pressure intensification. Thus, the cubic sample assembly (Figs. 1c and 6, right) can contain eight tungsten carbide blocks acting as a second stage. Each block has one corner removed and is oriented in such a way as to make an octahedrally shaped space between the blocks (Fig. 7a). This space is mainly filled with a solid pressure medium (e.g., poly-

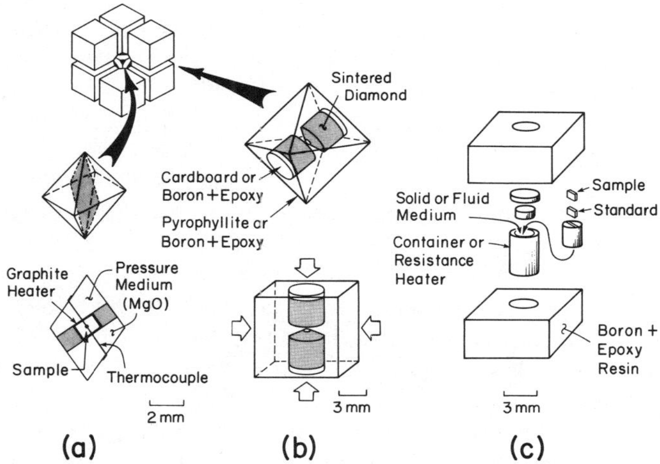

FIGURE 7. (a) Second stage in a large-volume press may consist of eight WC blocks with truncated inner corners that can accommodate an octahedrally shaped sample volume (*top;* cf. Fig. 6). The octahedral sample volume (*middle*) consists of a pressure medium (e.g., pressed MgO) and an inner sample chamber surrounded by a resistance heater (*bottom:* cross section along shaded plane in middle figure). Temperature in the sample is monitored by means of a conventional thermocouple (e.g., Pt/Pt–13% Rh or W/W–26% Re) (after Ohtani *et al.*[56]). (b) Sintered diamond anvils have been placed inside an eight-block second stage of anvils (*top*) or directly inside a cubic anvil press (*bottom*) (after Endo and Ito[57]). Using a boron–epoxy medium, the bottom configuration has been used with the MAX-80 press at the Tsukuba, Japan, synchrotron (Photon Factory) to obtain X-ray diffraction patterns at 300 K to pressures exceeding 40 GPa and at high temperatures but at lower pressures.[31,60] (c) Conventional sample configuration inside a cubic anvil press includes a resistance heater or a capsule (or both) around the sample volume, which may also contain a calibration standard. X rays passed diagonally through the boron–epoxy block (cf. Fig. 1c) can be used to measure the P–V equation of state of the sample in a hydrostatic medium. This arrangement has been used inside the MAX-80 press to obtain X-ray shadowgraphs of Pt and Mo spheres falling through silicate melts at pressures of 2–5 GPa and at temperatures near 1500 K (Kanzaki *et al.*[59]).

crystalline MgO) that compresses the sample and thermocouple assembly inside a resistance heater. Graphite, which has been used as a heating element in the past, is not useful as a heating element above 7–10 GPa because it transforms to diamond. Thus, it is being replaced by other materials, such as lanthanum chromate with titanium carbide electrodes. This combination is stable to at least 28 GPa and nearly 3000 K simultaneous pressures and temperatures.

A further intensification with a third stage has led to the highest pressures yet achieved in a large-volume press. Pressures in excess of 50 GPa have been attained using a system in which sintered diamond opposed anvils are compressed by a set of octahedral anvils with a cube press[57] (Fig. 7b, top). Similarly, sintered diamond anvils have been used directly as a second stage (Fig. 7b, bottom). In both cases, however, the sample volume has been enormously reduced in order to generate high pressures (cf. Fig. 2). It is within this ultra-high-pressure regime that "large-volume" technology becomes similar in size, design, and overall intent to the diamond anvil cell that is described below.

As with the resistance heaters, the increased $P–T$ range achieved in large-volume devices has required changes in pressure media and gasketing material. In this context, the gasketing does not refer to any of the metallic parts (e.g., the girdle in Fig. 6, left) but to the material between the metallic anvils and blocks. The gasketing is intended to compress and extrude plastically, thereby sealing the sample volume or inner stages of the apparatus. Thus, pyrophyllite has been traditionally used because it is easily machined, yet extrudes readily under pressure. At high temperatures and pressures above 8–10 GPa, however, it dehydrates and transforms to a hard, dense mixture of Al_2O_3 and SiO_2-stishovite which extrudes less readily. More importantly, the volume decrease caused by the density change causes the pressure to drop, and it is difficult to increase pressure further. Therefore, pure polycrystalline MgO is used as an ultra-high-$P–T$ pressure medium.[55] At low temperatures, a polycrystalline boron medium (with a small amount of epoxy binder) has been used effectively at high pressures (Fig. 7). This is also readily machinable and is relatively stable under pressure. Finally, the outer gasketing material (e.g., between the WC blocks or between the girdle and anvils in Fig. 6) can be a simple, easily available material that is even unstable at pressure: pyrophyllite, cardboard, and paper have all been used.

In order to achieve truly hydrostatic conditions, a fluid medium must be contained inside the large-volume press. A design consisting of a boron and epoxy second stage that fits inside a cubic anvil press is shown in Fig. 7c. This configuration has been used for isothermal compression measurements by X-ray diffraction at room temperature,[31] the advantage to using boron and epoxy being that they are relatively transparent to high-energy X rays (above ~10 keV). Alternatively, if the container is replaced by a resistance heater, the sample volume can be sufficiently heated to melt the sample entirely. It is in this manner that a relatively large volume (~10 mm³) of liquid can be held at up to 1800 K and 4 GPa (higher pressures with smaller volumes; Ref. 50).

4.3. X-Ray Diffraction and Imaging in Large-Volume Presses

Because of the opacity of large-volume press anvils, the uses of such presses until the mid-part of this decade were largely limited to synthesis of materials, determinations of phase equilibria derived from samples quenched in temperature and pressure, or electrical conductivity measurements. In conjunction with high-intensity laboratory (rotating-anode) and synchrotron X-ray sources, the ability has been developed to conduct X-ray diffraction experiments on samples inside large-volume presses at simultaneous high pressures and temperatures. The particular configuration used in the MAX-80 cube anvil press has already been shown in Figs. 1c and 7c. In either the cubic anvil or split-sphere geometry, the X-ray beam follows the same path, diagonally traversing between the anvils and through the gasketing material, pressure medium, and sample. The diffracted beam can be analyzed with an energy-dispersive detector or a conventional (angle-dispersive) diffractometer. X-ray diffraction has been conducted up to about 40 GPa using this technique at ambient temperature, and to between 5 and 10 GPa at 1300 K.[31,59,60] Additionally, shadow radiography has been applied to measure the rates at which chemically inert spheres sink through melts. That is, the sample is loaded with a sphere, typically made of platinum or molybdenum, at the top of the charge, the sample is heated to above the melting point, and the position of the falling sphere is imaged as a function of time.[59] From the settling velocities of two spheres of different materials in the same composition melt, both the viscosity (η) and the density of the melt are calculated by Stokes' law:

$$\eta = 2r^2 g \Delta\rho / 9v \tag{14}$$

Here, r is the radius of the sphere, g is the gravitational acceleration, $\Delta\rho$ is the density difference between the sphere and the melt, and v is the velocity with which the sphere falls. The position of the sphere as it sinks, and hence its velocity, is monitored as a function of time.

5. The Diamond Anvil Cell

The diamond anvil cell is capable of static pressures of 550 GPa (5.5 Mbar)[61] and, when coupled with laser heating, simultaneous temperatures as high as 7000 K.[62] As such, the diamond cell offers not only the highest pressure range, but also the largest temperature range of any method for generating sustained pressures. Although sample size is generally quite small (typically less than 0.02 mm^3), optic access to the sample inside the diamond cell makes a wide variety of *in situ* spectroscopic probes possible at high pressures and temperatures. Specifically, Raman,[63] Brillouin,[64] infrared,[65] Mössbauer,[66]

N.M.R,[67] fluorescence,[68] and optical absorption[69] and reflectance spectroscopy[70] have all been conducted at pressures in excess of 5 GPa, as have X-ray diffraction,[71] EXAFS,[72,73] positron annihilation,[74] and electrical conductivity[75–77] experiments.

As an excellent review of the technical details of diamond alignment and sample preparation exists,[78] we shall only briefly describe these techniques, and we instead emphasize detailed descriptions of actual high-temperature generation at high pressures within the diamond cell. Furthermore, there are several reviews of the application of the diamond cell to a variety of high-pressure studies.[52,79]

5.1. Diamonds

Because of its high strength and transparency, diamond represents the ideal material for use in opposed-anvil high-pressure cells. Although devices using tungsten carbide opposed anvils enjoyed considerable success in the early and mid-1960s, attaining maximum pressures in excess of 50 GPa,[80,81] the difficulties of using these devices for measurements other than electrical conductivity and occasionally X-ray diffraction limited their applications. As pressure is defined as force per unit area, the small area and high yield strength of the diamond anvil are an ideal combination for generation of ultra-high pressures. In this regard, gem-quality, inclusion-free, single-crystal diamonds are critical for success: any small flaws present within the anvil represent possible nucleation points for fracture at high pressures. Moreover, with clear, impurity-free diamonds, the optical transparency of the anvils is excellent, with the only absorption mechanisms between far-infrared and ultraviolet wavelengths being multiphonon bands at 3 to 6 μm.

Typical anvil sizes vary from 0.2 to 0.75 carat, although synthetic diamonds as small as 0.10 carat have been used.[82] The diamonds are usually cut in the brilliant style, or with a modification of this cut, with the culet point of the diamond removed in order to produce a flat pressure-generating surface having a diameter typically between 150 μm and 1 mm. Smaller culet sizes can generally attain higher pressures. Additionally, the culet may be beveled, making the overall shape close to hemispherical, in which case the central flat may be as small as 20 μm in diameter. Such beveled diamonds are generally used for work at pressures greater than 120 GPa, and they are generally not recoverable without fracturing upon decompression. In contrast, well-aligned diamonds with small flat culets can be reused many times up to pressures of about 130 GPa.

Anvils come in two major varieties, those bearing nitrogen as an impurity (Type I) and those without (Type II). These are further subdivided into Types Ia, Ib, IIa, and IIb. Type Ia diamonds are the most common type of naturally occurring diamond and have approximately 0.1% nitrogen present in small aggregates, including platelets. By contrast, nitrogen in Type Ib diamonds is dispersed substitutionally. Of the two Type II diamond types, Type IIb is semiconducting due to small amounts of boron impurities, whereas Type IIa diamonds are comparatively pure (for a set of comprehensive reviews of diamond material properties, see Ref. 83). Although Type I diamonds are appropriate for many high-pressure experiments, absorption due to the nitrogen impurities makes the region of the infrared between wavelengths of 6 and 13 μm difficult to access. Furthermore, nitrogen centers may be responsible for anvil fluorescence, which is sometimes observed under laser excitation at pressures greater than 100 GPa. Therefore, although Type II diamonds are approximately twice as costly as Type I diamonds of equivalent size, they are necessary for some spectroscopic and some ultra-high-pressure experiments. We note, however, that the presence of the nitrogen impurities is thought to strengthen Type I diamonds relative to Type II anvils at pressures above 1 Mbar.[84]

5.2. Design of the Diamond Cell

A variety of cell designs exist, yet all designs must fulfill two principal criteria. First, it must be possible to precisely align the diamond anvils; and second, the cell must maintain that alignment under high stresses. These two criteria are necessitated by the brittleness of diamond single crystals; if the two culets are initially misaligned, or become misaligned at pressure, a nonuniform stress is placed on the anvils. Such misalignment can produce a single crack or fracture or, in cases of extreme failure, crushing of one or both anvils into powder.

One example of a cell fulfilling both criteria, the Mao–Bell "Megabar" design,[85] is shown in Fig. 1d. The two diamonds are aligned by a combination of translation and tilting of tungsten carbide half-cylinders onto which each diamond is glued (not shown). The piston and cylinder of the diamond cell have either a small angled hole or a slit over which each diamond is attached; these allow optic access to the sample. The two diamond culets are optically aligned to better than 1 μm in translation. Similarly, tilt alignment is monitored by observing when the optical fringes (Newton's rings) between the opposed diamond culets vanish. Maximum deviations from parallelism by this procedure are less than 300 nm across the diameter of the culet. An apparatus of the type shown in Fig. 1d has been used to generate pressures in excess of 500 GPa (5 Mbar). However, more modest cell designs exist: the Merrill–Bassett cell[86] has attained pressures of up to 28 GPa.[87] Such cells rely simply on the simultaneous tightening of three bolts and use no rotational alignment, as the diamond is usually seated on a steel, beryllium, beryllium–copper, or tungsten carbide disk which can only be translationally aligned by set screws.

5.3. Sample Preparation

Many early experiments using the diamond cell simply compressed a sample between the two anvils, producing large pressure gradients across the sample. More recently, usage of a metal gasket, a foil with a hole containing the sample, has become accepted procedure (Fig. 8).[68] This sample configuration tends to minimize the pressure gradients across the sample and to allow the enclosure of hydrostatic and quasihydrostatic media between the diamonds. These gaskets, between 100 and 500 μm in initial thickness, are generally made of steel, such as hardened T-301 stainless steel and spring steel, or high-temperature alloys (e.g., Inconel X-750) or, alternatively, metals chosen for their incompressibility such as rhenium, tungsten, or tantalum. These foils are often preindented in the diamond cell by being pressurized up to as high as 50 GPa: this process produces an imprint of the diamonds on the foil and makes the drilling of a hole centered on the diamond imprint possible with a micro-drill press.[78] The size of the holes is, depending on the experiment, between 50 and 500 μm.

5.4. Pressure Media and Measurement

The most widely accepted technique for pressure calibration is to measure the fluorescence of a small amount of ruby (Al_2O_3:Cr^{3+}) included between the diamonds along with the sample.[68,88,89] The particular transition involved is from the split ^2E_g excited state to the ground state ($^4A_{2g}$) levels of trivalent chromium in ruby, which produces the R_1 and R_2 fluorescence lines at 694.2 and

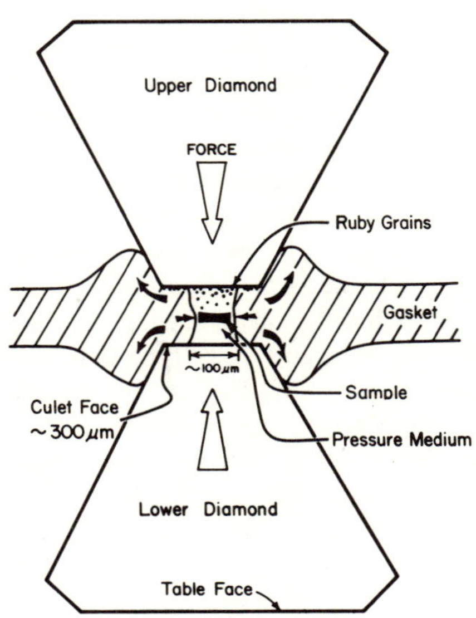

FIGURE 8. Schematic cross section of the diamond-anvil high-pressure cell, illustrating the way in which a sample is contained by a metal gasket while being squeezed between the points (culet faces) of two single-crystal diamonds. The gasket pushes inward on the pressure medium surrounding the sample and also oozes outward to provide support on the sides of the diamond. Ruby grains spread over the sample area are used for pressure calibration. Typical sample and culet dimensions are indicated (from Jeanloz[11]).

692.9 nm, respectively, at zero pressure and 300 K. The energy by which the two lines are separated is nearly independent of hydrostatic pressure.[90] At high levels of differential stress, dramatic line broadening of the ruby fluorescence is observed: from an initial width of the R_1 line of about 0.7 nm at ambient pressure and 300 K, widths of as much as about 50 nm have been reported under extremely nonhydrostatic conditions[91] (although under these conditions the R_1 and R_2 lines are indistinguishable). The shift of the R_1 line has been well calibrated both against the equations of state of metals and against fixed points determined by other techniques.[88,89,92–95] (Fig. 9). Such ruby fluorescence may be excited by a variety of light sources, with argon-ion and He–Cd lasers being among the most common (Fig. 10). Unfortunately, the intensity of the fluorescence declines dramatically at pressures greater than about 70 GPa and is often obscured by fluorescence from the diamonds at pressures above 120 GPa. Also, at high temperatures (greater than about 700 K) the intensity of the R_1 fluorescent line becomes difficult to distinguish from background fluorescence: this effect is produced by the thermally generated increase in intensity of broad vibronic bands surrounding the R_1 and R_2 lines.[96,97] In the temperature range to 700 K, however, the lifetime of the ruby fluorescence, which decreases markedly at high temperatures, may be used in combination with the spectral line position as an internal calibrant of both the temperature and pressure within the sample.[98]

An important difficulty encountered in high-pressure experiments at sub-

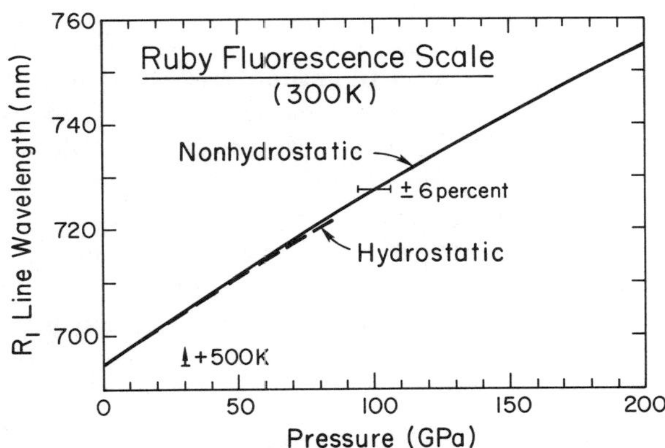

FIGURE 9. Pressure dependence of the R_1 fluorescence line of ruby at 300 K.[92,93] Multiple calibrations against shock-wave and ultrasonic measurements indicate that the ruby fluorescence wavelength can be used as an ultra-high-pressure gauge with an uncertainty of less than 6% and that deviations from hydrostaticity have relatively small effects on the calibration.[92–95] The effect on the wavelength of the ruby fluorescence induced by increasing the temperature by 500 K is shown at the lower left.[97] Note that increasing either pressure or temperature shifts the fluorescence to longer wavelengths.

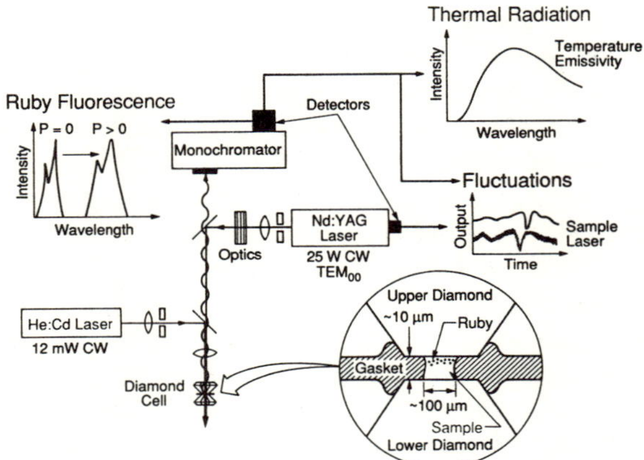

FIGURE 10. Schematic of pressure and temperature measurements in the laser-heated diamond cell. A He–Cd laser beam (*lower left*) is focused into the diamond cell to excite the ruby fluorescence at 300 K (before and after heating). The fluorescence signal (wavy line) is collected through the same microscope that is used to focus the laser and is passed into a monochromator and detector which resolve the ruby fluorescence doublet (*upper left:* the stronger R_1 line accompanied by the weaker R_2 line at shorter wavelengths). For heating, a high-power, continuous-wave Nd:YAG laser beam (*center right*) is filtered both spectrally and spatially and directed via a dichroic mirror through a microscope which focuses the beam onto the sample. Upon absorbing the laser radiation, the sample is heated and emits thermal radiation (wavy line) that is collected through the microscope optics. The emission spectrum at visible and near-infrared wavelengths is obtained through the monochromator and detector, and its shape and amplitude yield the temperature and emissivity of the sample inside the diamond cell (*upper right*). Fluctuations in laser output can be correlated with fluctuations in sample temperature as a function of time.[112]

solidus temperatures is the existence of pressure gradients across the sample. At pressures to 15 GPa, a 16:3:1 mixture of methanol, ethanol, and water remains hydrostatic, while a simple 4:1 mixture of methanol and ethanol is liquid to pressures of 10.4 GPa.[53,99] For work at higher pressures, noble-gas solids such as argon, xenon, or neon are sufficiently weak that they are nearly hydrostatic to pressures of 70 GPa or more.[92,100] Although pressure gradients due to non-hydrostaticity are often present, they are readily characterized if ruby powder is distributed across the sample. The degree of nonhydrostaticity is then quantified by the measured pressure distribution.[101]

5.5. High Temperatures in the Diamond Anvil Cell

Three different methods have been developed for achieving high temperatures within the diamond anvil cell: external heating, internal resistance heating, and laser heating. The first involves placing a furnace either around the diamonds and sample chamber or around the entire cell itself. For example, temperatures exceeding 800 K can be obtained with a circularly wound, resistance-wire furnace approximately 2 cm in diameter placed around the diamonds. The principal limitation of this technique lies in the metastability of diamond, with diamonds heated above 1000 K in air graphitizing. In vacuum, temperatures as high as 1500 K have been attained using this technique.[102] The less common external heating technique involves placing the entire cell within a furnace. The limitations in this case are not only in the metastability of diamond, but also in maintaining the force on the diamonds once the diamond cell is heated. In general, temperatures in the externally heated diamond cell are measured with a thermocouple fixed to the diamonds near the sample chamber.

Although comparatively little work other than visual observations of melting/freezing equilibria has been done on high-temperature liquids using external heating, X-ray diffraction experiments have been performed at pressures to 20 GPa and temperatures of about 1200 K on solid oxides and metals.[102] Also, Raman spectroscopy has been conducted on the liquid diatomics N_2 and O_2 at temperatures to 900 K at 18 GPa.[103,104]

Internal resistance heating involves placing a wire through the diamond cell sample chamber using techniques similar to those described in the next section for electrical conductivity experiments. The wire is typically insulated from the diamonds by a layer of either ruby or of silica glass.[105–107] High current is then run through the wire until it is heated to incandescent temperatures, which can then be measured spectroradiometrically. Temperatures of up to 2400 K have been reported at 40 GPa using this method.[106] One advantage of this technique is that the conductivity of the sample can be monitored at simultaneous high-pressure and high-temperature conditions. However, such resistive heating is only readily conducted on metals. Also, the temperature at the core of the wire is likely to be significantly higher than that observed radiatively at the wire surface, due to the closer proximity of the wire surface to the highly thermally conductive diamond anvils.

Laser heating is capable of temperatures much higher than either external or internal resistive heating, in excess of 7000 K at high pressures.[62,77] The sole requirement is that the radiation, usually from a Nd:YAG or CO_2 laser, be absorbed by the sample. The reason that high temperatures can be attained without decomposing the diamond is due to the extremely high thermal conductivity of diamond, and the comparatively massive volume of the diamond relative to that of the sample. The incident laser beam is absorbed only by the sample, within which thermal gradients develop (as much as 10^3 degrees per micron) in both the directions perpendicular and parallel to the anvil surfaces. Laser heating is achieved with the TEM_{00} laser mode (a Gaussian intensity profile of the beam) so that it is radially symmetric in the plane parallel to the anvil surfaces.

Temperatures under laser heating are measured using spectroradiometry, in which a gray-body model of emissivity is assumed. That is, the Planck radiation function is used to invert the thermal radiation observed from the sample, with the sample emissivity assumed to be constant with wavelength. A laser heating system including an apparatus designed to measure sample temperatures is schematically shown in Fig. 10.[108] Any reflected laser light from the sample is filtered out in a microscope, which focuses the thermal emission from the sample into a remotely controlled scanning monochromator. In these measurements, stability of the incident laser is extremely important: variations of a few percent in laser power can produce changes of hundreds of degrees kelvin in sample temperature.

The spatial variation of temperature across the laser-heated spot within the sample is also of critical importance. One method for characterizing these thermal gradients is by moving a slit across the sample image several times, with the monochromator of Fig. 10 tuned to a different wavelength for each scan. The resulting spectra as a function of slit location are inverted for the temperature distribution across the sample using the formalism of the Radon transform, which treats the slit as a line probe of sample emission.[108,109] Thus, the slit width must be narrow relative to the size of the laser-heated spot. Results of such a measurement are shown at the top of Fig. 11, for an experiment with gradients of the order of 700 K/μm.[109] Notably, finite difference calculations of the

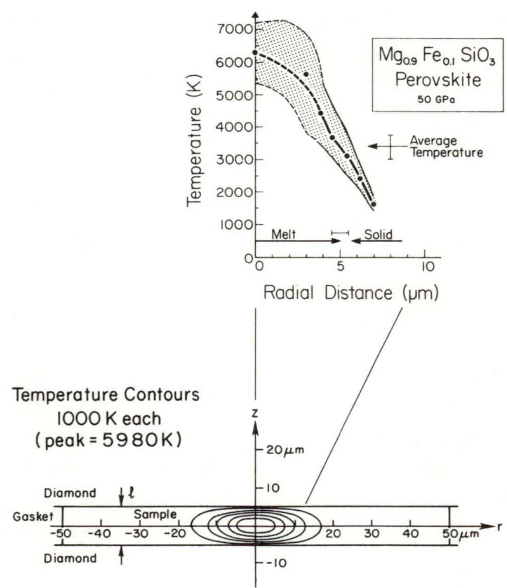

FIGURE 11. Temperature distribution in the laser-heated diamond cell. Results of a model calculation[110] (*bottom:* contours superimposed on a cross section of the sample configuration between the diamond anvils) is compared with the radial temperature distribution observed in an experiment on a silicate melt at 50 GPa[112] (*top:* shading indicates the uncertainty in the values shown by closed circles). The experimental measurement involves the use of a tomographic technique for determining the spatial variation of the temperature in the sample.[108,109]

temperature distribution within the laser-heated diamond cell samples yield results that are in accord with experimental observations[110] (Fig. 11, bottom). In this particular calculation, temperature at the diamond–sample interface was found to nearly equal ambient temperature due to the high thermal conductivity and large thermal inertia of the anvils relative to the sample.

Thus far, the primary applications of laser heating at very high pressures have involved either the synthesis of high-pressure phases or the determination of phase boundaries and, in particular, the variations of the melting temperatures of materials with pressure. For example, the fusion curves of iron,[62] iron sulfide,[111] iron oxide,[77] and $MgSiO_3$ perovskite[112] (Fig. 11) have all been measured to pressures in excess of 60 GPa. The manner in which melting of the sample is determined to have occurred is either by actual observation of fluid motion within the laser-heated spot or by changes in texture (e.g., relief or grain-boundary patterns) in the sample. In many cases, the sample (particularly those that are metallic) reacts with diamond at high temperatures. Thus, samples must be separated from the diamond interface by a layer of nonreactive material. Ruby is, in general, a suitable material for this purpose (it also functions as the pressure calibrant). Other possibilities include magnesium oxide and other refractory ceramics, including the sample material itself under the appropriate conditions.

Finally, we note the difficulties posed by thermal pressure.[112] If the sample remains at constant volume while being heated to high temperatures (i.e., there is no expansion of the sample chamber and no stress relaxation within the sample), then the rise in pressure induced by the increase in temperature is equal to

$$\Delta P = \alpha K_T \Delta T \tag{15}$$

where α is the average thermal expansion, and K_T is the mean bulk modulus over the ΔT interval that the temperature is increased. For typical material properties at 100 GPa, the increase in pressure for a 3000 K increase in temperature would be of the order of 10–20 GPa at constant volume. In reality, stress relaxation (viscous flow) at high temperatures reduces the actual thermal pressure considerably. At present, it is difficult to measure the magnitude of this effect. However, experiments suggest that the amount of stress relaxation (drop in peak pressure) on laser heating is comparable to the thermal pressure and that these represent competing and compensating effects.[112] Therefore, the pressure distribution prior to any laser heating yields the pressure at high temperatures, and the resulting uncertainty is less than 10 GPa.

5.6. Electrical Conductivity in the Diamond Anvil Cell

To date, most measurements of electrical properties in the diamond anvil cell have focused on the absolute value of sample resistivity. Less commonly, the activation energy for electrical conduction has been obtained by examining the effect of temperature on the conductivity. The introduction of electrical leads into

the diamond cell is conceptually straightforward, yet fairly sophisticated in practice.[75-77] This difficulty is due to a combination of both the small size of the
leads (and thus a lack of mechanical strength) and the changing geometry of
the sample as it is compressed. To prevent short-circuiting of the wires through
the metal gasket, the gasket material can be coated with a mixture of epoxy and
MgO or Al_2O_3. In addition, the indentation of the diamond in the metal gasket
can be punched out. This leaves an open hole in which a layer of an insulating
powder with high shear strength, often MgO or Al_2O_3, is compacted prior to
placing the sample in the center of the compacted powder. Either two separate
wires, usually between 10 and 25 μm in diameter, are then placed over the
sample and compressed between the anvils or leads are vapor-deposited onto the
diamond surface and placed in contact with the sample. Because the geometry of
the leads and the sample dimensions can be measured *in situ* inside the diamond
cell, the absolute value of resistivity (not just the relative resistance) is obtained.
For a resistance R measured across the sample, the sample resistivity is

$$\rho_e = Rwh/l \qquad (16)$$

where w is the distance between the leads, h the sample thickness, and l the
length over which the leads impinge on the sample. In general, the sample
thickness is most easily determined by examination of the sample after decompression, with small corrections being applied for the effect of compression.

5.7. Spectroscopic Techniques in the Diamond Anvil Cell

Although a wide variety of spectroscopic measurements are possible within
the diamond anvil cell, there are several primary methods involved in most such
experiments. First, the diamond anvil cell is particularly amenable to measurements made in a transmission geometry, such as infrared and visible absorption,[65,69] Mössbauer spectroscopy,[66] or extended X-ray absorption fine structure[72,73] (EXAFS). Second, scattering and excitation experiments are readily
performed at pressure when the diamond cell is coupled with laser radiation. The
major varieties of scattering experiments conducted have involved Raman[63]
(including resonance Raman[113] and hyper-Raman[114]) and Brillouin[64] spectroscopies, although impulsive stimulated scattering (ISS), a nonlinear optical
technique, has also been recently applied at pressure.[115] Among excitation
experiments, luminescence spectroscopy has been conducted extensively within
the diamond anvil cell, particularly in regard to the ruby fluorescence pressure
scale.[68,88-90,97-99,116,117] Many of these laser-mediated techniques, with il-
lustrative applications, are reviewed in Ref. 117.

Infrared transmission experiments have been performed at pressure on a
wide variety of materials, including many alkali and alkaline earth halides and
chalcogenides. A review of much of this work, combined with a discussion of the
experimental techniques involved, is given by Ferraro.[118] Infrared emission

studies are notably absent from techniques utilized within the diamond cell: although diamond windows have been conventionally used in emission studies of molten salts, they have not been used in a pressure-generating anvil arrangement.[119] Raman scattering measurements are conducted at pressure in a variety of geometries, varying from nearly back-scattering to scattering at the more common 135° angle. In nonhydrostatic experiments, microfocusing of the incident laser beam to 10–30 μm diameter, or less, is common in order to minimize the pressure gradients across the scattering volume in the sample. For these measurements, diamonds exhibiting the least amount of impurity-induced fluorescence are naturally preferable. It remains difficult to obtain polarized information on samples held at pressure due to the strain birefringence present within the anvils.

Finally, nuclear magnetic resonance measurements have recently been conducted successfully with a diamond anvil cell.[67] An rf coil surrounds the diamonds, generating radio frequency magnetic field lines that run parallel to the gasket. By using Be–Cu as a gasket, the magnetic field lines are not shielded in this direction, and they penetrate by dipping into the sample chamber, despite the "shadowing" effect of the hole edges. With this technique, both the T_1 and T_2 relaxation times of hydrogen-bearing liquids have been measured. We further note that electron spin resonance measurements have been performed at pressure in opposed-anvil apparatuses using sapphire anvils, in which the sapphire anvil also serves as a microwave cavity.[120]

6. Illustrative Applications

In this section, we examine some representative applications of the high-pressure, high-temperature techniques described above. Our choices for illustration are intended not only to emphasize results of scientific interest on a variety of different materials, but also to indicate the usefulness of the overlapping pressure and temperature ranges of the various techniques. For example, the phase diagram of NaCl is now known to pressures and temperatures exceeding 60 GPa and 3500 K by combining the results of piston–cylinder, shock-wave, and diamond-cell experiments (Fig. 12). Sodium chloride is the prototypical ionic material, and the topology of its high-pressure phase relations is similar to those of most other alkali halides. The melting curve is derived from piston–cylinder experiments[121] and from temperature measurements under shock loading.[122–124] In contrast, the transition from the B1 (NaCl structured) to the B2 (CsCl structured) phase is determined entirely from diamond-cell experiments, which also show that the B2 phase is stable to over 70 GPa.[125–127]

The B1–B2–liquid triple point has recently been inferred to occur at about 23.5 GPa and 2250 K, based on anomalously high temperatures observed between 20 and 35 GPa along the Hugoniot. However, measurements of the slope

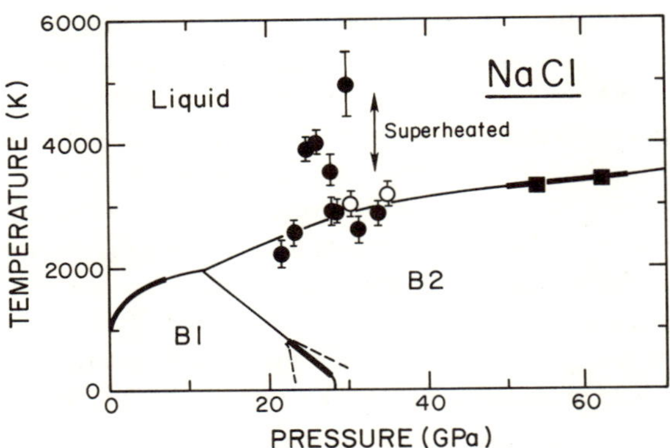

FIGURE 12. Phase diagram of NaCl at pressure and temperature. Bold segments indicate *P–T* ranges of data (not shown except at 50–60 GPa), and light curves indicate interpolated phase boundaries. The melting curve was determined to 6.5 GPa in a piston–cylinder apparatus[121] and at 50–60 GPa from measurements of temperature under shock loading.[123] Recent measurements of shock temperatures (open[122] and closed[124] circles) are interpreted as being at and above the melting temperature of the B2 phase, as described in the text. The B1–B2 transition is obtained from diamond-cell experiments,[125,127] with the dashed curves indicating the kinetically induced hysteresis in the transition.[125]

of the B1–B2 transition within the diamond cell to 700 K indicate that the triple point should be near 12 GPa and 2000 K,[125] as shown in Fig. 12. Thus, we attribute the highest temperatures observed under shock loading (between 24 and 30 GPa) to superheating, with the lower temperatures giving the onset of melting within shock-induced heterogeneities in B2-structured NaCl. This conclusion is in accord with the earlier analysis of Ahrens *et al.*,[128] and it brings the measurements by all of the techniques into mutual agreement. The rapid change in slope of the B1 melting curve and the shallow cusp at the triple point are associated with changes in melt structure: at pressures near the triple point, there is likely to be a distribution of coordination numbers within the liquid, from the sixfold coordination of the low-pressure structure to the eightfold coordination of the B2 structure.[4–6] These pressure-induced changes in melt coordination are clearly evident in molecular dynamics simulations.[6] The appearance of eightfold-coordinated ions in the liquid, prior to the pressure of the B1–B2 phase transition, increases the density of the liquid relative to that of the solid. Similarly, at pressures greater than that of the B1–B2 transition, some sixfold-coordinated ions remain in the liquid, with the opposite effect on its density. Therefore, the decrease in volume of fusion prior to the crystalline phase transition produces a shallowing of the melting curve on the low-pressure side of the triple point, possibly even generating a cusp, as in the phase diagram of KCl.[5,129] Above the triple point, the initial fusion curve is steeper than at higher pressures, due to the effects of compression and the final loss of ions in lower coordination.

Figure 13 shows the sound velocity of CsI measured under shock load-

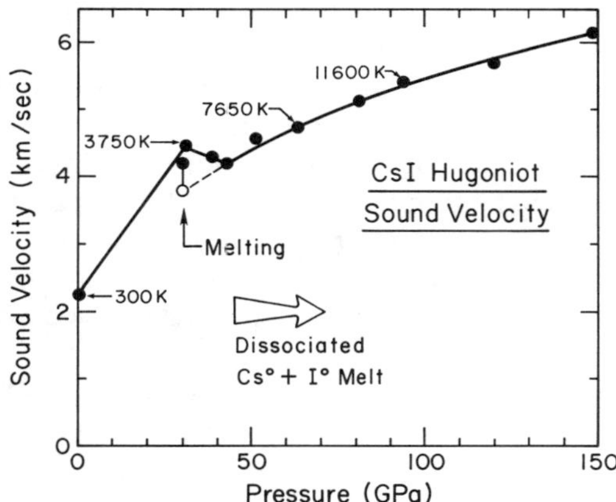

FIGURE 13. Measurements of sound velocity in CsI under shock loading.[130] At pressures above 30 GPa, the sound velocity of CsI decreases, indicating that melting has occurred. Linked open and closed symbols at 30.3 GPa represent the presence of a two-phase mixture at this pressure, with the open symbol representing the apparent sound velocity in the liquid. In the pressure range between 30 and 45 GPa, CsI has been shown to dissociate into its constituent elements at high temperatures.[10] The shock-loaded liquid is thus likely to be a mixture of neutral cesium and iodine atoms, forming a metallic liquid under these conditions, along with undissociated CsI [measurements of electrical conductivity under shock loading document that the molten mixture shows metallic (and apparently time-dependent) conductivity at pressures between 40 and 80 GPa[142]]. This instability of the ionized configurations is predominantly due to the collapse of the $5d$ shell of Cs under compression, making it volumetrically preferable at high pressures for Cs to retain its outermost electron in this shell.[10]

ing.[130] The drop in sound velocity above 30 GPa documents a lowering of the elastic modulus, from the longitudinal value in the solid to the bulk modulus in the melt (see Eq. 10). These results demonstrate that CsI melts above 30 GPa along the Hugoniot (shock-compression curve). However, compression measurements in the diamond cell show that between 30 and 50 GPa the ionized configuration of Cs^+ and I^- becomes volumetrically unstable relative to the elemental electronic configurations of Cs^0 and I^0.[10,131] This instability is driven by electronic changes in cesium (primarily) and in iodine. Specifically, elemental Cs undergoes a volume collapse between 0 and 10 GPa, with a volume at 10 GPa which is 27% of its zero pressure value.[132,133] This collapse has been attributed to a change in the electronic character of the outermost valence orbital of Cs from predominantly $6s$ to $5d$[134−138]; because of the smaller volume of the $5d$ orbital, it is energetically less advantageous to ionize cesium at high pressures than at low pressures.[10] The breakdown of CsI to its constituent elements has been achieved directly with the laser-heated diamond anvil cell at pressures exceeding 45 GPa.[10] The precise pressure range of this reaction is difficult to determine under shock loading. Nevertheless, as the density of the shock-loaded melt[9] is anomalously high relative to that determined for metastable CsI,[139−141] the transition to a metallic, elemental melt of Cs + I is inferred to occur between about 40 and 60 GPa under shock[10,131] (Fig. 13).

Although the vast majority of high-pressure experiments are concerned with thermodynamic properties and phase equilibria, there has also been significant interest in the effect of pressure on rheological properties. In particular, a change in viscosity is expected to be correlated with structural changes induced either in liquids or in solids under pressure. For melts that are characterized at zero pressure by a framework structure of tetrahedrally coordinated Al and Si, the pressure-induced shift to sixfold coordination should produce a more ionic melt, with lower viscosity.[143,144]

Figure 14 demonstrates an application of radiography to imaging spheres falling in a melt of albite composition ($NaAlSi_3O_8$) at a pressure of 2 GPa and temperature of 1100 ($\pm$ 200) °C. The experiment was carried out in the MAX-80

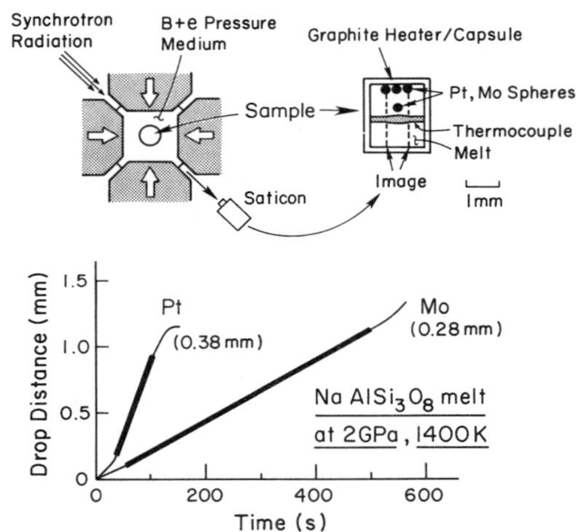

FIGURE 14. Schematic diagram of melt viscosity measurements inside the MAX-80 cubic-anvil press at high pressures and temperatures[59] (*top*). Synchrotron X rays pass through the boron and epoxy (B + e) pressure medium and cylindrical sample and are recorded with a Saticon television camera (shown in plan view: *top left*). In cross section (*top right*), the sample consists of Pt and Mo spheres, and a thermocouple, inside the melt that is contained in a cylindrical resistance heater. The heater acts as a capsule for the melt. Position as a function of time is shown for platinum and molybdenum spheres sinking in an albite ($NaAlSi_3O_8$) composition melt at 2 GPa and 1100 ($\pm$ 200)°C (*bottom:* after Kanzaki *et al.*[59]). The linear regions (bold segments) are those used for determining the steady-state velocities of the spheres. Nonlinear behavior near the top and bottom of the charge is probably due to wall effects, which can be quantitatively modeled.[147] Results of this study are presently regarded as semiquantitative (see text; also, the fusion temperature of albite at 3 GPa, the maximum pressure of these measurements, is near 1350°C[148]: spheres were, however, observed to sink at 1100 ($\pm$ 200)°C. Thus, freezing-point depression of the liquid by a contaminant appears to have taken place. Also, the results derived from these experiments are discrepant with those of Ref. 146).

press with synchrotron X rays and an X-ray-sensitive television camera.[59] The velocities of the falling spheres are derived from measurements of sphere positions versus time, and because the density of molybdenum is roughly half that of platinum (both values can be accurately estimated at these pressure–temperature conditions), both viscosity and density of the melt can be calculated from Eq. (14). Notably, it is in the 2–3-GPa pressure range of these experiments that aluminum is expected to change coordination from fourfold to sixfold, based on phase-equilibrium results (at pressures of about 3.3 GPa, $NaAlSi_3O_8$ begins to melt incongruently to $NaAlSi_2O_6$-jadeite, with aluminum in octahedral coordination, plus liquid[145]). The measurements shown in Fig. 14 yield a melt viscosity of 77 Pa·s, in comparison to a 1-bar value of 11,300 Pa·s,[146] thus supporting the view that the melt structure (Al coordination) is changing. However, this result is presently regarded as preliminary because it is likely that the sample becomes contaminated by the breakdown of the epoxy in the pressure medium inside the large-volume press.[59]

Acknowledgments

This work was supported by the NSF and NASA. This is contribution number 70 of the Institute of Tectonics (Mineral Physics Lab) at the University of California, Santa Cruz. Work was partially supported by the W. M. Keck Foundation.

References

1. P. W. Bridgman, *The Physics of High Pressure,* G. Bell and Sons, London (1949).
2. C. W. F. T. Pistorius, *Prog. Solid State Chem.* **11,** 1 (1976).
3. Y. Sato-Sorenson, *J. Geophys. Res.* **88,** 3543 (1983).
4. N. Kawai and Y. Inokuti, *Jpn. J. Appl. Phys.* **9,** 31 (1970).
5. J. L. Tallon, *Phys. Lett. A* **72,** 150 (1979).
6. M. Ross and F. J. Rogers, *Phys. Rev. B* **31,** 1463 (1986).
7. Q. Williams and R. Jeanloz, *Phys. Rev. Lett.* **56,** 163 (1986).
8. R. Reichlin, M. Ross, S. Martin, and K. A. Goettel, *Phys. Rev. Lett.* **56,** 2858 (1986).
9. H. B. Radousky, M. Ross, A. C. Mitchell, and W. J. Nellis, *Phys. Rev. B* **31,** 1457 (1985).
10. Q. Williams and R. Jeanloz, *Phys. Rev. Lett.* **59,** 1132 (1987).
11. R. Jeanloz, *Ann. Rev. Phys. Chem.,* **40,** 237–259 (1989).
12. I. L. Spain and J. Paauwe (eds.), *High Pressure Technology,* Vols. 1–3, Marcel Dekker, New York (1977).
13. G. C. Ulmer (ed.), *Research Techniques for High Pressure and High Temperature,* Springer-Verlag, New York (1971).
14. A. D. Edgar, *Experimental Petrology, Basic Principles and Techniques,* Clarendon Press, Oxford (1973).
15. G. C. Ulmer and H. E. Barnes (eds.), *Hydrothermal Experimental Techniques,* Wiley-Interscience, New York (1987).

16. J. Holloway and B. Wood, *Simulating the Earth*, Unwin-Allen, New York (1988).
17. B. Vodar and P. Marteau, *High Pressure Science and Technology*, Vols. 1–3, Pergamon Press, Oxford (1980).
18. C.-M. Backman, T. Johannisson, and L. Tegner (eds.), *High Pressure Research in Science and Industry*, Arkitektkopia, Uppsala (1982).
19. C. Homan, R. K. MacCrone, and E. Whalley (eds.), *High Pressure Science and Technology*, Parts 1–3, North-Holland, New York (1984).
20. N. J. Trappeniers *et al.* (eds.), *Proceedings of the Xth AIRAPT International High Pressure Conference, 1985*, North-Holland, Amsterdam (also in *Physica B + C* **139/140**) (1986).
21. *J. Phys., Colloq.* **C8** (1984).
22. H. Vollstädt (ed.), *High Pressure Geosciences and Material Synthesis*, Akademie-Verlag, Berlin (1987).
23. M. H. Manghnani and S. Akimoto (eds.), *High Pressure Research: Applications in Geophysics*, Academic Press, New York (1977).
24. S. Akimoto and M. H. Manghnani (eds.), *High Pressure Research in Geophysics*, Center for Academic Publications, Tokyo (1982).
25. M. H. Manghnani and Y. Syono (eds.), *High-Pressure Research in Mineral Physics*, American Geophysical Union, Washington, D.C. (1987).
26. J. R. Asay, R. A. Graham, and G. K. Straub (eds.), *Shock Waves in Condensed Matter—1983*, Elsevier, Amsterdam (1984).
27. Y. Gupta (ed.), *Shock Waves in Condensed Matter—'85*, Plenum Press, New York (1986).
28. S. C. Schmidt and N. C. Holmes, *Proceedings of the APS Topical Conference on Shock Waves in Condensed Matter, 1978*, Elsevier, Amsterdam (1988).
29. M. Kumazawa, in: *High-Pressure Research: Applications in Geophysics* (M. H. Manghnani and S. Akimoto, eds.), pp. 563–572, Academic Press, New York (1977).
30. O. Fukunaga, S. Yamaoka, T. Endoh, M. Akaishi, and H. Kanda, in: *High Pressure Science and Technology*, Vol. 1 (K. D. Timmerhaus and M. S. Barber, eds.), pp. 846–852, Plenum Press, New York (1979).
31. O. Shimomura, S. Yamaoka, T. Yagi, M. Wakatsuki, K. Tsuji, H. Kawamura, N. Hamaya, O., Fukunaga, K. Aoki, and S. Akimoto, in: *Solid State Physics under Pressure* (S. Minomura, ed.), pp. 351–356, Reidel, Dordrecht (1985).
32. E. K. Graham, *J. Geophys. Res.* **91**, 4630 (1986).
33. L. V. Al'tshuler, *Sov. Phys. Usp.* (Engl. Transl.), **85**, 52 (1965).
34. Y. B. Zel'dovitch and Y. P. Raiser, *Physics of Shock Waves and High Temperature Hydrodynamic Phenomena*, 2 vols., Academic Press, New York (1966).
35. M. H. Rice, R. G. McQueen, and J. M. Walsh, in: *Solid State Physics*, Vol. 6 (F. Seitz and D. Turnbull, eds.), pp. 1–63, Academic Press, New York (1958).
36. T. J. Ahrens, in: *Methods of Experimental Physics*, Vol. 24A, pp. 185–235, Academic Press, New York (1987).
37. P. A. Urtiew and R. Grover, *J. Appl. Phys.* **45**, 140 (1974).
38. B. Svendsen, T. J. Ahrens, and J. D. Bass, in: *High-Pressure Research in Mineral Physics* (M. H. Manghnani and Y. Syono, eds.), pp. 403–423, American Geophysical Union, Washington, D.C. (1987).
39. R. Jeanloz and T. J. Ahrens, *J. Geophys. Res.* **84**, 7545 (1979).
40. R. G. McQueen, J. W. Hopson, and J. N. Fritz, *Rev. Sci. Instrum.* **53**, 245 (1982).
41. J. M. Brown, J. W. Shaner, and C. A. Swenson, *Phys. Rev. B* **32**, 4507 (1985).
42. Q. Johnson and A. C. Mitchell, *Phys. Rev. Lett.* **29**, 1369 (1972).
43. Q. Johnson, A. Mitchell, R. N. Keeler, and L. Evans, *Phys. Rev. Lett.* **25**, 1099 (1970).
44. T. Goto, T. J. Ahrens, and G. Rossman, *Phys. Chem. Miner.* **4**, 253 (1979).
45. N. C. Holmes, W. J. Nellis, W. B. Graham, and G. E. Walrafen, *Phys. Rev. Lett* **55**, 2433 (1985).

46. P. D. Horn and Y. M. Gupta, *Appl. Phys. Lett.* **49**, 856 (1986).
47. P. D. Horn and Y. M. Gupta, *Phys. Rev. B* **39**, 973 (1989).
48. T. J. Ahrens, *J. Appl. Phys.* **37**, 2532 (1966).
49. S. D. Hamann and M. Linton, *Trans. Faraday Soc.* **62**, 2234 (1966).
50. H. Carslaw and J. C. Jaeger, *Conduction of Heat in Solids*, Clarendon Press, Oxford (1959).
51. H. T. Hall, *Rev. Sci. Instrum.* **33**, 1278 (1962).
52. A. Jayaraman, *Rev. Mod. Phys.* **55**, 65 (1983).
53. I. Fujishiro, G. J. Piermarini, S. Block, and R. G. Munro, in: *High Pressure Research in Science and Industry*, C. M. Backman, T. Johannison, and L. Tegner, eds., pp. 608–611, Arkitektkopia, Uppsala (1982).
54. O. Fukunaga, S. Yamaoka, M. Akaishi, H. Kanda, T. Osawa, O. Shimomura, T. Nagashima, and M. Yoshikawa, in: *High-Pressure Research in Mineral Physics* (M. H. Manghnani and Y. Syono, eds.), pp. 17–28, American Geophysical Union, Washington, D.C. (1987).
55. K. Suito, in: *High-Pressure Research: Applications in Geophysics* (M. H. Manghnani and S. Akimoto, eds.), pp. 255–266, Academic Press, New York (1977).
56. E. Ohtani, M. Kumazawa, T. Kato, and T. Irifune, in: *High-Pressure Research in Geophysics* (S. Akimoto and M. H. Manghnani, eds.), pp. 259–270, Center for Academic Publications, Tokyo (1982).
57. S. Endo and K. Ito, in: *High-Pressure Research in Geophysics* (S. Akimoto and M. H. Manghnani, eds.), pp. 3–12, Center for Academic Publications, Tokyo (1982).
58. E. Ito and E. Takahashi, in: *High-Pressure Research in Mineral Physics* (M. H. Manghnani and Y. Syono, eds.), pp. 221–229, American Geophysical Union, Washington, D.C. (1987).
59. M. Kanzaki, K. Kurita, T. Fujii, T. Kato, O. Shimomura, and S. Akimoto, in: *High-Pressure Research in Mineral Physics* (M. H. Manghnani and Y. Syono, eds.), pp. 195–200, American Geophysical Union, Washington, D.C. (1987).
60. S. Akimoto, T. Suzuki, T. Yagi, and O. Shimomura in: *High-Pressure Research in Mineral Physics* (M. H. Manghnani and Y. Syono, eds.), pp. 149–154, American Geophysical Union, Washington, D.C. (1987).
61. J. Xu, H. K. Mao, and P. M. Bell, *Science* **232**, 1401 (1987).
62. Q. Williams, R. Jeanloz, J. Bass, B. Svendsen, and T. J. Ahrens, *Science*, **236**, 181 (1987).
63. H. K. Mao, P. M. Bell, and R. J. Hemley, *Phys. Rev. Lett.* **55**, 99 (1985).
64. A. Polian and M. Grimsditch, *Phys. Rev. B* **27**, 6409 (1983).
65. Q. Williams. R. Jeanloz, and P. McMillan, *J. Geophys. Res.* **92**, 8116 (1987).
66. K. Kurimoto, S. Nasu, S. Nagatomo, S. Endo, and F. E. Fujita, *Physica B* 139, 140, 495 (1986).
67. S.-H. Lee, K. Luszczynski, R. E. Norberg, and M. S. Conradi, *Rev. Sci. Instrum.* **58**, 415 (1987).
68. J. D. Barnett, S. Block, and G. J. Piermarini, *Rev. Sci. Instrum.* **44**, 1 (1973).
69. H. K. Mao, in: *The Physics and Chemistry of Minerals and Rocks* (R. G. J. Strens, ed.), pp. 573–581, Wiley, New York (1976).
70. K. Syassen and R. Sonnenschein, *Rev. Sci. Instrum.* **53**, 644 (1982).
71. W. A. Bassett, T. Takahashi, and P. W. Stook, *Rev. Sci. Instrum.* **38**, 37 (1967).
72. R. Ingalls, G. A. Garcia, and E. A. Stern, *Phys. Rev. Lett.* **40**, 334 (1978).
73. O. Shimomura and T. Kawamura, in: *High-Pressure Research in Mineral Physics* (M. H. Manghnani and Y. Syono, eds.), pp. 187–193, American Geophysical Union, Washington, D.C. (1987).
74. H. B. Radousky, R. L. Reichlin, and R. H. Howell, *J. Phys. Colloq.* **C8**, 369 (1984).
75. H. K. Mao and P. M. Bell, *Rev. Sci. Instrum.* **52**, 615 (1981).
76. R. Reichlin, *J. Phys. Colloq.* **C8**, 399 (1984).
77. E. Knittle and R. Jeanloz, *J. Geophys. Res.*, in press (1990).

78. A. P. Jephcoat, H. K. Mao, and P. M. Bell, in: *Hydrothermal Experimental Techniques* (H. E. Barnes and G. C. Ulmer, eds.), pp. 469–506 Wiley-Interscience, New York (1987).

79. A. Jayaraman, *Rev. Sci. Instrum.* **57,** 1013 (1986).

80. A. S. Balchan and H. G. Drickamer, *Rev. Sci. Instrum.* **32,** 308 (1961).

81. H. G. Drickamer, R. W. Lynch, R. L. Clendenen, and E. A. Perez-Albuerne, in: *Solid State Physics,* Vol. 19 (F. Seitz and D. Turnbull, eds.), pp. 135–228, Academic Press, New York (1966).

82. A. Onodera, K. Furuno, and S. Yazu, *Science* **232,** 1419 (1986).

83. J. E. Field (ed.), *The Properties of Diamond,* Academic Press, New York (1979).

84. K. A. Goettel, H. K. Mao, and P. M. Bell, *Rev. Sci. Instrum.* **56,** 1420 (1985).

85. H.-K. Mao, P. M. Bell, K. J. Dunn, R. M. Chrenko, and R. C. DeVries, *Rev. Sci. Instrum.* **50,** 1002 (1979).

86. L. Merrill and W. A. Bassett, *Rev. Sci. Instrum.* **45,** 290 (1974).

87. M. Pasternak, J. N. Farrell, and R. F. Taylor, *Phys. Rev. Lett.* **58,** 575 (1987).

88. G. J. Piermarini and S. Block, *Rev. Sci. Instrum.* **46,** 973 (1975).

89. G. J. Piermarini, S. Block, J. D. Barnett, and R. A. Forman, *J. Appl. Phys.* **46,** 2774 (1975).

90. J. M. Besson and J. P. Pinceaux, *Rev. Sci. Instrum.* **50,** 541 (1979).

91. W. C. Moss, J. O. Hallquist, R. Reichlin, K. A. Goettel, and S. Martin, *Appl. Phys. Lett.* **48,** 1258 (1986).

92. H.-K. Mao, J. Xu, and P. M. Bell, *J. Geophys. Res.* **91,** 4673 (1986).

93. P. M. Bell, J. Xu, and H.-K. Mao, in: *Shock Waves in Condensed Matter—'85* (Y. Gupta, ed.), pp. 125–130, Plenum Press, New York (1986).

94. H.-K. Mao, P. M. Bell, J. Shaner, and D. Steinberg, *J. Appl. Phys.* **49,** 3276 (1978).

95. D. L. Heinz and R. Jeanloz, *J. Appl. Phys.* **55,** 885 (1984).

96. R. C. Powell, B. DiBartolo, B. Birang, and C. S. Naiman, *Phys. Rev.* **155,** 296 (1967).

97. O. Shimomura, S. Yamaoka, N. Nakazawa, and O. Fukunaga, in: *High-Pressure Research in Geophysics* (S. Akimoto and M. H. Manghnani, eds.), pp. 49–60, Center for Academic Publications, Tokyo (1982).

98. Y. Sato-Sorenson, in: *High-Pressure Research in Mineral Physics* (M. H. Manghnani and Y. Syono, eds.), pp. 53–59, American Geophysical Union, Washington, D. C. (1987).

99. G. J. Piermarini, S. Block, and J. D. Barnett, *J. Appl. Phys.* **44,** 5377 (1973).

100. A. P. Jephcoat, H. K. Mao, and P. M. Bell, *J. Geophys. Res.* **91,** 4677 (1986).

101. C. M. Sung, C. Goetze, and H. K. Mao, *Rev. Sci. Instrum.* **48,** 1386 (1977).

102. D. Schiferl, J. N. Fritz, A. I. Katz, M. Schaefer, E. F. Skelton, S. B. Qadri, L. C. Ming, and M. H. Manghnani, in: *High-Pressure Research in Mineral Physics* (M. H. Manghnani and Y. Syono, eds.), pp. 75–83, American Geophysical Union, Washington, D.C. (1987).

103. J. Yen and M. Nicol, *J. Phys. Chem.* **91,** 3336 (1987).

104. A. S. Zinn, D. Schiferl, and M. F. Nicol, *J. Chem. Phys.* **87,** 1267 (1987).

105. L. Liu and W. A. Bassett, *J. Geophys. Res.* **80,** 3777 (1975).

106. R. Boehler, *Geophys. Res. Lett.* **13,** 1153 (1986).

107. R. Boehler, M. Nicol, C. S. Zha, and M. L. Johnson, *Physica B* **139–140,** 916 (1986).

108. D. L. Heinz and R. Jeanloz, in: *High-Pressure Research in Mineral Physics* (M. H. Manghnani and Y. Syono, eds.), pp. 113–127, American Geophysical Union, Washington, D.C. (1987).

109. R. Jeanloz and D. Heinz, *J. Phys. Colloq.* **C8,** 83–92 (1984).

110. S. Bodea and R. Jeanloz, *J. Appl. Phys.* **65,** 4688 (1989).

111. Q. Williams and R. Jeanloz, *J. Geophys. Res.,* in press (1990).

112. D. L. Heinz and R. Jeanloz, *J. Geophys. Res.* **92,** 11437 (1987).

113. B. Khodadoost, S. A. Lee, J. B. Page, and R. C. Hanson, *Phys. Rev. B* **38,** 5288 (1988).

114. K. Inoue, S. Akimoto, and T. Ishidate in: *Solid State Physics under Pressure: Recent Advances with Anvil Devices* (S. Minomura, ed.), pp. 87–92, Reidel, Dordrecht (1985).

115. J. M. Brown, L. J. Slutsky, K. A. Nelson, and L. T. Cheng, *Science* **241,** 65 (1988).

116. Q. Williams and R. Jeanloz, *Phys. Rev. B* **31**, 7449 (1985).
117. R. J. Hemley, H. K. Mao, and P. M. Bell, *Science* **237**, 605 (1987).
118. J. R. Ferraro, *Vibrational Spectroscopy at High Pressures*, Wiley-Interscience, New York (1986).
119. J. Hvistendahl, P. Klaeboe, E. Rytter, and H. A. Øye, *Inorg. Chem.* **23**, 706 (1984).
120. J. D. Barnett, S. D. Tyagi, and H. M. Nelson, *Rev. Sci. Instrum.* **49**, 348 (1978).
121. J. Akella, S. N. Vaidya, and G. C. Kennedy, *Phys. Rev.* **185**, 1135 (1969).
122. D. R. Schmitt, T. J. Ahrens, and B. Svendsen, *J. Appl. Phys.* **63**, 99 (1987).
123. S. B. Kormer, *Sov. Phys. Usp.* **11**, 229 (1968).
124. K. Kondo and T. J. Ahrens, *Phys. Chem. Miner.* **9**, 173 (1983).
125. X. Li and R. Jeanloz, *Phys. Rev. B* **36**, 474 (1987).
126. D. L. Heinz and R. Jeanloz, *Phys. Rev. B* **30**, 6045 (1984).
127. W. A. Bassett, T. Takahashi, H. K. Mao, and J. S. Weaver, *J. Appl. Phys.* **39**, 319 (1968).
128. T. J. Ahrens, G. A. Lyzenga, and A. C. Mitchell, in: *High-Pressure Research in Geophysics* (S. Akimoto and M. H. Manghnani, eds.), pp. 579–594, Center for Academic Publications, Tokyo (1982).
129. S. M. Stishov, *Sov. Phys. Usp.* **11**, 816 (1969).
130. C. A. Swenson, J. W. Shaner, and J. M. Brown, *Phys. Rev. B* **34**, 7924 (1986).
131. Q. Williams and R. Jeanloz, in: *Shock Waves in Condensed Matter 1987* (S. C. Schmidt and N. C. Holmes, eds.), pp. 73–75, Elsevier, New York (1988).
132. K. Takemura and K. Syassen, *Phys. Rev. B* **32**, 2213 (1985).
133. K. Takemura, S. Minomura, and O. Shimomura, *Phys. Rev. Lett.* **49**, 1772 (1982).
134. R. Sternheimer, *Phys. Rev.* **78**, 235 (1950).
135. A. Jayaraman, R. C. Newton, and J. M. McDonough, *Phys. Rev.* **159**, 527 (1967).
136. D. B. McWhan, G. Parisot, and D. Bloch, *J. Phys. F* **4**, L69 (1974).
137. A. K. McMahan, *Phys. Rev. B* **17**, 1521 (1978).
138. D. Glötzel and A. K. McMahan, *Phys. Rev. B* **20**, 3210 (1979).
139. E. Knittle and R. Jeanloz, *Science* **223**, 53 (1984).
140. K. Asaumi, *Phys. Rev. B* **29**, 1118 (1984).
141. Y. K. Vohra, K. E. Brister, S. T. Weir, S. J. Duclos, and A. L. Ruoff, *Science* **231**, 1136 (1986).
142. L. A. Gatilov and L. V. Kuleshova, *Sov. Phys. Solid State* **23**, 1663 (1981).
143. H. S. Waff, *Geophys. Res. Lett.* **2**, 193 (1975).
144. C. A. Angell, P. Cheeseman, and S. Tamaddon, *Science* **218**, 885 (1982).
145. P. M. Bell and E. H. Roseboom, Jr., *Miner. Soc. Amer. Special Paper* **2**, pp. 151–162 (1969).
146. I. Kushiro, in: *Physics of Magmatic Processes* (R. B. Hargraves, ed.), pp. 93–120, Princeton University Press, Princeton (1980).
147. H. R. Shaw, *J. Geophys. Res.* **68**, 6337 (1963).
148. V. C. Kress, Q. Williams, and I. S. E. Carmichael, *Geochim. Cosmochim. Acta* **52**, 283 (1988).
149. F. P. Bundy, *Physics Reports* **167**, 133 (1988).

7

Battery Construction, Testing, and Materials

Paul A. Nelson and Thomas D. Kaun

1. Introduction

Batteries having molten salt electrolytes are of interest because the ionic conductivity of molten salts is high; lightweight, reactive elements with alloying materials can be used at the high temperatures and oxygen-free conditions appropriate for molten salt electrolytes. For example, primary cells with molten salt electrolytes have been developed for the couples $Li-Al/FeS_2$, $Li-Si/FeS_2$, and $Li-Al/FeS$. In general, these primary cells are formed into bipolar stacks of very thin electrodes, providing high power and high specific energy for various military or space power applications. Secondary batteries based on molten salt electrolytes also have favorable characteristics, including high specific energy and long cycle life. Research and development work has been carried out on such rechargeable batteries having (1) negative electrodes of lithium or lithium alloys and calcium or calcium alloys and (2) positive electrodes of tellurium, selenium, sulfur, and metal sulfides such as FeS, FeS_2, Ni_2S_3, NiS_2, CoS_2, and TiS_2. Molten halide electrolytes and separators of boron nitride fibers or magnesia powder typically have been used.

In this chapter, discussion will be focused on $Li-alloy/FeS$ and $Li-alloy/FeS_2$ cells and batteries being developed in programs under sponsorship of the Department of Energy. The applications for batteries developed in these programs are for powering electric vehicles and for stationary energy storage, such as electric utility load leveling. In the discussions that follow, methods of

Paul A. Nelson and Thomas D. Kaun • Argonne National Laboratory, Chemical Technology Division, Argonne, Illinois 60439. Work supported by the U.S. Department of Energy, Office of Energy Storage and Distribution under Contract W-31-109-Eng-38. The U.S. Government retains a nonexclusive, royalty-free license to publish or reproduce the published form of this contribution, or allow others to do so, for U.S. Government purposes.

229

construction and battery testing will be emphasized. Other summary articles have been written on the chemistry of the cell,[1–4] and in this chapter only sufficient information will be given on cell chemistry to provide the necessary background for the subjects to be discussed.

Work on the Li/S cell with an electrolyte of LiCl–KCl (mp 352°C) was begun at Argonne National Laboratory in the late 1960s.[1] At the cell operating temperature of about 400°C, both electrodes of this system are molten, and serious problems arise because of (1) the escape of liquid lithium from the negative electrode, causing the cell to be short-circuited, and (2) the high solubility of both sulfur and lithium in the electrolyte. The solutions for these problems, which were first introduced in the period 1972–1973, are the use of Li–Al or Li–Si negative electrodes and FeS or FeS_2 positive electrodes.[2,5] These electrode materials are solids at the operating temperature of the cell, and thus their use avoids the problems of the liquid electrodes and makes practical cells possible. Also, the use of lithium alloys decreases the lithium activity with respect to that of elemental lithium, and this makes possible the use of boron nitride fiber separators.

The theoretical specific energies of the lithium/sulfide cells and the competing sodium/sulfur, sodium/iron chloride, and sodium/nickel chloride cells are compared in Table I. As shown in this table, the substitution of alloys and compounds for the elemental materials reduces the theoretical specific energy from about 2555 Wh/kg for the Li/S cell to 460 Wh/kg for the Li–Al/FeS cell and to 475 to 630 Wh/kg for the Li–Al/FeS_2 cell. The value for the Li–Al/FeS_2 cell depends upon the extent of reaction of the positive electrode. The sodium-electrode cells are discussed briefly at the end of this chapter.

Some examples of overall cell reactions are the following:

TABLE I. Specific Energies of Selected Cell Couples

System	Positive electrode discharge product	Average open-circuit voltage (V)	Theoretical specific energy (Wh/kg)
Li/S	Li_2S	2.14	2555
Li–Al/FeS	Fe, Li_2S	1.34	460[a]
Li–Al/FeS_2	Li_2FeS_2	1.66	475[a]
Li–Al/FeS_2	Li_2S, Fe	1.50	630[a]
$Li_{3.25}$–Si/FeS_2	Li_2S, Fe	1.58	930
Na/S	$Na–S_{1.5}$	2.01	760
Na/$FeCl_2$	NaCl, Fe	2.35	728
Na/$NiCl_2$	NaCl, Ni	2.59	794

[a]Based on Li–50 atom % Li–Al.

$$2Li-Al + FeS \rightarrow Li_2S + Fe + 2Al \qquad (1)$$
$$2Li-Al + FeS_2 \rightarrow Li_2FeS_2 + 2Al \qquad (2)$$
$$4Li-Al + FeS_2 \rightarrow 2Li_2S + Fe + 4Al \qquad (3)$$

The fraction of the theoretical specific energy that can be achieved for practical cells differs markedly for the couples shown in Table I. The important factors are (1) the operating voltages of the cells, (2) the extent of utilization of the active materials, and (3) the densities of the reactants and the discharge products. The densities of the materials have a surprisingly large effect on the achievable specific energy.[6] For instance, the third system shown in Table I, $Li-Al/FeS_2$ discharged to Li_2FeS_2, has a theoretical specific energy of only 475 Wh/kg but has promise of achieving a higher actual specific energy than any of the other couples shown in the table because both FeS_2 and the reaction product, Li_2FeS_2, have high densities (4.8 g/cm^3 and 3.0 g/cm^3, respectively) and, thus, small volumes. The small active material volumes in all states of charge result in the need for less electrolyte to fill the voids in comparison to other electrodes. It is apparent, as will be discussed in more detail below, that the theoretical specific energy is a poor relative measure of the achievable specific energy for these molten salt cells.

Most of the discussion below will be concerned with the development of a prismatic cell design, comprised of either $Li-Al$ or $Li-Si$ negative electrodes, FeS or FeS_2 positive electrodes, a molten salt electrolyte (typically LiCl–KCl), and a separator of BN or MgO.

2. Lithium/Monosulfide Cells

As noted above, cells having monosulfide positive electrodes have lower specific energy than those having disulfide positive electrodes. However, the theoretical specific energy of 460 Wh/kg for a representative monosulfide cell, $Li-Al/FeS$, is considerably higher than that of cells in ambient-temperature secondary batteries such as the lead–acid battery and the nickel–iron battery. Also, the monosulfide cell has advantages over some of the other high-temperature cells listed in Table I, such as lithium/disulfide and sodium/sulfur, that at least partially compensate for the lower specific energy of the $Li-Al/FeS$ cells. The most notable advantage is that the positive electrode in all states of discharge is compatible with iron and nickel, which may be used as construction materials or current collectors. Both FeS_2 and sulfur rapidly attack these construction materials in $Li-Al/FeS_2$ and Na/S cells. The $Li-Al/FeS$ cell has a virtually constant open-circuit potential over the entire discharge of 1.34 to 1.33 V, resulting in an average operating voltage during a 4-h discharge of approximately 1.2 V. Thus, the monosulfide cells are charged and discharged between voltages of approximately 1.5 and 1.0 V.

2.1. Negative Electrodes

A variety of lithium alloys have been tested for use in secondary Li-alloy/iron sulfide cells, including those shown in Table II. This table illustrates the effects of the density and usable capacity of the alloys upon the effective capacity density of the electrode. The effective capacity density, as defined for this table, involves the total volume of the alloys, but not the volume of the electrolyte associated with the electrode.

Most of the lithium/sulfide cells tested have had lithium–aluminum negative electrodes. The phase diagram for the Li–Al system is shown in Fig. 1, and the voltage for a Li–Al electrode as lithium is titrated into the alloy is shown in Fig. 2.[7] Most of the operation of the Li–Al electrode takes place on the $\alpha + \beta$ voltage plateau (300 mV relative to lithium), which extends from about 8 atom % lithium to about 48 atom % lithium. The electrode can be operated in the β-phase region and in the $\beta + \gamma$ region, but the high lithium activity in these regions, as shown in Fig. 2, requires the use of MgO separators rather than BN separators.

In early studies of the Li–Al/FeS cell, the rate of capacity decline was unacceptably high, up to 0.25% per cycle. Metallographic examinations identified agglomeration of the porous Li–Al electrode as the main cause of this rapid capacity decline.[8] These electrodes had been cold-pressed to high loading densities and discharged to about 85% lithium utilization. Under these conditions, the electrodes sintered at the center to dense structures high in aluminum concentration and did not react readily with lithium ions on charge. One method developed to alleviate this problem was to fabricate cells with a negative-to-positive capacity ratio of about 1.3 rather than 1.0 and thus maintain adequate lithium concentration in the negative electrodes to avoid sintering. This sacrificed specific energy and resulted in poor power near the end of discharge because of the poor performance of the FeS electrode at deep discharge. However, the capacity decline of the Li–Al electrode was reduced to about 0.02% per

TABLE II. Capacity Densities of Secondary Li Electrodes

Composition	Density (g/cm³)	Theoretical capacity (Ah/cm³)	Usable capacity[a] (% of Theoretical capacity)	Effective capacity density (Ah/cm³)
$Li_{3.2}Si$	1.05	1.80	65	1.17
Li–Al (50 atom % Li)	1.75	1.39	82	1.14
Li–Al (53 atom % Li)	1.65	1.47	83	1.22
50 atom % Li–Al + 50 atom % $Li_{3.2}Si$	1.28	1.59	70	1.11
88 atom % Li–Al + 12 atom % $Li_{3.2}Si$	1.66	1.44	80	1.15

[a]Electrode capacity having lithium activity of ≤ 0.45 V vs. lithium.

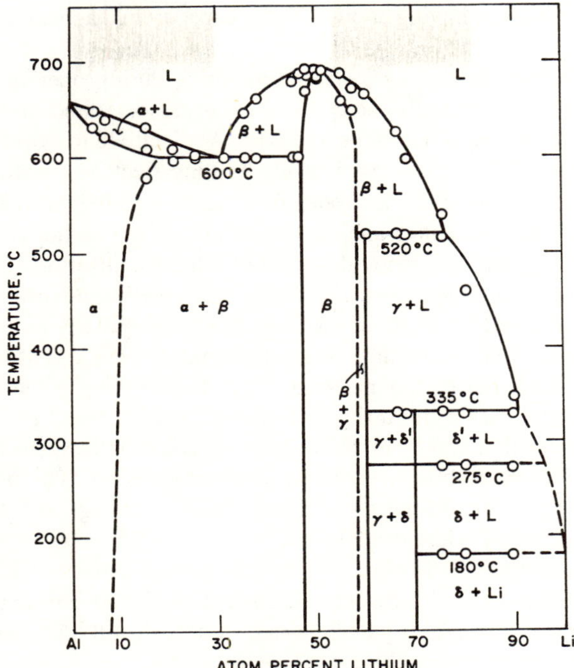

FIGURE 1. Li–Al phase diagram.

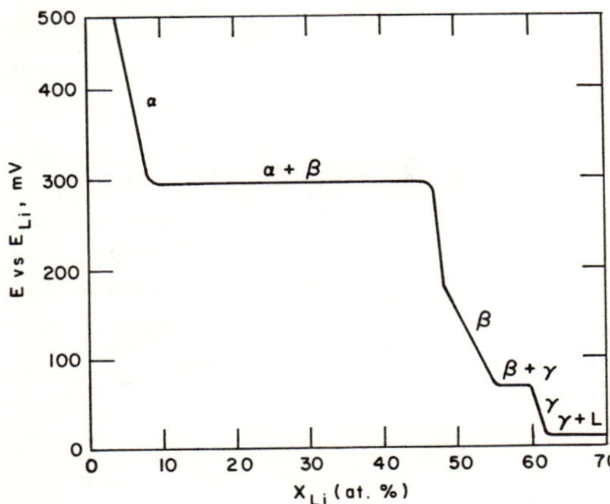

FIGURE 2. Coulometric titration curve for lithium–aluminum system at 423°C.[7]

cycle. Cells of this type having about 300-Ah capacity had a mean time to failure of about 1000 cycles.[9]

Methods of circumventing electrode sintering during operation have been studied extensively in this laboratory. This effort concentrated primarily on the examination of addition of additives to the Li–Al electrode to avoid sintering. In these cell tests, the Li–Al electrode was the limiting electrode, having a capacity of about 50% that of the FeS electrode. As seen in Fig. 3, cell WR-1 exhibited Li–Al electrode characteristics analogous to those of the earlier cells, that is, a 0.093% rate of capacity decline with an initial lithium utilization of 69% (earlier work had shown even higher rates of decline for higher lithium utilizations). The addition of carbon and MgO additives was effective in greatly reducing the rate of capacity decline, but the method of electrode fabrication was also found to be important. Whereas the earlier electrodes had been cold-pressed and sintered without an additive, electrodes made by forming a slurry of the Li–Al powders in an organic solvent that was later removed by evacuation resulted in electrodes having longer life than those formed by powder pressing. The use of carbon powder or MgO powder additives was an additional advantage over the use of the slurry forming method without additives.

For a Li–Al electrode loading density (theoretical capacity per unit volume of electrode) of 0.9 Ah/cm^3, powder additions of MgO and carbon in 3–5 vol % amounts suppressed the capacity decline rates to less than 0.02% per cycle, with the lithium utilization approaching 80%. This performance was accomplished under harsh operating conditions involving high polarization of the negative electrode at the end of the discharge. In developmental cells, the improved Li–Al electrode demonstrated stable cell capacity, and its higher utilization yielded a 9% increase in specific energy from that of previous cells.

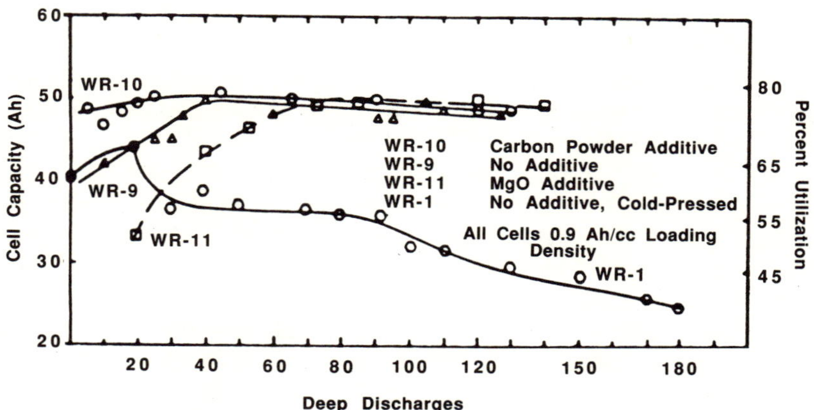

FIGURE 3. Capacity stability with cycling for slurry-formed Li–Al electrodes compared with conventional (cold-pressed) Li–Al electrodes.

TABLE III. Characteristics of Li–Si Alloy Electrode

Plateau voltage observed[a] (mV)	Proposed phase transition		Lithium reacted (ΔLi/Si)	
	Wen and Huggins[b]	Schnering *et al.*[c]	Wen and Huggins[b]	Schnering *et al.*[c]
-250	$Li_{4.41}Si \rightarrow Li_{3.317}Si$	$Li_{3.85}Si \rightarrow Li_{3.25}Si$	1.09	0.65
-142	$Li_{3.317}Si \rightarrow Li_{2.367}Si$	$Li_{3.25}Si \rightarrow Li_{2.33}Si$	0.95	0.92
-15	$Li_{2.367}Si \rightarrow Li_{1.731}Si$	$Li_{2.33}Si \rightarrow Li_{1.71}Si$	0.64	0.62
$+34$	$Li_{1.731}Si \rightarrow Si$	$Li_{1.71}Si \rightarrow Si$	1.73	1.71

[a]At 450°C vs. Li–Al ($\alpha + \beta$).
[b]Reference 10.
[c]Reference 11.

The lithium–silicon alloy electrode has a higher theoretical specific energy than Li–Al because silicon is light and will accept several atoms of lithium per atom of silicon, as shown in Table III.[10–12] However, the use of silicon as an alloying agent for lithium is not as advantageous as it would first appear.[13] First, much of the lithium is unusable because it is released at too low an operating cell voltage. This additional lithium increases the volume of the electrode, decreasing the effective capacity density compared to that of Li–Al (Table II), and results in a cost penalty for the expensive unused lithium. Second, the rate of reaction for the Li–Si electrode is lower (particularly on the lowest-voltage plateau) than that for Li–Al. This results in a higher specific impedance for the Li–Si electrode than for the Li–Al electrode during a 15-s high-rate discharge, such as would be needed for acceleration of an electric vehicle. Finally, silicon is corrosive to the current collector; nickel may not be acceptable for Li–Si electrodes, whereas steel current collectors show negligible corrosion in contact with Li–Al. Improved designs may result in the Li–Si alloy being favored over Li–Al, but in current designs, the Li–Al alloy appears to have a slight overall advantage in specific energy, power, and cost.

Ternary lithium alloys such as Li–Al–Si have also been examined (Table II). The effective capacity densities of ternary alloy electrodes are similar to that of Li–Al. Generally, higher lithium activity than is afforded with Li–Al at the beginning of discharge is exchanged for lower lithium activity at 80% depth of discharge. All of the alloys investigated appear to have about the same effective capacity density, approximately 1.17 ± 0.05 Ah of delivered capacity per cubic centimeter of alloy in electrodes containing 35 to 40 vol % electrolyte at full charge.

An objective of continued research on lithium–alloy electrodes is to increase the effective capacity density and to increase the lithium activity and cell voltage, especially at 80% depth of discharge. A further objective is to avoid the need for unavailable lithium, which is quite considerable for the Li–Si electrode as noted above.

2.2. Monosulfide Electrodes

The metal sulfides, such as sulfides of Fe, Co, Ni, and Ti, have proved to be effective and practical replacements for elemental sulfur electrodes. They provide high specific energy and, as was the case for the lithium-alloy electrode, constitute a solid-state electrode material at the high operating temperature of the cell. Of the sulfides, FeS is of particular interest because it is chemically stable in contact with metallic iron, and, thus, low-cost iron current collectors are practical. Long cycle life for the Li–Al/FeS cell was also identified early in the development effort. For these reasons, initial battery development focused on cells having FeS rather than FeS_2 electrodes.

The FeS electrode consists of a compressed bed of FeS particles containing a molten salt electrolyte. Contact between the electronically conductive FeS particles provides an electronic path to the current collector, which may be a continuous sheet of metal located at the center of the electrode or perforated sheets on the electrode faces. Detailed investigations of the chemistry and electrochemistry of the FeS electrode indicate that its reactions are much more complex than shown by the overall reaction of Eq. (1). As deduced from the Li–Fe–S phase diagram in Fig. 4,[12] the reactions for lithium ions on discharge of the FeS electrode are as follows.

$$2FeS + 2Li^+ + 2e^- \rightarrow Li_2FeS_2 + Fe \tag{4}$$
$$Li_2FeS_2 + 2Li^+ + 2e^- \rightarrow 2Li_2S + Fe \tag{5}$$

Both of these reactions take place at 1.34 to 1.33 V. However, in LiCl–KCl electrolyte, the formation of $LiK_6Fe_{24}S_{26}Cl$, which has been denoted the "J-

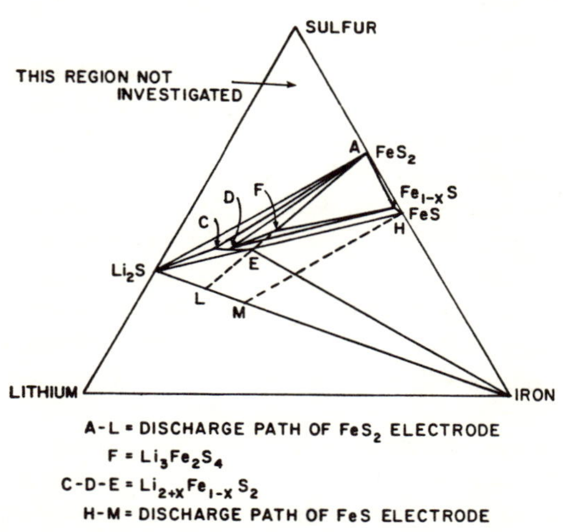

FIGURE 4. Phases in the Li–Fe–S system at 450°C.

phase," kinetically hinders the discharge of the cell. In early investigations, addition of Cu_2S to the FeS was found to inhibit J-phase formation, but, unfortunately, premature cell failures resulted from copper deposition within the separator region. Additional chemical studies indicated that J-phase formation could be suppressed by increasing the LiCl content of the electrolyte and by increasing the operating temperature of the cell.[14] Subsequent investigations of the FeS-electrode chemistry concentrated on the effectiveness of the molten salt composition. Compositions that were found to avoid J-phase formations, if the cells are operated above about 450°C, are LiCl-rich LiCl–KCl (68 mol % LiCl–32 mol % KCl) and an all-Li-cation electrolyte (22 mol % LiF–32 mol % LiCl–46 Mol% LiBr).

The objective of FeS electrode development is to optimize the specific energy and power, that is, to achieve high electrode capacity density and low internal electrode resistance. Secondary cells must also have long cycle life. As the FeS electrode discharges, it increases in volume; FeS has a density of 4.8 g/cm^3, and Li_2S, the most voluminous discharge product, has a density of only 1.6 g/cm^3. Thus, the FeS electrode must be designed with an initially high volume fraction of electrolyte, usually about 50%. To maintain dimensional stability, some of the electrolyte is forced out of the electrode during discharge to accommodate the higher volume of the discharge products. Also, the entire cell must be constrained with a pressure of about 100 kPa (15 psi) during cycling to avoid swelling and to force the transfer of electrolyte between electrodes. The amount of electrolyte to be transferred between electrodes on cycling can be reduced by the use of electrolyte-starved electrodes. This also reduces the required restraining pressure.

The FeS electrode requires higher overvoltage to maintain the current as the cell discharges. Ionic transport diminishes as the molten salt electrolyte is displaced by the increasing volume of the reaction products. Also, the electronic conductivity of the electrode bed diminishes, while the volume fraction of Li_2S (an insulator) increases as a result of the reaction given in Eq. (5). The FeS electrode exhibits an increasing rate of diminuition in conductivity at about 60% utilization of theoretical capacity.

In the optimization of FeS electrode performance, a number of engineering parameters have been considered. The capacity density of FeS per unit of electrode volume was optimized at 1.4 Ah/cm^3, as illustrated in Figs. 5 and 6. The capacity density affects the balance of electronic and ionic conductivity within the FeS electrode bed. Although the utilization of theoretical capacity diminishes as FeS volume replaces electrolyte volume, the cell specific energy is still improved. The addition of electronically conductive additives to the FeS electrode, such as carbon powder or fiber and Ni_3S_2, improves the specific power by augmenting the current collection. For high-density electrodes (1.4 Ah/cm^3), which are already very conductive, the use of additives is less advantageous. Similarly, the use of porous metal and carbon matrices to provide a distributed

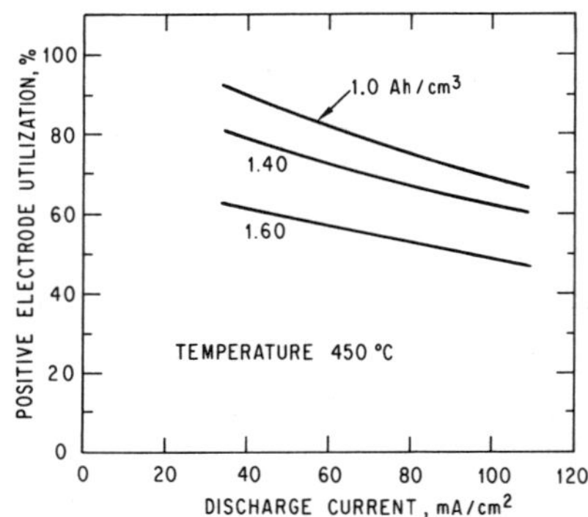

FIGURE 5. Effect of initial capacity density on FeS electrode utilization.

current collector for the FeS electrode yielded only marginal improvements in performance, and investigation of these additives has not been continued. Electrode impedance is not significantly affected by electrode thickness in the range of 0.5 to 1.0 cm for an electrode reacted from both faces, provided the FeS electrode has good current collection. Variation of the electrode thickness provides an approach for trading off specific power and energy. Greater specific power is achieved with more electrode area and lower cell weight, but with a slight decrease in specific energy.

The drawbacks of the FeS electrode performance, particularly low power at the end of the discharge, can be alleviated through limiting the depth of discharge to about 70% of the theoretical capacity. The development of cells having the capacity limited by the lithium electrode resulted in higher specific energy and power. An FeS capacity density of 1.4 Ah/cm³, an electrode thickness of 0.4 to 0.8 cm (reacted from both faces), and fabrication by cold-pressing a mixture of FeS and electrolyte powders resulted in near-optimum performance. Slurry meth-

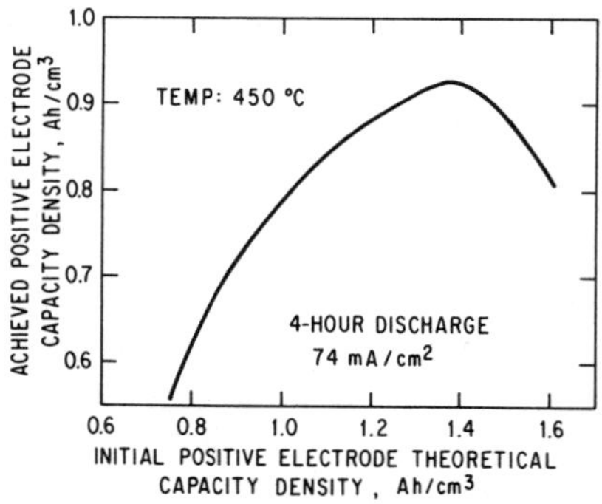

FIGURE 6. Effect of initial capacity density on achieved electrode capacity.

ods have also been used for fabrication, but only in laboratory studies at Argonne National Laboratory and not in engineering efforts by commercial firms.

The cycle life of Li-alloy/FeS cells was limited by the Li–Al electrode in early development efforts, as noted above, but later the FeS electrode became a limiting factor in achieving long cycle life. A cell shorting mechanism, which occurs at about 400 cycles for Li–Al–Si/LiF–LiCl–LiBr/FeS cells (developed by Gould, Inc.), is the deposition of iron from the positive electrode into the separator region. Contacting between the deposited particles electrically short-circuits the cell. Earlier Li–Al/FeS cells had operated for more than 1000 cycles without this short-circuiting problem. Peculiarities of the Gould design, which apparently cause repeated mild overcharging of the FeS electrode, are under study. Such overcharging results in formation of dissolved $FeCl_2$, which probably accounts for transfer of iron to the separator region. To further bolster the cell longevity, design changes that enable overcharge tolerance of these cells have been identified (see Section 4.1).

2.3. Separators

The separator for these molten salt cells provides (1) a porous medium that can be filled with molten salt for ionic conductivity and (2) an effective electronic insulator between the electrodes. The candidate materials are limited to those that withstand the high lithium activity of lithium-alloy electrodes, the high sulfur activity of the metal sulfide electrode, and the corrosiveness of molten salt electrolyte.[15–17]

The initial choice of separator material, boron nitride fiber, was first used in the form of a woven fabric and later as a nonwoven felt.[18,19] The felt has good strength and is 70% open space when compressed between the electrodes. It has demonstrated a long life of 1000 cycles of operation with Li–Al electrodes over two years. However, it is difficult to remove all the oxygen from the BN fiber during manufacture, and, as a result, the BN is attacked by lithium at high lithium activities near that of elemental lithium and a conductive film is formed.

Magnesia powder was later considered for the separator because it is more widely available and is less costly.[17,20,21] It also has greater chemical stability than BN fiber in contact with lithium at high activity. An MgO powder separator with very high surface area provides a separator having plasticlike properties with 60 to 80 vol % molten salt. However, cells having MgO powder separators must be operated in an electrolyte-starved state to avoid formation of a low-viscosity suspension of the MgO in the electrolyte, which would permit the electrodes to touch and form a short circuit. A composite of MgO powder/BN fiber structure is now being considered and may provide a good mix of physical and chemical properties.

The MgO powder separator is fabricated as a cold-pressed plaque from finely ground particles of electrolyte salt and high-surface-area MgO. These plaques have been made in rather large sizes of 12 × 17 cm sheets of 1- to 2-mm thickness. They are of sufficient strength for routine cell assembly. Separators of boron nitride felt sheets are handled similarly to the MgO powder plaques, except that the electrolyte is not incorporated into the separator during forming. Boron nitride is not readily wetted by the molten salt, and the wetting treatment (surface deposition of MgO) transforms the BN felt into a stiff, boardlike material. When BN felt is used, the electrolyte is vacuum-infiltrated after cell assembly. No electrolyte is added in cells having MgO powder separators.

Generally, the separator is physically supported by perforated metal sheets. Metal screens have also aided in particle retention. Cost considerations are driving separator development toward application of MgO, with approaches which rely on the resiliency and tortuosity of the separator for particle retention rather than the use of perforated metal sheets. The conditions of fabrication (MgO powder size and shape, pressing pressure) play an important role in the separator operating properties.

2.4. Electrolytes

The lithium-ion-containing molten salts used as electrolytes in these high-temperature cells have outstanding ionic conductance of 1 to 10 Ω^{-1} cm^{-1} for lithium ions. In the initial development work, the electrolyte was almost always the LiCl–KCl eutectic. Greater understanding of electrochemical interactions of the electrodes with electrolytes resulted in use of different electrolyte compositions that improved electrode kinetics and cell capacity stability.

For the FeS-electrode cells having MgO powder separators that are operated with limited electrolyte content (electrolyte starved), the LiF–LiCl–LiBr eutectic composition has been preferred. Since this electrolyte contains only lithium as a cation constituent, it will not undergo local freezing, as has sometimes occurred with LiCl–KCl electrolytes that develop large composition variations at high lithium-ion fluxes. Also the high lithium-ion conductivity of the all-lithium-cation electrolytes is important for the electrolyte-starved condition, which is required for MgO separators. The avoidance of J-phase formation resulting from potassium-ion reaction in the FeS electrodes is an added attraction. However, the cost of the LiF–LiCl–LiBr electrolyte is much higher than that of other electrolytes. In the future, it is expected that attention will be given to the use of lithium-rich LiCl–KCl electrolytes for FeS electrode cells having MgO powder separators.

2.5. Materials of Construction

For Li-alloy/FeS cells, the contemporary design is based upon a rectangular plate-type electrode and a prismatic cell design. This design was selected because of the ease of cell assembly and compactness of the battery. Corrosion studies have indicated that both low carbon and stainless steel are suitable for the cell housing. Cell housings can be formed by welding sheet metal or stamped forms and by deep drawing. Internal parts such as current collectors, electrode encapsulations, and electrode facing sheets have been made from both low-carbon steel and stainless steel. Above 500°C at the negative electrode, steel undergoes a slow reaction with aluminum to form aluminum–iron intermetallic compounds. This can generally be avoided by restricting the temperature during operation to below 500°C and recharging immediately after discharge. At the positive electrode, the current collector can be attacked galvanically by over-charging as the electrical potential and sulfur activity exceed those of the FeS electrode (above 1.5 V vs. Li–Al reference). Again, proper cell operation has avoided this problem, and, as discussed later, a chemical overcharge mechanism has been developed to provide additional protection. Low-carbon steel undergoes a low rate of intergranular corrosion, which is mitigated by the presence of a small excess of iron powder.[17] Under normal operating conditions, steel is expected to have an operating life of about five years based on a 10-μm/year corrosion rate. Nickel also has been used as an FeS electrode structural material. The disadvantage of the higher cost of nickel is at least partially offset by its increased electrical conductivity and extended corrosion life.

The prismatic cell design requires only one electrical feedthrough since one set of electrodes, usually the negative set, is grounded to the cell housing. The feedthrough is composed of annular ceramic insulating components[22] formed of BeO, MgO, or BN, which are positioned around a rod-shaped electrode terminal of about 6- to 12-mm diameter. The ceramic feedthrough components must satisfy the same chemical compatibility requirements as the separator. For most of the cells tested, the seal has been formed by compressing a preshaped annular ring of BN powder between a lower ceramic ring and an upper ceramic ring. This arrangement forces the BN powder against the rod-shaped terminal and the feedthrough housing.[23] In more recent proprietary developments at Electrofuel Corporation, Toronto, Canada, a hermetic, high-temperature seal is formed by means of a glass-to-metal seal. Based on current designs, the battery is intended to be operated under an inert blanketing gas, and a hermetic seal may not be required, but would be advantageous for long life. Lithium-alloy/FeS cells with hermetically sealed feedthroughs have been successfully operated for periods of about one year in air atmospheres. Feedthrough development efforts continue to address the issues of cost, stability, and design improvement.

2.6. Cell Design

Simplicity, reproducibility, and dependability are criteria to be met for low-cost fabrication of Li-alloy/FeS cells. Cells must be amenable to assembly into a high-density battery configuration for optimum results; the volume of the cell modules has a marked effect on the weight of the thermal insulating jacket. The specific energy and power of the cells are increased by minimizing nonactive material weights (hardware and electrolyte). The multiplate cell design with thick electrodes (<6 mm thick) has resulted in 200-Ah-capacity cells attaining about 20 to 25% of their theoretical specific energy (100 to 120 Wh/kg). These multiplate cells consist of rectangular, vertically oriented electrodes having electrode bus-bar connections welded together above the electrodes. The electrode faces have areas of 150 to 225 cm^2, with 15×15 cm and 12×18 cm being the most often used sizes. The battery power requirements usually dictate the cell engineering specifications: the steady-state power delivered by the battery is limited by the electrode current density, which should fall in the range of 50 to 150 mA/cm^2 for high utilization of active materials. The maximum cell power is determined by the internal impedance.

The capacity of the positive electrode relative to that of the negative electrode influences the internal cell impedance as a function of depth of discharge. Generally, for power-demanding applications (e.g., electric vehicles), usable capacity is limited by the need for the short power pulse required at the end of battery discharge for a satisfactory rate of vehicle acceleration. When the cell can no longer provide the required power for a 30-s acceleration, the cell is considered fully discharged. For that application, the lithium-limited cell design has demonstrated higher specific energy and higher specific power than the FeS-limited cell design. In particular, the FeS electrode impedance increases by 100% at about 60 to 80% of theoretical capacity, whereas the Li–Al impedance is relatively unchanged in this range of capacity. Therefore, the of the Li–Al electrode capacity is sized relative to that of the FeS electrode so that it is exhausted before the FeS electrode begins its upswing in electrode impedance. The cell capacity, stability, and performance are optimal at about 80% utilization of the Li–Al electrode and about 70% utilization of the FeS electrode.

The current collector structures for a Li–Al/FeS cell are important not only because they affect specific energy and power, but also because they encapsulate the electrodes. The negative and positive electrodes are each enclosed in a picture-frame type of structure, which prevents active material from being forced out of the edges of electrodes by internal pressure and short-circuiting the cell. The front faces of the electrodes typically have been fabricated of perforated metal (usually 5-mil-thick stainless steel) to retain the electrode particles. Sometimes, this perforated sheet has been backed with a fine screen on the electrode side of the sheet. In some designs, the perforated sheet has been the main current collector, but usually the main current collector is a flat sheet located at the center of the electrode and welded to a bar across the top edge of the electrode. This type of structure requires that the positive electrode be fabricated as two thin plaques that are placed one on each side of the central sheet. In other designs, the active material has been pressed into a grid-type structure, which is then enclosed by a picture-frame structure that contains perforated sheets on the electrode faces. This type of structure requires close spacing of the grid structure in the positive electrode to limit the electronic resistivity of the FeS, which otherwise would be a major portion of the cell internal resistance. However, this design is not favorably considered now because it is difficult to press FeS into a closely spaced grid structure without damaging the structure.

Each set of electrodes is interconnected by a bus-bar structure above the electrodes. The welding of the electrode tabs to the bus bar is done after assembling all the electrodes and separators into a single unit, as described below.

Inappropriate design or poor connection of the electrode tabs, interelectrode bus bar, or the cell terminal may be a source of high internal cell resistance. For instance, an insufficient cross section for the welds has led to much higher resistance than expected. To minimize internal cell resistance, the best designs have been based on broad electrode tabs, substantial interelectrode bus bars, and large-diameter cell terminals (6 to 12 mm). In properly designed cells of approximately 250-Ah capacity, the resistance from the electrode terminals to the cell terminals has been 25 to 50 μ Ω. For cells of that size, the resistance is in the range of 0.5 to 0.7 Ω/cm^2 of separator area.

2.7. Cell Assembly

Cell assembly is designed to be simple and to require only a few steps,[24] as illustrated in Fig. 7. As indicated previously, the electrodes are assembled from cold-pressed plaques with the brittle electrode particles held together by an electrolyte matrix. Plaques are positioned on each side of the sheet current collector. Framelike enclosures which have solid edge caps and perforated metal faces are positioned over this subassembly. Spot welding (or a crimping operation) secures the enclosures. The electrodes are then stacked in a fixture in the following order: outside negative electrode, separator/electrolyte plaque, positive electrode, inside negative electrode, positive electrode, and so on. A compressive force is applied to the stack to maintain positioning. In some cell designs, separator material is positioned at the side and top edges of this stack. The stack is then slid into the top of a properly sized cell housing, which retains the compressive force on the stack. The cell terminal and interelectrode bus bar assemblies are positioned and welded into place. In some cell designs, a preassembled feedthrough has been used. The base of the terminal in this unit is welded to the internal bus bar assembly before the cell top is welded closed.

Earlier cells, particularly with BN fiber separators, required vacuum infiltration of electrolyte through a fill tube, which was welded closed after the filling. The contemporary cells with MgO powder separators contain the required amount of electrolyte within the cell components. Hermetically sealed cells are assembled in an air atmosphere. The small amount of air trapped in the cell reacts with the lithium on heating, thus avoiding pressure buildup from the thermal expansion of trapped gas.

After assembly, the cells are submitted to a quality-control procedure. Cells

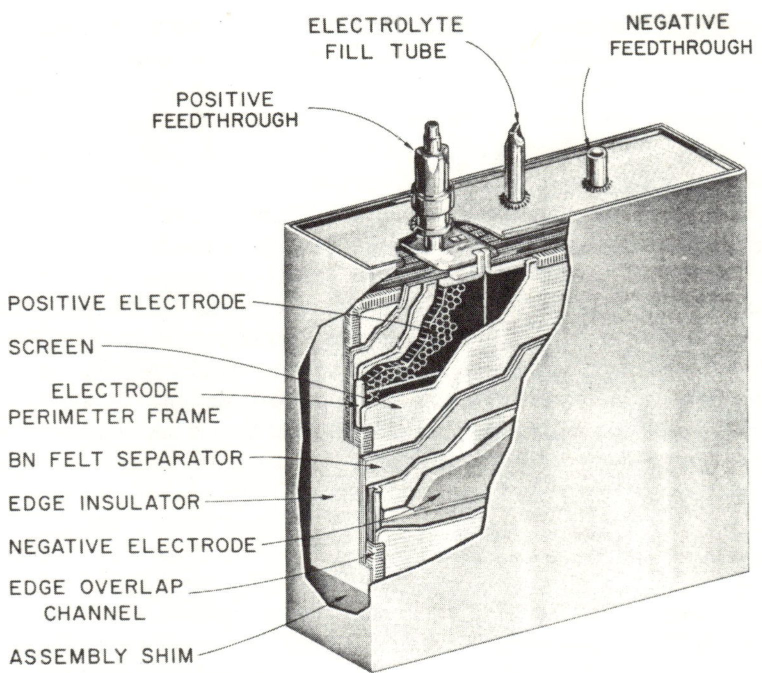

FIGURE 7. Li–Al/FeS multiplate cell design.

are checked for cold resistance (~ 1 MΩ) and gas tightness (which varies relative to feedthrough design). Cells for use in developmental batteries are typically given break-in and performance tests before being assembled into a battery.

The cold-pressing electrode fabrication and the simple cell assembly procedure enable a single technician to build cells at a rate of about four per day. A battery production facility would require automated handling procedures, but this appears to be achievable; cold-pressed primary batteries of the Li–Al/FeS$_2$ type are already being made with semiautomated equipment.

In the above procedure, the cells are fabricated in the fully charged state because of long experience with this method and the ease of fabrication. However, it may eventually be more cost effective to fabricate cells in the uncharged state (porous Al negative electrodes and Li$_2$S + Fe positive electrodes) or even a partially charged state, which mixes discharged materials with charged materials. Cost projections indicate substantial savings can be achieved by using Li$_2$S as the source of lithium for the cell, but the difficulties in manufacturing the cells with the chalklike Li$_2$S have not been fully assessed. It is likely that fabrication methods other than cold-pressing would be required for uncharged cell manufacture.

2.8. Cell Testing

The purpose of cell testing is to evaluate the status of cell performance (i.e., specific energy and power) and cell stability (i.e., cycle life and capacity retention) at set operating conditions. The most common testing consists of constant-current discharge (25 to 200 mA/cm^2) to a voltage cutoff (1.0 to 0.9 V) and constant-current charge (15 to 50 mA/cm^2) to a voltage cutoff (1.45 to 1.55 V). The charging scheme has been receiving attention as an approach to further extend the cycle life of FeS cells (discussed in Section 4.1). Computer-simulated driving profile tests are conducted as a means of more accurately projecting cell performance under vehicle driving conditions. Cells are prepared for testing by securing them between $\frac{1}{4}$-in.-thick stainless steel plates to simulate the degree of dimensional restraint expected from the confinement of cells in the battery. Separate current and voltage leads are attached before the test. The cells are brought up to operating temperature within one half hour, and operation is initiated after an additional half hour.

A number of diagnostic tests have been used. For peak power evaluation, cells were submitted to periodic pulse power sequences of 15- to 30-s duration with 2-min intervening rests. The power-pulse testing ranged from 100 to 1000 mA/cm^2 at depths of discharge of 5, 50, and 80% of the determined cell capacity. From these measurements, the internal impedance of the cell and the power at any current can be determined. Impedance measurements have also been made by interrupting the cell current for short periods during discharge of the cell. Varying the length of the period for voltage relaxation provides information on the values of diffusion overvoltage and resistance losses.

Determinations of cell electrode potentials by use of reference electrodes have provided better understanding of cell degradation processes and helped in designing new components that improve performance. The reference electrodes developed for this use have long-term stability and are isolated with a high-impedance operational amplifier. The coordinated use of reference electrode monitoring with other cell diagnostic tests, such as the current interruption technique, provides information on the impedance of separate cell components. The reference electrode is also used extensively for investigation of cycle-life-related phenomena. Specifically, cell capacity decline has been caused by degradation of only the positive or negative electrode; capacity decline and cell short-circuiting in recent tests are believed to be related to polarization of the positive electrode at the end of charging, which causes formation of dissolved FeCl$_2$, followed by precipitation of iron in the separator. These phenomena were readily studied by means of reference electrodes. More recently, overcharge tolerance tests have been conducted routinely with the use of reference electrodes in cells specifically designed for these studies. The need for overcharge tolerance to permit equalization of the state of charge among cells becomes apparent at battery-level testing, and this is discussed in Section 4.

Tests to evaluate cell durability have been conducted on a nonroutine basis. Freeze/thaw cycling has been studied in detail, and little impact on cell performance of this thermal cycling is apparent. During the course of normal testing, cells have been frozen, repaired, and rapidly reheated without concern for the effect on cell performance. Also, vibration and shock tests have been conducted on individual cells without effect on cell performance.

Another important area of the testing program has been postoperative evaluation. Extensive metallographic, analytical chemistry, and scanning-electron microscopy analyses have provided valuable information for understanding cell degradation phenomena. In the earlier work, cell mechanical faults contributed to cell short-circuiting. Also, the agglomeration in the Li–Al electrodes and the deposition of iron from the positive electrode into the separator region were identified, by postoperative examination, as contributing to cell degradation.

At this stage, lithium-limited FeS cells with MgO powder separators (Westinghouse Oceanic Division, Cleveland, Ohio) are generally perceived as being furthest along in development. These lithium-limited FeS cells have good power at 80% depth of discharge and provide a usable capacity of about 235 Ah under driving profile testing at a specific energy of about 115 Wh/kg. The cycle-life expectancy of these cells is about 400 deep discharge cycles. Another developer, Electrofuel of Canada, uses a BN fiber-based separator, sometimes with the addition of MgO powder, and a LiCl–KCl electrolyte. Less is known about the performance of these somewhat smaller cells, but cycle life is reported to be greater than 700 cycles.

3. Lithium/Disulfide Cells

The Li-alloy/FeS_2 cell is similar to the Li-alloy/FeS cell in many respects, but there are important differences. Some similarities are the use of essentially the same negative electrode, the same type of steel cell container (which is at negative potential in both cases), and the same separator. The most important difference between the two types of cells is the sulfur activity in the positive electrode.

Early versions of the FeS_2-electrode cell had only marginal success; these cells had specific energies only about 20% greater than those of Li–Al/FeS cells, and they suffered unacceptably high capacity decline rates of 0.1 to 0.25% per cycle. The decline in capacity was especially severe for the upper voltage plateau and resulted from sulfide species migrating from the disulfide electrode into the separator region and depositing there as Li_2S.[25]

A new electrolyte and a new approach to the design of the FeS_2 electrode, called the dense upper-plateau (U.P.) FeS_2 electrode, dramatically improved cell performance and capacity stability.[6] The new electrolyte,[26] 25 mol % LiCl–37 mol % LiBr–38 mol % KBr (mp 310°C), rather than 58 mol % LiCl–42 mol % KCl (mp 352°C), permitted operation of the cell at lower temperature. This solved the problem of sulfide species migration and resulted in longer life. The improved electrode design involved the use of a higher loading density (2.4 vs. 1.5 Ah/cm^3) for the FeS_2 electrode, which was then operated only on its higher voltage plateau (1.66 V vs. Li–Al).[27]

As indicated in Table I, the limitation of the discharge to the formation of Li_2FeS_2 results in a theoretical specific energy of only 475 Wh/kg. However, the higher average discharge voltage and the reduction in electrolyte weight resulting from the high loading density more than compensate for the higher specific energy of cells designed to operate on both voltage plateaus.

3.1. Lithium–Aluminum Electrodes

The Li-alloy electrodes that have been developed for the FeS cells have direct application to cells with FeS_2 positive electrodes. Lithium–aluminum electrodes having safeguards against sintering (carbon or MgO powder additions) appear to be most appropriate. In advanced cell designs, the Li-alloy electrode is called upon to prevent overcharge or overdischarge of the positive electrode. As a result, the Li-alloy electrode must remain stable, even under severe potential swings.

Limiting the discharge of the positive electrode is especially important for practical operation of the Li–Al/U.P. FeS_2 cell. With the high loading density of the FeS_2 electrode, the discharge must be restricted to the capacity of the upper voltage plateau of the FeS_2 electrode.[28] Excessive expansion of the FeS_2 electrode would result from discharge at the lower FeS_2 voltage plateau because of low-density Li_2S formation. It is impractical to limit the discharge of each cell in a battery by voltage controls. The most straightforward method of preventing overdischarge of the positive electrode is to restrict the capacity of the negative electrode. Typically, this is accomplished by designing for 90% utilization of the U.P. FeS_2 capacity and 80% utilization of the Li–Al electrode. This results in a marked reduction in cell voltage toward the end of discharge without appreciable formation of Li_2S in the positive electrode. The Li–Al electrode is polarized by at least 300 mV at the end of each cycle. A concern with operating the Li–Al electrodes under such severe conditions was that aluminum might deposit in the separator. This concern was mitigated by the operation of a Li–Al/U.P. FeS_2 cell with discharge cutoff voltages ranging from 1.0 to 1.3 V (cell 5 in Table IV). This cell maintained 98+% coulombic efficiency for over 900 cycles. In the operation of a battery of such cells that are well matched in state of charge, the voltage and power of the entire battery will decline markedly at the end of discharge, and extended overdischarge of any individual cell would be avoided. The means of maintaining the matching of state of charge of cells in the battery is discussed in Section 4.1.

TABLE IV. Cycle-Life Tests of Upper-Plateau (U.P.) FeS_2 Cells[a]

		Performance		Lifetime	
		U.P. FeS_2		Capacity	
Cell		utilized,	Discharge	Change	
no.	Composition of limiting electrode (mol %)	(%)	rate (h)	(%)	Cycles
1	FeS_2:CoS_2 (85:15)	89	4	1	400
3	FeS_2:CoS (85:15)	88	4	4	260
4	FeS_2:CoS_2 (85:15)[b]	80	2	18	1020
5	Li–Al, carbon	55	1.3	10	900
6	FeS_2:NiS_2:Li_2S (50:50:5)	93	4	5	250
7	FeS_2:CoS_2 (85:15), overcharge tolerant	86	4	1	>250

[a]Prismatic bicells use LiCl–LiBr–KBr (mp 310°C) with BN-felt separator (100 cm^2) and 400°C operation.
[b]Cell capacity was limited by the negative electrode after 500 cylces, as determined by the reference electrode.

3.2. Disulfide Electrodes

The disulfide electrode discharges through several stages as it reacts along the path shown in the three-component diagram of Fig. 4.[12] The sulfide phases observed during charging of Li–Al/FeS_2 cells are shown in Table V[29,30] and the voltage versus capacity curve of a Li–Al/FeS_2 cell discharged at a low rate is shown in Fig. 8.[31] Experimental results and calculations of cell weight have indicated that the best combination of high power and high specific energy can be obtained in a practical Li–Al/FeS_2 cell by limiting the discharge to the "A",

TABLE V. Sulfide Phases Observed during Charge of Li–Al/FeS_2 cells Operated at 410°C Using Eutectic Electrolyte[a]

| Charge potential[b] vs. Li–Al ($\alpha + \beta$) | X-ray findings | | Metallographic findings |
	Major phase	Minor phase	
1.53	Li_2FeS_2	$LiK_6Fe_{24}S_{26}Cl$	Li_2FeS_2 + trace of $LiK_6Fe_{24}S_{26}Cl$
1.64	$Li_{2.33}Fe_{0.67}S_2$ and	$Fe_{1-x}S$	Li_2FeS_2 + $Fe_{1-x}S$ + $Li_{2.23}Fe_{0.67}S_2$
1.72	Li_2FeS_2	None detected	$Li_3Fe_2S_4$ only
1.79	$Li_3Fe_2S_4$	None detected	$Li_3Fe_2S_4$ + 5% $Fe_{1-x}S$
1.82	$Li_3Fe_2S_4$	FeS_2 + $Fe_{1-x}S$	Not examined
1.85	$Li_3Fe_2S_4$ FeS_2	$Fe_{1-x}S$	Fe_2S + $Fe_{1-x}S$

[a]Li_2S and Fe react at lower potential (1.33 V) to form Li_2FeS_2.
[b]Constant-current charge at ~18 mA/cm^2 followed by >18-h constant-voltage charge (final current density, >2 mA/cm^2).

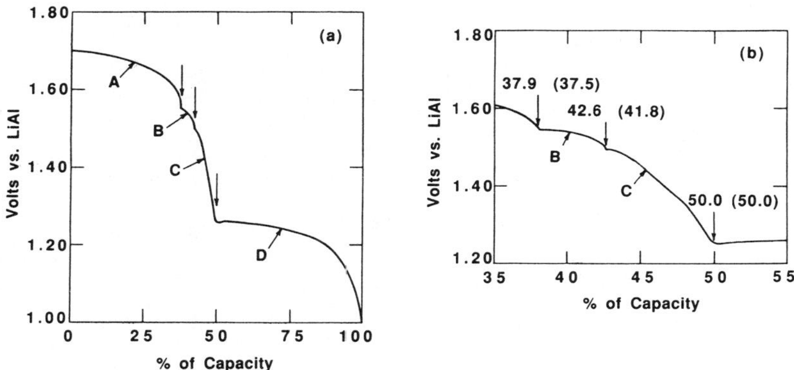

$$FeS_2 \xrightarrow{A} Li_3Fe_2S_4 \xrightarrow{B} Li_{2-x}Fe_{1-x}S_2 + Fe_{1-x}S \xrightarrow{C} Li_2FeS_2 \xrightarrow{D} Li_2S + Fe$$

FIGURE 8. Voltage curves on discharge for Li–Al/FeS$_2$ cell. Electrolyte: LiCl–KCl eutectic; temperature: 417°C; current density: 12 mA/cm^2.

"B," and "C" reactions shown in Fig. 8, or continuing only slightly beyond these.[6] Reaction on the "D" part of the discharge curve requires a larger Li–Al electrode and a higher volume fraction of electrolyte in the positive electrode to accommodate the low-density Li$_2$S formed on discharge on the "D" part of the curve. These larger weights coupled with the lower voltage for the "D" part of the curve result in a lower specific energy, lower power near the end of the discharge, and the requirement that the power conversion system be able to accommodate a large voltage drop on discharge. Therefore, recent experimental work has been confined to cells operating on the upper plateau of the voltage curve.[28]

The use of various additives in the FeS$_2$ electrode to improve stability or to increase electronic conductivity has been evaluated. Additives evaluated include NiS$_2$, CoS$_2$, TiS$_2$, and Li$_2$S.[13,32,33] The most successful additive has been CoS$_2$.

In tests of the FeS$_2$ electrode, cells were prepared having 100-cm^2 separators, 24-Ah FeS$_2$ electrodes, and 35-Ah negative electrodes containing 53 atom % Li–Al alloy. The excess capacity of the negative electrodes provided an opportunity to test the positive electrodes in deep discharges. The positive electrodes contained 15 mol % CoS$_2$ and were fabricated either as dense (2.4 Ah/cm^2) U.P. electrodes for discharge to Li$_2$FeS$_2$ or as a looser compaction (1.5 Ah/cm^2) for discharge on both voltage plateaus to iron and Li$_2$S.

The discharge curves for cells with these two different positive electrodes at 50 mA/cm^2 are shown in Fig. 9. The U.P. FeS$_2$ electrode clearly outperformed the two-plateau electrode, providing approximately 50% higher utilization of the theoretical capacity at a 10% higher average discharge voltage. The capacity

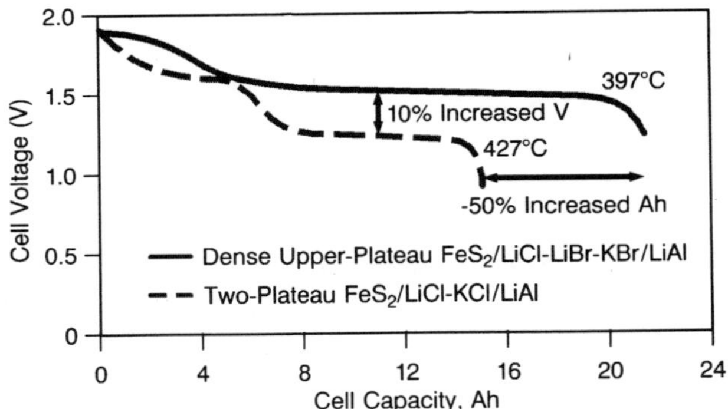

FIGURE 9. Voltage/capacity curves at a discharge rate of 50 mA/cm². FeS_2 cell energy density improved by 75% and power doubled at 80% DOD for U.P. FeS_2 electrode as compared to two-plateau electrode [cells of same volume and separator area (100 cm²)].

utilization of the U.P. FeS_2 electrode was even better at higher discharge rates, with utilizations of the FeS_2 capacity on the upper plateau of 82 and 75% for 100 and 150 mA/cm² current densities, respectively, at 397°C. The two-plateau cell achieved only 45% utilization at 150 mA/cm² and 427°C. The utilization achieved for the U.P. FeS_2 electrode at a discharge current density of 100 mA/cm² only decreased slightly as temperature was decreased from 427 to 388°C.

The improved power and energy densities of the iron disulfide electrode at the reduced operating temperatures of 388 to 427°C are believed to have resulted primarily from improved electronic conductivity of the electrode. According to the Bruggeman equation,[34]

$$Km = (1 - f)^{3/2} \tag{6}$$

where Km is the bed conductivity/dense body conductivity ratio, and f is the electrolyte fraction. Thus, the increased electrode loading density (from 32 to 50 vol %) doubles the electronic conductivity of the iron disulfide electrode.

3.3. Electrolytes

The performance and stability of the U.P. FeS_2 electrode appears highly dependent upon the molten salt electrolyte composition. The electrolyte of choice, 25 mol % LiCl–37 mol % LiBr–38 mol % KBr (mp 310°C), has many attractive properties, which give it preference over the commonly used 58 mol % LiCl–42 mol % KCl (mp 352°C).

Its liquidus range of 1.25 to 2.6 Li^+/K^+ ratio at 400°C is of particular benefit to the dynamic conditions of electrode operation at high current density. For LiCl–KCl, the liquidus range at 400°C is only 1.25 to 1.81 Li^+/K^+ ratio. The broader liquidus for LiCl–LiBr–KBr would tend to prevent salt crystallization with variation in the Li^+/K^+ ratio resulting from Li^+ ion transport. Melting point and cooling curve determinations indicate a melting point of 321 ± 1°C, but, once liquid, this molten salt exhibits a lower freezing point of 296 ± 4°C. Accordingly, test cells began to exhibit an output voltage during initial heating when the temperature reached 310 ± 10°C.[35]

The specific resistivity of the molten salt and the solubility of the metal disulfide electrode materials are linked to Li^+ ion activity and temperature. An added consideration is the thermal stability of FeS_2. It has been determined[36] that thermal decomposition of FeS_2 at 400°C is rather slow and that thousands of hours would pass before a significant amount of decomposition would occur, whereas at 500°C decomposition would occur within a few hours. Earlier studies have documented the increasing solubility of Li_2S in molten salts with increasing Li^+ ion content and increasing temperature.[37] The solubility of Li_2S at 450°C is 1860 ppm in 65 mol % LiCl–35 mol % KCl and 840 ppm in 55 mol % LiCl–45 mol % KCl. The all-Li^+-cat ion salt, LiCl–LiBr–LiF, at 465°C has a much higher Li_2S solubility[38] of 6840 ppm. Generally, a 50°C temperature rise in the range of 400 to 500°C will approximately double Li_2S solubility in all of these molten salts. The specific resistivity of the LiCl–KCl eutectic at 400°C is approximately 0.8 $\Omega \cdot cm$, and it decreases as the LiCl mole fraction is increased. The liquidus point increases more rapidly for the LiCl-rich compositions of LiCl–KCl than for LiCl–LiBr–KBr. It is apparent that, compared with the LiCl–KCl eutectic, the LiCl–LiBr–KBr electrolyte allows a lower temperature operation, which minimizes the thermal decomposition and solubility of the electrode material. It also provides a higher Li^+ ion concentration, which improves FeS_2 electrode performance by increasing the Li^+ ion conductivity.

Cyclic voltammetry experiments[35] have indicated good stability of FeS_2 in LiCl–LiBr–KBr. At 425°C or less, soluble polysulfides are not generated, as evidenced by the lack of activity in the voltage region 1.95 to 2.10 V vs. Li–Al. Generation of soluble polysulfides would lead to electrode capacity loss. Unlike FeS_2, both CoS_2 and NiS_2, in charging to their higher voltage reaction, generate Li_2S, which can be further oxidized to soluble Li_2S_2.[39]

Cyclic voltammograms indicated more complete reversibility of FeS_2 electrodes in LiCl–LiBr–KBr than in LiCl–KCl, as determined by the smaller peak separation between the higher voltage reaction peaks.[35] For the LiCl–KCl electrolyte, separation in the peaks on charge and discharge indicated a slight irreversibility, with formation of a chemical species near the top of charge that is probably associated with capacity loss on cycling.

3.4. Positive-Electrode Current Collectors

The higher sulfur and chlorine activities encountered in cells with the FeS_2 electrode relative to those for cells with the FeS electrode require current collector materials having higher corrosion resistance. For most experimental studies, the current collector material for the FeS_2 electrode has been molybdenum, rather than iron or nickel as used for FeS electrodes.

Other materials that have been considered are carbon, tungsten, titanium nitride, and titanium carbide, which are difficult to work with and have lower conductivities. Molybdenum contributes to the long-cycle-life capability of the U.P. FeS_2 cell. Its high oxidation potential of 2.4 V vs. Li–Al is 0.35 V above the normal charge voltage cutoff of 2.05 V. The corrosion rate has been found to be sufficiently low such that a 0.025-mm-thick molybdenum layer on a metallic substrate would provide a 10-year corrosion life[16] under ideal conditions, with no cracking or spalling of the coating.

Welding of molybdenum parts requires special attention. The metal recrystallizes to form a somewhat brittle connection. Molybdenum current collectors have been designed such that little mechanical stress upon the weld area results from cell assembly and operation. For these designs, the welds serve mainly to improve electrical conductivity.

There is considerable promise for developing coated current-collector structures[40] from lower-cost base metals such as Fe or Ni. Coatings of Mo, MoC, and TiN[41] can be obtained by a number of methods. Success with a coated current collector for the disulfide electrode would increase design flexibility, as well as lead to lower cell costs.

3.5. Cell Configuration and Assembly

The assembly of the Li-alloy/FeS_2 cell is similar to that of the Li-alloy/FeS cells (see Section 2.7) except that the positive current collector requires a different design. Multiplate cells of Li–Al/FeS typically have been assembled with welded connections between the current collectors and a bus bar within the cell housing.[42] Suitable welding procedures have not been developed for multiplate FeS_2 cells having molybdenum current collectors. Therefore, the FeS_2 cells have been assembled with a single positive electrode and a central molybdenum current collector sheet welded to a vertical rod that extends through the feedthrough in the top of the cell housing. This type of design would limit the cell size to about 100–150 Ah.

3.6. Disulfide Cell Testing

3.6.1. Cycle-Life Stability

In early tests of cells having LiCl–KCl electrolyte and FeS_2 electrodes operated on both voltage plateaus, special care was taken in the charging procedures to enhance the cycle life. The cells were charged at low current density (15 mA/cm^2) and to a charge voltage cutoff of no greater than 1.9 V. By these procedures, generation of soluble polysulfide, which would migrate into the separator and deposit there as Li_2S, was diminished. Although this resulted in improved cycle life, it was achieved at the cost of not fully charging the cell and suffering the associated capacity loss. Voltammetric experiments discussed in Section 3.3 showed the means for improving cycle life without loss of capacity. These experiments on FeS_2 and LiCl–LiBr–KBr electrolyte indicated no generation of polysulfide during the charge half cycle.[35] Tests of Li–Al/U.P. FeS_2 cells having LiCl–LiBr–KBr electrolyte demonstrated the usefulness of this approach. Such cells were routinely charged at higher rates (25–30 mA/cm^2) to a higher voltage cutoff (2.05 V vs. 1.95 V) than those having LiCl–KCl electrolyte and exhibited a much lower rate of capacity decline on cycling.

An example of the excellent cycle-life stability achieved for Li–Al/U.P. FeS_2 cells[28] is shown in Fig. 10. This cell showed only moderate loss of capacity after 1000 cycles and 7000 h of continuous operation. In contrast, Li–Al/FeS_2 cells having LiCl–KCl electrolyte and operated on both voltage plateaus had a mean life of only about 100 cycles. The Li–Al/U.P. FeS_2 cell capacity remained constant in the first 500 cycles; thereafter, test facility equipment (e.g., glovebox failures) resulted in 18% decline in capacity after 1000 cycles (Fig. 10). A reference electrode indicated that cell capacity had become limited by the Li-alloy electrode. The U.P. FeS_2 electrode retained at least 90% of its initial capacity, as indicated by the reference electrode, throughout the 1000-cycle test. The loss of Li-alloy electrode capacity resulted in more abrupt cell voltage swings at the end of charge and discharge. During the 1000 cycles of operation, the high specific power of the cell did not change, the area-specific impedance remained at about 0.7 $\Omega \cdot cm^2$ over the range of 5 to 80% depth of discharge, and the coulombic efficiency was 96+%.

In five of six U.P. FeS_2 cells (Table IV), the utilization of the capacity on the

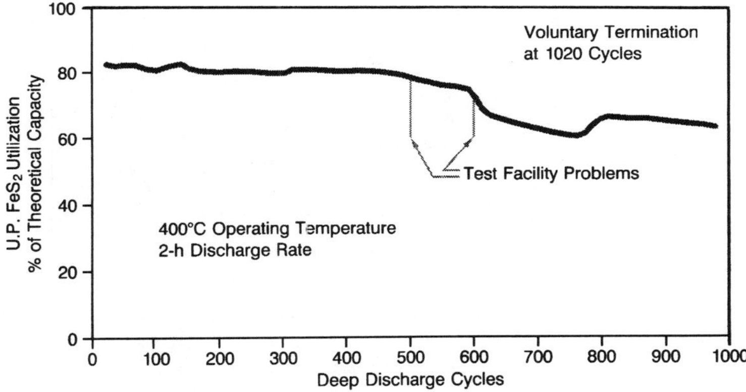

FIGURE 10. 1000-cycle test of Li-alloy/U.P. FeS_2 cell (cell capacity limited by the Li–Al electrode after cycle 500; <5% decline in 500 cycles, <20% decline in 1000 cycles).

upper voltage plateau was 80–90% at discharge rates of 2–4 h. For cell no. 5, utilization was limited by the capacity in the Li-alloy electrode as a test of that electrode. In testing of cell no. 4, the lithium electrode became limiting after 500 cycles. The results on the other cells indicated greater than 95% capacity retention for the U.P. FeS_2 electrodes. The coulombic efficiencies during these tests were 96–99%.

3.6.2. Performance

Performance tests have demonstrated that the power of Li–Al/FeS_2 U.P. cells is appreciably higher than that of cells designed to use both voltage plateaus (T.P. cells). Tests of a Li–Al/LiCl–LiBr–KBr/dense U.P. FeS_2–15 mol % CoS_2 cell at 400°C demonstrated at least a 50% improved energy density in comparison to that of a Li–Al/LiCl–KCl/T.P. FeS_2–15 mol % CoS_2 cell of equal theoretical capacity operated at 427°C.[43] As seen in Fig. 9, capacity utilization is almost 50% greater, and the average discharge voltage is 10% higher. Cell power density was improved by at least 100% because of two factors. First, the cell voltage at 80% depth of discharge (DOD) was about 0.3 V higher than that of a T.P. FeS_2 cell. Second, cell impedance was lower for the dense U.P. FeS_2 cell at 400°C, ranging from 0.65 to 0.85 $\Omega \cdot cm^2$ for 5–80% DOD compared with 1.2 to 1.6 $\Omega \cdot cm^2$ for the T.P. FeS_2 cell at 427°C. The higher voltage and lower resistivity of the U.P. FeS_2 cell increased the power of the cell from 0.3 W/cm^2 to 0.8 W/cm^2 (based on separator area) at 80% DOD.

High capacity utilizations of about 80% were achieved for cells having U.P. FeS_2 electrodes and LiCl–LiBr–KBr electrolyte, when operated at current densities of 50–150 mA/cm^2 and at a temperature of 397°C (Fig. 11). The results attest to the excellent electrode kinetics of this system. At a current density of

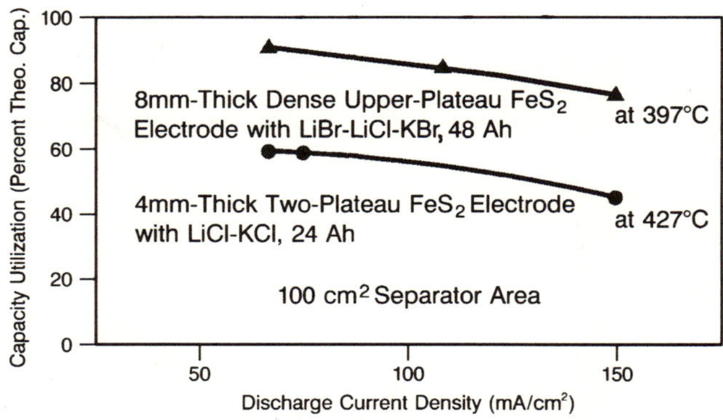

FIGURE 11. Utilization versus discharge current density for Li–Al/FeS_2 cells with 100-cm^2 separator area. The electrode thickness of the U.P. FeS_2 cell was double that of the two-plateau FeS_2 cell. High performance thick electrodes increased area specific capacity fourfold.

150 mA/cm^2, the utilization was 70% greater than that for FeS_2 electrodes operated on both voltage plateaus. Cells with U.P. FeS_2 electrodes have been operated at temperatures as low as 380°C, but the sensitivity of the cell to operating temperature becomes appreciable at temperatures less than 400°C. At a 1-h discharge rate (150 mA/cm^2), 78% utilization is achieved at 400°C.

In summary, the performance achieved in these tests indicates that a 250-Ah prismatic cell having two U.P. dense FeS_2 electrodes and operated at 400°C would be expected to have a specific energy of 175 Wh/kg at the 4-h rate and a specific power of 200 W/kg at 80% DOD.[42]

4. Prismatic-Cell Batteries

The assembly of batteries of cells that operate at high temperatures is obviously complicated by the need for a well-insulated jacket with temperature control and for bus bars and sensors within the jacket. These would seem to be engineering problems, not necessarily appropriate for a chapter dealing primarily with the materials and construction of molten salt batteries. However, some of the problems associated with battery operation can best be solved by fundamental changes in the cell chemistry. For this reason, attention is given here to battery development. The problems associated with batteries of prismatic cells also encourage strong consideration of bipolar batteries (see Section 5).

4.1. Effect of Battery Operations on Cell Performance

In lithium/iron sulfide battery development programs, most of the cell testing has been carried out with individual cells. As noted above, much emphasis is placed on determining the energy and power per unit weight and the cycle life of the cells. In carrying out these tests, cells are usually discharged until the voltage decreases to a designated level, and then they are recharged until the voltage reaches a safe upper limit. By this means, the cells are protected from overdischarging or overcharging.

When cells are connected in a battery arrangement, it is more complicated and, perhaps, impractical to provide electronic controls that cycle each individual cell between specified voltage limits. Ordinarily, the cells are connected in series so that each cell (or each group of parallel-connected cells in a series-parallel configuration) conducts the same current on charge and discharge. If the cells are mismatched in capacity or, worse yet, have slightly different coulombic efficiencies, the voltage of the cells at the end of charge and discharge may vary considerably. For instance, if two cells connected in series receive and deliver the same coulombic charge but have coulombic efficiencies that differ by 1% (e.g., 99% vs. 98%), these cells will be mismatched in state of charge by 50% of their capacity after 50 cycles. If some mechanism or action does not alleviate this problem, the battery would probably be inoperable before reaching 50 cycles because of problems arising from overcharge or overdischarge of individual cells.

For the lithium/sulfide batteries, overcharging generally results in oxidation of the current collector in the positive electrode. For example, in the monosulfide system, which has an iron current collector, overcharging results in dissolution of the current collector to form Fe^{2+} ions in increasing concentrations in the molten salt as the voltage is raised during overcharge, becoming a serious problem at about 1.7 V vs. Li–Al. For an FeS_2 electrode having a molybdenum current collector, Mo^{2+} ions would form in significant quantities starting at about 2.4 V. These reactions on overcharge result in destruction of the positive current collectors and short-circuiting of the cells by deposition of metallic particles in the separator.

If a sensing wire is provided for each voltage stage in the battery, the mismatching of cells can be adjusted by a small trickle charge through this sensing wire. However, this is a complex system. A vehicle battery may require more than 100 cells, and the introduction of a wire for every cell through the insulated casing provides an opportunity for short circuits through these wires and adds to the cost. Also, an electronic charger/equalizer is more expensive than a simple charger.

In aqueous electrolyte batteries, such as the lead–acid battery, overcharge results in hydrolysis of the electrolyte to form hydrogen at the negative electrode and oxygen at the positive electrode, rather than attack of the current-collector structure. Because the gases so formed recombine within the cell, the excess current through charged cells is harmlessly dissipated as cells that are undercharged gradually approach full charge. This mechanism for repeatedly rematching the state of charge of cells in the battery cannot be effectively used in molten salt batteries because practical current-collector materials are attacked before the electrolyte decomposes and generates chlorine gas at the positive electrode. It has been found, however, that there are means of designing lithium/iron sulfide cells that are tolerant to a trickle charge for an essentially indefinite period after the cell has reached full charge without deleterious effects to the current collector. The mechanism for this overcharge tolerance is an increase in the self-discharge rate as the cell reaches full charge. The mechanism involves a "lithium shuttle"[44,45]: (1) dissolution of Li^0 (possibly as Li_2^+) at the Li-alloy electrode surface, (2) diffusion of the Li^0 across the electrolyte/separator to the positive electrode, (3) electrochemical discharge of the Li^0 to produce Li^+ ion at the positive electrode, and (4) transfer of Li^+ ion to the negative electrode at the trickle charge rate.

If the diffusion of Li^0 in the above mechanism matches the trickle charge rate, the lithium shuttle mechanism will maintain the cell unharmed at full charge while undercharged cells continue to charge. The parameters that control this mechanism are the solubility of Li^0 in the electrolyte, which is a function of the voltage at the electrode relative to Li^0, and the rate of diffusion of Li^0 through the electrolyte/separator structure.

The amounts of the constituents in the electrodes must be selected so that polarization of the negative electrode occurs prior to polarization of the positive electrode on charging. Polarization of the positive electrode would cause (1) formation of soluble sulfur species with rapid loss of cell capacity and (2) attack of the positive current collector. To ensure polarization at the negative electrode, excess lithium should be added to the cell (as metal in the negative electrode or Li_2S in the positive electrode), the alloying elements in the negative electrode must be limited in amount, and the iron-to-sulfur ratio must be balanced.

To demonstrate the feasibility of this mechanism, potentiometric experiments have been carried out in which a solid-particulate Li–Al electrode was polarized in molten LiF–LiCl–LiBr (22:31:47 mol %) at 511°C. As shown in Fig. 12, self-discharge rates as high as 18 mA/cm² were achieved at voltages below that at which molten lithium alloys would form (Figs. 1 and 2). Other experiments showed that temperature has a significant effect on the self-discharge rate; a 50°C change doubled the self-discharge rate at a set potential in the range of −175 to −225 mV relative to a reference electrode of αAl + βLi–Al.

Although achieving the desirable effect with Li–Al electrodes appears feasible, other lithium-alloy electrodes such as Li–Al–Si and Li–Al–Fe have been considered. Variation in the amount of the additional alloying elements permits variation in the length of the voltage plateau at the overcharge potential and the level of the potential relative to lithium.

Tests of cells designed for overcharge tolerance have verified that the lithium shuttle mechanism is effective.[44] In Fig. 13, the electrode potentials of a U.P. FeS₂ cell [LiAl + 10 mol % Li₅Al₅Fe₂/LiCl–LiBr–KBr (MgO/BN)/ U.P.–FeS₂] are plotted versus capacity for a charge/discharge cycle. After charging at a rate of 25 mA/cm² to a voltage of 2.03 V, the cell was trickle

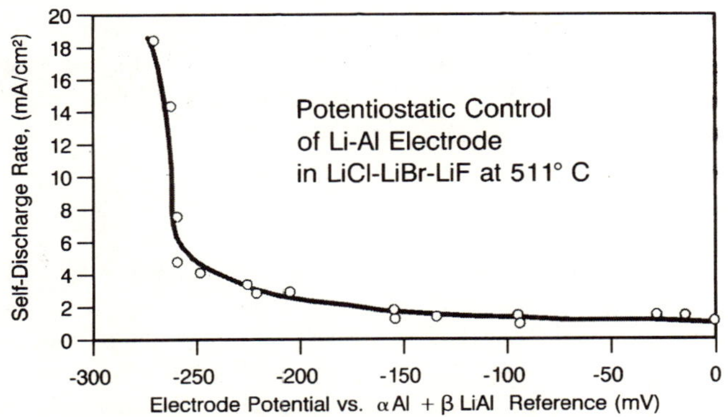

FIGURE 12. Self-discharge rate as function of Li-alloy electrode potential.

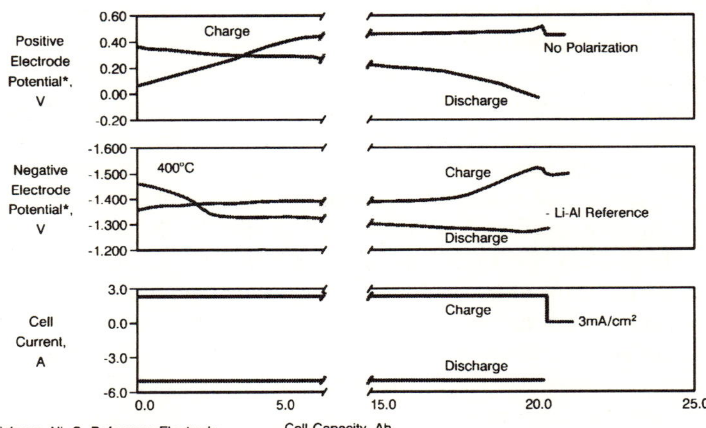

FIGURE 13. Demonstration of trickle-charge tolerance for U.P. FeS$_2$ cell in which 5% additional capacity is charged without polarizing the positive electrode.

charged at a rate of 3.0 mA/cm^2. During the trickle-charge period, the U.P. FeS$_2$ electrode maintained a sufficiently low potential to avoid the deleterious effects of polarization, namely, generation of soluble sulfur species or attack of the molybdenum current collector. Also, the potential of the lithium-alloy electrode remained at about -220 mV vs. Li$-$Al during the trickle-charge period. During this test, 5% more coulombic charge was consumed than the rated capacity. This additional charge did not contribute to the additional capacity of the cell in the subsequent discharge. Repeated tests involving cells having either FeS or FeS$_2$ electrodes have shown that the cells will accept 5% overcharge at rates of 2.0–3.0 mA/cm^2 frequently repeated during operation of over 200 cycles without deleterious effects to the cell capacity.

The use of the lithium shuttle mechanism shows promise for eliminating the need for a wire to each cell in a battery for the purpose of charge equalization. Further tests must be conducted to demonstrate the cycle life and power of cells designed to accommodate overcharge to the extent needed for the differences in coulombic efficiencies of the cells in a practical battery.

4.2. Thermal Control Systems

The high-temperature operation of molten salt batteries restricts their use to applications requiring batteries large enough to make practical the maintenance of elevated operating temperature of 400°C or more. Thus, low-cost civilian applications that require batteries of more than 10 kWh of storage capacity are favored, although applications involving smaller sizes can be considered, especially if the battery has high usage. Some possible practical applications of these batteries are for powering electric vans and cars, electric airport vehicles, and forklift trucks and for storage of off-peak energy at electric utilities. For year-round use for these applications, in most regions any battery system would require a thermal control system involving the use of an insulating jacket and dissipation of waste heat. The thermal control system for the compact lithium/sulfide batteries is somewhat more complex than for conventional batteries, but probably not much costlier.

For an electric utility van, a battery of approximately 60 kWh is needed. Ideally, this battery should be no more than about 25 cm in height and between 300 and 500 liters in volume. The thermal jacket for a lithium-alloy/iron-sulfide battery for this application is a double-walled vacuum structure fabricated of sheet metal. The evacuated space between the walls is about 2 to 3 cm in thickness and is filled with an insulating material that is effective in blocking both conductive and radiative heat transfer. One of the main means of heat transfer through conventional insulation is by conduction and convection through air spaces, and this means of heat transfer is eliminated by evacuation of the thermal jacket. Various insulating materials for use in the evacuated space have been investigated, including rigid fiber materials and alternating layers of aluminum foil and fiberglass mats. The latter type has been demonstrated to be the better insulator. The jacket may be evacuated by standard vacuum equipment and then sealed, or it may be evacuated after sealing by the use of a getter material on the high-temperature side of the jacket. The latter approach has the advantage in manufacturing of requiring less attention and of continuing to maintain a vacuum during slow outgassing. Experimental work and design studies with this type of jacket at Argonne National Laboratory[46] have indicated the feasibility of achieving heat loss rates of about 200 to 300 W for a 60-kWh battery. Most of this heat could be supplied by heat generated during operation of the battery, as indicated below.

Cooling of electric vehicle batteries during high rates of operation could be accomplished by forced air flow through isolated air passages within the hot region of the battery. The judicious use of thin insulating layers on the coolant surfaces would improve the uniformity of cooling.

For such a thermal system, the battery temperature should be permitted to fluctuate during operation over a range of at least 50°C. This feature would allow storage of heat and would greatly reduce the amount of energy required from outside sources to maintain the temperature. Most of the heat to compensate for the 200-W heat loss through the jacket wall would be generated during operation of the battery. The battery would cool at a rate at about 1°C per hour. Only after periods longer than two or three days of nonoperation would additional energy be required from an outside power source (perhaps that used for recharging the battery) to maintain the battery above the melting point of the electrolyte or at a temperature that would provide sufficient power on initial operation. If it is necessary to maintain the battery at nearly constant temperature, the amount of heat required might amount to as much as 25% or more of the energy supplied to the battery, depending on the duty cycle.

For application at electric utilities (off-peak energy storage), the batteries would be very large. At first, this would seem to decrease the problem of providing an insulated housing. However, evaluation of batteries for this application indicated problems concerned with creep within the support structures for the heavy cells in such a battery. Also, bus bars for operation at high temperature would be expensive. These are problems that are common to all high-temperature batteries for utility application.

5. Bipolar Batteries

The physical configuration of a bipolar battery differs from that of a conventional battery. The electrodes of cells in conventional batteries contain current-collector structures. In multiple-electrode cells, the positive current collectors are connected in parallel to the positive terminal, and the negative current collectors are connected to the negative terminal. For cells capable of rapid discharge, the current collectors, which run the full length of the cell, add considerably to the weight of the battery. On the other hand, the bipolar battery is formed of bipolar plates having a negative electrode on one side of a nonporous metal plate and a positive electrode on the other side. Adjacent bipolar plates are sealed at the edges with an insulator and separated from each other by a separator filled with electrolyte. The current flow through the battery is in a direction normal to the electrode faces and the bipolar plate. Little, if any, current-collector structure is required, and the bipolar plate also serves as the cell container. In addition to this weight saving, the bipolar design results in high power that is limited only by electrode kinetics and not by the current-collector structure. However, the cell capacity is limited by the practical size of a single electrode reacted from one side, whereas a conventional cell may contain many electrodes of each polarity, and all the electrodes, except those at the ends of the cell, are reacted from both sides. Therefore, paralleled bipolar battery designs need to be considered.

An example of a bipolar battery is the thermal primary battery, which is designed for military applications requiring high power. Most of the thermal batteries now utilize the lithium-alloy/FeS_2 couple. These primary batteries are designed for discharge periods ranging from a few seconds to about 30 minutes. Long shelf life, an important feature, is achieved by storing the batteries frozen at ambient temperature and then initiating discharge by rapid heating. The shelf life of these batteries exceeds 25 years. In fabricating these batteries, pellets are formed of the electrode materials and the MgO powder separators by mixing these materials with LiCl–KCl electrolyte and pressing. Bipolar batteries are assembled by stacking these pellets in the proper order with bipolar plates that incorporate a heat pellet. Operation is initiated by activating the heat pellet with a pyrotechnic charge. These thermal batteries are raised to stack operating temperature of 500 to 550°C in less than one second.

The low weight and high power of the bipolar design are attractive for secondary batteries of the Li–Al/FeS$_2$ couple. As noted earlier, the cost of molybdenum current collectors is too high for many applications, and the bipolar design reduces the amount of current collectors required. Also, the use of cost-effective coatings on the flat surface of the positive side of the bipolar plate is more feasible than on the complex current-collector structure of the electrodes in cells of conventional design.

The major problems to be solved in developing a bipolar Li–Al/FeS$_2$ battery of long operating life (five years) are the development of an appropriate insulating seal between the outer edges of adjacent bipolar plates and the achievement of overcharge tolerance.

The equalization of state of charge during charging by external control involving a wire to each cell would be especially impractical for the small cells required in a bipolar array. A modest effort has been made in the past at General Motors and Argonne National Laboratory to seal the edges of plates with a gasket material, for instance, boron nitride fiber and powder, that is not easily wetted by the molten salt electrolyte. This approach has met with only limited success and does not show promise for long life.

More recently, the development of overcharge tolerance in cells by the lithium shuttle mechanism described above has inspired new interest in the bipolar design. Work is under way at Argonne National Laboratory to develop hermetically sealed cells[47] with new glass-ceramic composite materials. If successful, a fully insulated Li–Al/FeS$_2$ secondary battery is projected to have a specific power greater than 500 W/kg and a specific energy greater than 200 Wh/kg. Such an achievement would be notable in that this projected specific energy is 42% of the theoretical value of 475 Wh/kg (Table I).

6. Other Molten Salt Battery Types

The lithium/sulfide batteries described above are the only extensively studied secondary batteries that have molten salt electrolytes for transport of ions between the electrodes. This is because few combinations of negative and positive electrodes are compatible with a low–melting molten salt, especially if only inorganic salts (the most conductive) are considered.

One type of cell that has been given limited attention is the calcium/iron sulfide cell.[48] In those studies the calcium was usually alloyed with aluminum and silicon, and the electrolyte was the BaCl$_2$–CaCl$_2$–NaCl–LiCl eutectic (mp 383°C). This system appeared to be power–limited in the configuration tested and required a high operating temperature of 475°C, but it showed promise for calcium-based negative electrode materials, which would be inexpensive.

Aluminum has also been studied as a negative electrode material in evaluation of cells of the type $Al/NaCl-AlCl_3/MeS_x$, where MeS_x is a transition metal sulfide.[49] The most satisfactory sulfides as positive electrode materials for operation of cells at 175°C over repeated cycles were FeS, $NaFeS_2$, CoS_2, Ni_3S_2, and MoS_3. The cells with these materials had only moderate open circuit voltages (0.85 to 1.38 V), but some of the combinations would involve only low cost materials and further consideration seems warranted.

Sodium is not compatible with a known low-melting salt suitable for use with a cell having a sufficiently high voltage. For instance, the eutectic of $NaCl-KCl$ melts at the high temperature of 660°C, and $NaCl-AlCl_3$, which melts below 160°C, reacts with sodium to form $NaCl$ and aluminum. To circumvent this problem, a solid electrolyte has been used to separate the electrodes. The most successful such electrolyte is β''–alumina, which contains 9 weight % Na_2O and low concentrations of Li_2O or MgO.[50,51] Two types of cells have been studied that utilize this electrolyte: the sodium/sulfur cell,[52] which operates at about 330–350°C, and the sodium/metal chloride cells,[53–55] which operate at 250–350°C. Both types of cells involve molten salts; molten sodium polysulfides form at the positive electrode on discharge of sodium/sulfur cells, and $NaCl$–$AlCl_3$ accumulates at the positive electrode of sodium/metal chloride cells during discharge. For this reason, it is appropriate to provide a brief description of these systems, although they are not usually considered to be "molten salt batteries," a term normally reserved for those having molten salt electrolytes in the interelectrode space.

Most studies of sodium/sulfur cells have employed tubular β''–alumina electrolytes with the positive (sulfur) electrode outside the electrolyte tube. This configuration provides higher power for a given electrolyte surface area than that for cells having the positive electrode on the inside of the tube. The positive electrode is usually formed by molding sulfur with a graphite felt matrix, which acts as a current collector. The negative (sodium) electrode typically contains a metal sleeve that is slightly smaller than the electrolyte tube. This provides a wicking action to maintain sodium over the entire electrolyte surface, and it also limits the amount of sodium available for immediate reaction in the event of electrolyte breakage. The cells are hermetically sealed by means of glass sealants and alumina/aluminum/container metal bonds formed under heat and pressure.

The sodium/sulfur cell has been studied more extensively than any other high–temperature secondary battery system, and it has some obvious advantages over other systems. The raw materials for the cells are plentiful and inexpensive. The coulombic efficiency is virtually 100%, and the theoretical specific energy is very high (760 Wh/kg, Table I). The mean time to failure for individual cells is thousands of cycles.

A shortcoming of sodium/sulfur cells is that failure of the ceramic separator usually occurs first as a short circuit (an electronic path between the electrodes through a crack in the electrolyte), followed by an open circuit on formation of

Na_2S_2, which is an insulator. The cells also fail on overdischarge by formation of Na_2S_2. Achieving a reliable battery with these cell failure mechanisms requires that the battery have either a large number of cells in parallel (at least 30), or a device to short circuit a failed cell in a long series string of cells.

Steady progress has been made in the development of sodium/sulfur batteries in work that is ongoing at Asea Brown Boveri, Ltd. in Germany, Chloride Silent Power, Ltd. in the United Kingdom, Hitachi, Ltd. in Japan, and Yuasa Battery Co. in Japan. The state-of-the-art performance for a 25 kWh battery is a specific energy of about 90 Wh/kg, a specific power of about 120 W/kg at 80% depth of discharge, and a life of several hundred cycles.

The sodium/metal chloride batteries have the same negative electrode material and the same type of solid electrolyte as the sodium/sulfur batteries. However, instead of a sulfur positive electrode they have either a $FeCl_2$ or a $NiCl_2$ positive electrode, which contains $NaCl$-$AlCl_3$ as a liquid electrolyte to conduct sodium ions from the solid electrolyte to the metal chloride, where it reacts to form $NaCl$ and iron or nickel. One version of the metal chloride electrode is fabricated in the uncharged state by sintering the metal powder (iron or nickel) in an intimate mixture with $NaCl$. The $NaCl$-$AlCl_3$ is injected as a liquid into the positive electrode during assembly of the cell. After assembly, the cell is charged to transfer sodium to the negative electrode and form $FeCl_2$ or $NiCl_2$ in the positive electrode by reaction with the porous metal matrix. Typically, the materials fractions are selected for reaction of 30% of the metal matrix, and the balance serves as current collector.

The sodium/metal chloride cells have high theoretical specific energy (728 Wh/kg for $Na/FeCl_2$ and 794 Wh/kg for $Na/NiCl_2$, Table I) and appear to have long cycle life similar to that for sodium/sulfur cells. However, the latter is based on less extensive data than available for the sodium/sulfur cell. On failure of the electrolyte, sodium reacts with the $NaCl$–$AlCl_3$ to form aluminum, which consistently results in a short circuit. Thus, large cells can be connected in series, and the battery so formed will operate with several failed cells, with the loss of energy restricted primarily to that associated with the loss of voltage for the failed cells. As a result of this failure mode, sodium/metal chloride batteries, when further developed, may have a higher specific energy than is achievable with practical sodium/sulfur batteries. As of 1989, the achieved specific energy and cycle life of sodium/metal chloride batteries were similar to that of sodium/sulfur batteries, but the specific power was slightly lower.

Disadvantages for sodium/metal chloride batteries include a relatively high area–specific resistance for the present large cells (40 to 100 Ah), especially beyond 70% depth of discharge. When operated at a high voltage drop to improve the power, a second disadvantage is evident: that of a high heat generation rate. It appears that these problems could be alleviated by designing for a high solid electrolyte area per unit capacity (e.g., use of many small diameter electrolyte tubes).

Development work is under way on the sodium/metal chloride batteries at Zebra Power Systems in South Africa, Beta R&D, Ltd. in the United Kingdom, and the United Kingdom Atomic Energy Agency, Harwell, in the United Kingdom. A small research effort began in the United States at Argonne National Laboratory in 1987 and at Beta Power, Inc. in 1989 to evaluate design concepts involving high solid electrolyte areas per unit of capacity.

References

1. E. J. Cairns and R. K. Steunenberg, in: *Progress in High Temperature Physics and Chemistry* (C. Rouse, ed.), Vol. 5, Pergamon Press, New York (1973); E. J. Cairns, in: *Molten Salt Technology* (D. G. Lovering, ed.), Plenum Press, New York (1982).
2. D. R. Vissers, Z. Tomczuk, and R. K. Steunenberg, *J. Electrochem. Soc.* **121**, 655 (1974).
3. J. R. Selman, in: *The Sulfur Electrode:* Fused Salts and Solid Electrolytes (R. P. Tischer, ed.), Academic Press, New York (1983).
4. D. R. Vissers, in: *Materials for Advanced Batteries* (J. Broadhead and B. C. H. Steel, eds.), pp. 47–89, Plenum Press, New York (1981).
5. J. Birk and R. K. Steunenberg, in: *New Uses of Sulfur* (J. R. West, ed.), *Adv. Chem. Ser.*, No. 140, pp. 186–202, American Chemical Society, Washington, D.C. (1975).
6. T. D. Kaun, *J. Electrochem. Soc.* **132**, 3063 (1985).
7. C. J. Wen, B. A. Boukamp, R. A. Huggins, and W. Weppner, *J. Electrochem. Soc.* **126**, 2258 (1979).
8. T. D. Kaun, in: Proceedings of the International Workshop on High-Temperature Molten Salt Batteries, Argonne National Laboratory Report ANL-86-40 (1986), pp. B101–B114.
9. T. D. Kaun, W. E. Miller, L. Redey, and J. D. Arntzen, in: *Proceedings of the Symposium on Lithium Batteries* (H. Venkatasetty, ed.), Vol. 81-4, pp. 421–428, The Electrochemical Society, Pennington, New Jersey (1981).
10. C. J. Wen and R. A. Huggins, *J. Solid State Chem.* **37**, 271 (1981).
11. H. G. V. Schnering, R. Nesper, K. F. Tebbe, and J. Curda, *Z. Met.* **71**, 357 (1980).
12. D. R. Vissers, Z. Tomczuk, L. Redey, and J. E. Battles, in: *Lithium* (O. Bock, ed.), Chapter 10, Wiley, New York (1985).
13. T. D. Kaun, L. Redey, and P. A. Nelson, in: Proceedings of 22nd Intersociety Energy Conversion Engineering Conference, Philadelphia, Pennsylvania, August 10–14, 1987, pp. 1085–1090.
14. D. R. Vissers, K. E. Anderson, C. K. Ho, and H. Shimotake, *Proceedings of the Symposium on Battery Design and Optimization* (S. Gross, ed.), Vol. 71-1, p. 416, The Electrochemical Society, Pennington, New Jersey (1979).
15. J. Battles, in: Proceedings of the International Workshop on High-Temperature Molten Salt Batteries, Argonne National Laboratory Report ANL-86-40 (1986), pp. B59–B66.
16. J. E. Battles, J. A. Smaga, and K. M. Myles, *Met. Trans.* **9A**, 183 (1978).
17. J. Smaga, in: Proceedings of the International Workshop on High-Temperature Molten Salt Batteries, Argonne National Laboratory Report ANL-86-40 (1986), pp. B72–B87.
18. J. A. Smaga, F. C. Mrazek, K. M. Myles, and J. E. Battles, in: Materials Requirements in $LiAl/LiCl\text{-}KCl/FeS_x$ Secondary Batteries, *Materials Considerations in Liquid Metals Systems in Power Generation* (N. J. Hoffman and G. A. Whitlow, eds.), pp. 52–64, National Association of Corrosion Engineers, Houston (1978).
19. R. B. Swaroop, J. E. Battles, and R. S. Hamilton, in: Proceedings of the Sixth Inter-American Conference on Materials Technology, San Francisco, California, August 12–15, 1980, pp. 67–71.
20. J. P. Mathers, C. W. Boquist, and T. W. Olszanski, *J. Electrochem. Soc.* **125**, 1913 (1978).

21. Gould Inc., in: Lithium/Iron Sulfide Batteries for Electric-Vehicle Propulsion and Other Applications, Progress Report for October 1980–September 1981, Argonne National Laboratory Report ANL-81-65 (1982), pp. 77–83.

22. J. A. Smaga and J. E. Battles, *J. Electrochem. Soc.* **129,** 496 (1982).

23. K. M. Myles and J. L. Settle, in: High-Performance Batteries for Stationary Energy Storage and Electric-Vehicle Propulsion, Progress Report for the Period October–December 1976, Argonne National Laboratory Report ANL-77-17 (1977), pp. 31–33.

24. E. C. Gay, R. K. Steunenberg, W. E. Miller, J. E. Battles, T. D. Kaun, F. J. Martino, J. A. Smaga, and A. A. Chilenskas, Lithium-Alloy/FeS Cell Design and Analysis Report, Argonne National Laboratory Report ANL-84-93 (1985).

25. J. E. Battles, F. C. Mrazek, and N. C. Otto, Post-Test Examinations of Li–Al/FeS$_x$ Secondary Cells, Argonne National Laboratory Report ANL-80-130 (1980).

26. A. G. Bergman and A. S. Arabadzhan, *Russ. J. Inorg. Chem.* (English Trans.) **8,** 369 (1963).

27. T. D. Kaun, U.S. Patent 4,764,437 (1988).

28. T. D. Kaun, T. F. Holifield, and W. H. DeLuca, Extended Abstracts, 174th Electrochemical Society Meeting, Chicago, Illinois, October 9–14, 1988, Vol. 88-2, p. 71; also, *Proceedings of the Materials and Processes for Lithium Batteries Symposium* (K. M. Abraham and B. B. Owens, eds.), Vol. 89-4, p. 373, The Electrochemical Society, Pennington, New Jersey (1989).

29. A. E. Martin, in: High Performance Batteries for Electric-Vehicle Propulsion and Stationary Energy Storage, Argonne National Laboratory Report ANL-78-94 (1980), p. 167.

30. A. E. Martin and Z. Tomczuk, in: High-Performance Batteries for Electric-Vehicle Propulsion and Stationary Energy Storage, Progress Report for the Period October 1978–March 1979, Argonne National Laboratory Report ANL-79-39 (1979), pp. 71–73.

31. Z. Tomczuk, B. Tani, N. C. Otto, M. F. Roche, and D. R. Vissers, *J. Electrochem. Soc.* **129,** 925 (1982).

32. L. Redey, in: Proceedings of the International Workshop on High-Temperature Molten Salt Batteries, Argonne National Laboratory Report ANL-86-40 (1986), pp. B131–B137.

33. C. C. McPheeters, W. W. Schertz, and N. P. Yao, *Extended Abstracts,* Battery Division, Electrochemical Society Meeting, Dallas, Texas, October 5–10, 1975, Vol. 75-2, p. 70, The Electrochemical Society, Pennington, New Jersey (1975).

34. D. A. Bruggeman, *Ann. Physik* **24,** 636 (1935).

35. T. D. Kaun, in: *Proceedings of Joint International Symposium on Molten Salts,* 172nd Electrochemical Society Meeting, Honolulu, Hawaii, October 18–23, 1987, Vol. 87-7, p. 621, The Electrochemical Society, Pennington, New Jersey (1987).

36. S. Dallek, Proceedings of the 32nd International Power Sources Symposium, Cherry Hill, New Jersey, June 9–12, 1986, p. 643.

37. D. Warin, Z. Tomczuk, and D. R. Vissers, *J. Electrochem. Soc.* **130,** 64 (1983).

38. Z. Tomczuk, D. R. Vissers, and M. L. Saboungi, in: *Proceedings of 4th International Symposium on Molten Salts* (M. Blander, ed.), Vol. 84-2, p. 352, The Electrochemical Society, Pennington, New Jersey (1984).

39. S. K. Preto, Z. Tomczuk, S. Von Winbush, and M. J. Roche, *J. Electrochem. Soc.* **130,** 264 (1983).

40. N. Kuora, J. E. Kincinas, and N. P. Yao, *J. Electrochem. Soc. Jpn.* **46,** 395 (1978).

41. G. Bandyopadhyay, *J. Electrochem. Soc.* **128,** 2545 (1981).

42. T. D. Kaun, in: Proceedings of 21st Intersociety Energy Conversion Engineering Conference, San Diego, California, August 25–29, 1986, p. 1048.

43. J. F. Martino, W. E. Moore, and E. C. Gay, in: Lithium/Iron Sulfide Batteries for Electric-Vehicle Propulsion and Other Applications, Progress Report for October 1981–September 1982, Argonne National Laboratory Report ANL-83-62 (1983), pp. 29–38.

44. T. D. Kaun, T. F. Holifield, M. F. Nigohosian, and P. A. Nelson, in: *Proceedings of the Materials and Processes Symposium* (K. M. Abraham and B. B. Owens, eds.), Vol. 89-4, p. 383, The Electrochemical Society, Pennington, New Jersey (1989).

45. L. Redey, *J. Electrochem. Soc.* **136,** (1989).
46. A. A. Chilenskas, R. F. Malecha, A. F. Tummillo, F. J. Meyer, and J. R. Missig, Development of Compressed Multifoil Insulation for High-Temperature Batteries and Other Applications, Argonne National Laboratory Report ANL-89-4 (1989).
47. T. D. Kaun and J. A. Smaga, U.S. Patent No. 4,687,717 (1987).
48. S. K. Preto, L. E. Ross, J. F. Lomax, N. C. Otto, and M. F. Roche, in: *Proceedings of 16th Intersociety Energy Conversion Engineering Conference,* Atlanta, Georgia, August 9–14, 1981, Vol. 1, pp. 765–768.
49. H. A. Hjulen, R. W. Berg, and N. J. Bjerrum, in: Proceedings of 14th International Power Sources Symposium, Brighton, U.K. September 1984, pp. 1–21.
50. J. T. Kummer and N. Weber, *SAE J.* **76,** 1003–1007 (1968).
51. B. J. McEntire, G. R. Miller, and R. S. Gordon, in: *Processing of Metal and Ceramic Powders* (R. M. German and K. W. Lay, Eds.), AIME Press, pp. 215–239, Warrendale, Pennsylvania (1982).
52. J. L. Sudworth and A. R. Tilley, *The Sodium Sulfur Battery,* Chapman and Hall Pub. Co., London (1985).
53. J. Coetzer, *J. Power Sources* **18,** 377–380 (1986).
54. R. C. Galloway, *J. Electrochem. Soc.* **134,** 256 (1987).
55. R. J. Bones, D. A. Teagle, S. D. Brooker, and F. L. Cullen, *J. Electrochem. Soc.* **136,** 1274 (1989).

Contents of Previous Volumes

Volume 1

Volume 2

Volume 3

Index